Computer and Intrusion Forensics

For quite a long time, computer security was a rather narrow field of study that was populated mainly by theoretical computer scientists, electrical engineers, and applied mathematicians. With the proliferation of open systems in general, and of the Internet and the World Wide Web (WWW) in particular, this situation has changed fundamentally. Today, computer and network practitioners are equally interested in computer security, since they require technologies and solutions that can be used to secure applications related to electronic commerce. Against this background, the field of computer security has become very broad and includes many topics of interest. The aim of this series is to publish state-of-the-art, high standard technical books on topics related to computer security. Further information about the series can be found on the WWW at the following URL:

http://www.esecurity.ch/serieseditor.html

Also, if you'd like to contribute to the series by writing a book about a topic related to computer security, feel free to contact either the Commissioning Editor or the Series Editor at Artech House.

Recent Titles in the Artech House Computer Security Series

Rolf Oppliger, Series Editor

Computer Forensics and Privacy, Michael A. Caloyannides

Computer and Intrusion Forensics, George Mohay, Alison Anderson, Byron Collie, Olivier de Vel, and Rodney McKemmish

Demystifying the IPsec Puzzle, Sheila Frankel

Developing Secure Distributed Systems with CORBA, Ulrich Lang and Rudolf Schreiner

Electronic Payment Systems for E-Commerce, Second Edition, Donal O'Mahony, Michael Pierce, and Hitesh Tewari

Implementing Electronic Card Payment Systems, Cristian Radu

Implementing Security for ATM Networks, Thomas Tarman and Edward Witzke

Information Hiding Techniques for Steganography and Digital Watermarking, Stefan Katzenbeisser and Fabien A. P. Petitcolas, editors

Internet and Intranet Security, Second Edition, Rolf Oppliger

Java Card for E-Payment Applications, Vesna Hassler, Martin Manninger, Mikhail Gordeev, and Christoph Müller

Non-repudiation in Electronic Commerce, Jianying Zhou

Secure Messaging with PGP and S/MIME, Rolf Oppliger

Security Fundamentals for E-Commerce, Vesna Hassler

Security Technologies for the World Wide Web, Second Edition, Rolf Oppliger

For a listing of recent titles in the *Artech House Computing Library*, turn to the back of this book.

Computer and Intrusion Forensics

George Mohay
Alison Anderson
Byron Collie
Olivier de Vel
Rodney McKemmish

Artech House
Boston • London
www.artechhouse.com

Library of Congress Cataloging-in-Publication Data
Computer and intrusion forensics / George Mohay...[et al.].
   p. cm.—(Artech House computer security series)
  Includes bibliographical references and index.
  ISBN 1-58053-369-8 (alk. paper)
    1. Computer security. 2. Data protection. 3. Forensic sciences.
    I. Mohay, George M., 1945–
  QA76.9.A25C628 2003
  005.8—dc21   2002044071

British Library Cataloguing in Publication Data
Computer and intrusion forensics—(Artech House computer security series)
    1. Computer security 2. Computer networks—Security measures 3. Forensic sciences
    4. Computing crimes—Investigation
  I. Mohay, George M., 1945–
  005.8

ISBN 1-58053-369-8

Cover design by Igor Valdman

© 2003 ARTECH HOUSE, INC.
685 Canton Street
Norwood, MA 02062

International Standard Book Number: 1-58053-369-8
Library of Congress Catalog Card Number: 2002044071

10 9 8 7 6 5 4 3 2 1

Contents

Foreword by Eugene Spafford . . . . *xi*

Preface . . . . *xvii*

Acknowledgments . . . . *xix*

Disclaimer . . . . *xxi*

1   Computer Crime, Computer Forensics, and
Computer Security 1

    1.1   Introduction 1

    1.2   Human behavior in the electronic age 4

    1.3   The nature of computer crime 6

    1.4   Establishing a case in computer forensics 12

         1.4.1   Computer forensic analysis within the forensic tradition 14

         1.4.2   The nature of digital evidence 21

         1.4.3   Retrieval and analysis of digital evidence 23

         1.4.4   Sources of digital evidence 27

    1.5   Legal considerations 29

    1.6   Computer security and its relationship
to computer forensics 31

         1.6.1   Basic communications on the Internet 32

         1.6.2   Computer security and computer forensics 35

v

1.7   Overview of the following chapters.   37

References.   39

2   Current Practice   41

2.1   Introduction   41

2.2   Electronic evidence   42

    2.2.1   Secure boot, write blockers and forensic platforms   44

    2.2.2   Disk file organization   46

    2.2.3   Disk and file imaging and analysis   49

    2.2.4   File deletion, media sanitization   57

    2.2.5   Mobile telephones, PDAs   59

    2.2.6   Discovery of electronic evidence   61

2.3   Forensic tools.   63

    2.3.1   EnCase.   67

    2.3.2   ILook Investigator.   69

    2.3.3   CFIT   72

2.4   Emerging procedures and standards   76

    2.4.1   Seizure and analysis of electronic evidence   77

    2.4.2   National and international standards.   86

2.5   Computer crime legislation and computer forensics   90

    *2.5.1   Council of Europe convention on cybercrime and
    other international activities*   90

    2.5.2   Carnivore and RIPA   94

    2.5.3   Antiterrorism legislation   98

2.6   Networks and intrusion forensics   103

References.   104

3   Computer Forensics in Law Enforcement and
National Security   113

3.1   The origins and history of computer forensics   113

3.2   The role of computer forensics in law enforcement   117

3.3   Principles of evidence.   121

      3.3.1   Jurisdictional issues   123

      3.3.2   Forensic principles and methodologies.   123

3.4   Computer forensics model for law enforcement.   128

      3.4.1   Computer forensic—secure, analyze,
      present (CFSAP) model   128

3.5   Forensic examination.   133

      3.5.1   Procedures.   133

      3.5.2   Analysis   143

      3.5.3   Presentation.   146

3.6   Forensic resources and tools   147

      3.6.1   Operating systems.   147

      3.6.2   Duplication   149

      3.6.3   Authentication   152

      3.6.4   Search   153

      3.6.5   Analysis   154

      3.6.6   File viewers.   159

3.7   Competencies and certification   160

      3.7.1   Training courses.   163

      3.7.2   Certification.   164

3.8   Computer forensics and national security   164

      3.8.1   National security   165

      3.8.2   Critical infrastructure protection   167

      3.8.3   National security computer forensic organizations   168

References.   169

4   **Computer Forensics in Forensic Accounting** **175**

4.1   Auditing and fraud detection   175

      4.1.1   Detecting fraud—the auditor and technology   176

4.2   Defining fraudulent activity   177

      4.2.1   What is fraud?.   178

4.2.2   Internal fraud versus external fraud.   180

4.2.3   Understanding fraudulent behavior   183

4.3   Technology and fraud detection   184

4.3.1   Data mining and fraud detection.   187

4.3.2   Digit analysis and fraud detection   188

4.3.3   Fraud detection tools.   189

4.4   Fraud detection techniques.   190

4.4.1   Fraud detection through statistical analysis   191

*4.4.2   Fraud detection through pattern
          and relationship analysis.*   200

4.4.3   Dealing with vagueness in fraud detection.   204

4.4.4   Signatures in fraud detection   205

4.5   Visual analysis techniques   206

4.5.1   Link or relationship analysis   207

4.5.2   Time-line analysis   209

4.5.3   Clustering   210

4.6   Building a fraud analysis model   211

4.6.1   Stage 1: Define objectives   212

4.6.2   Stage 2: Environmental scan.   214

4.6.3   Stage 3: Data acquisition   215

4.6.4   Stage 4: Define fraud rules.   216

4.6.5   Stage 5: Develop analysis methodology   217

4.6.6   Stage 6: Data analysis.   217

4.6.7   Stage 7: Review results   218

References.   219

Appendix 4A   221

5   Case Studies **223**

5.1   Introduction   223

5.2   The case of "Little Nicky" Scarfo.   223

5.2.1   The legal challenge   225

5.2.2   Keystroke logging system   226

5.3   The case of "El Griton"   229

   5.3.1   Surveillance on Harvard's computer network.   230

   5.3.2   Identification of the intruder: Julio Cesar Ardita.   231

   5.3.3   Targets of Ardita's activities   232

5.4   Melissa   236

   5.4.1   A word on macro viruses.   236

   5.4.2   The virus   237

   5.4.3   Tracking the author   239

5.5   The World Trade Center bombing (1993) and
      Operation Oplan Bojinka   242

5.6   Other cases   244

   5.6.1   Testing computer forensics in court.   244

   5.6.2   The case of the tender document   248

References.   253

6   Intrusion Detection and Intrusion Forensics   257

6.1   Intrusion detection, computer forensics, and
      information warfare.   257

6.2   Intrusion detection systems   264

   6.2.1   The evolution of IDS.   264

   6.2.2   IDS in practice.   267

   6.2.3   IDS interoperability and correlation   274

6.3   Analyzing computer intrusions.   276

   6.3.1   Event log analysis.   278

   6.3.2   Time-lining.   280

6.4   Network security   285

   6.4.1   Defense in depth.   285

   6.4.2   Monitoring of computer networks and systems   288

   6.4.3   Attack types, attacks, and system vulnerabilities   295

6.5   Intrusion forensics   303

   6.5.1   Incident response and investigation   303

    6.5.2   Analysis of an attack.   306

    6.5.3   A case study—security in cyberspace.   308

6.6   Future directions for IDS and intrusion forensics   310

References.   312

7   Research Directions and Future Developments.   319

7.1   Introduction   319

7.2   Forensic data mining—finding useful patterns
      in evidence   323

7.3   Text categorization.   327

7.4   Authorship attribution: identifying e-mail authors   331

7.5   Association rule mining—application to
      investigative profiling.   335

7.6   Evidence extraction, link analysis, and link discovery   339

      7.6.1   Evidence extraction and link analysis.   340

      7.6.2   Link discovery   343

7.7   Stegoforensic analysis.   345

7.8   Image mining   349

7.9   Cryptography and cryptanalysis   355

7.10   The future—society and technology   360

References.   364

Acronyms   369

About the Authors   379

Index   383

Foreword by Eugene Spafford

Computer science is a relatively new field, dating back about 60 years. The oldest computing society, the ACM, is almost 55 years old. The oldest degree-granting CS department in academia (the one at Purdue) is 40 years old. Compared to other sciences and engineering disciplines, computing is very young.

In its brief lifespan, the focus of the field has evolved and changed, with new branches forming to explore new problems. In particular, at a very high level of abstraction, we can see computing having several major phases of system understanding. In the first phase, starting in the 1940s, scientists and engineers were concerned with discovery of what could be computed. This included the development of new algorithms, theory, and hardware. This pursuit continues today. When systems did not work as expected (from hardware or software failures), debugging and system analysis tools were needed to discover why. The next major phase of computing started in the the 1960s with growing concern over how to minimize the cost and maximize the speed of computing. From this came software engineering, reliability, new work in language and OS development, and many new developments in hardware and networks. The testing and debugging technology of the prior phase continued to be improved, this time with more sophisticated trace facilities and data handling. Then in the 1980s, there was growing interest in how to make computations robust and reliable. This led to work in fault tolerance and an increasing focus on security. New tools for vulnerability testing and reverse engineering were developed, along with more complex visualization tools to understand network state.

Another 20 years later, and we are seeing another phase of interest develop: forensics. We are still interested in understanding what is happening on our computers and networks, but now we are trying to recreate behavior resulting from malicious acts. Rather than exploring faulty behavior, or probing efficiency, or disassembling viruses and Y2K code, we are now developing tools and methodologies to understand misbehavior given indirect evidence, and do so in a fashion that is legally acceptable. The problem is still one of understanding "what happened" using indirect evidence, but the evidence itself may be compromised or destroyed by an intelligent adversary. This context is very different from what came before.

The history of computer forensics goes back to the late 1980s and early 1990s. Disassembly of computer viruses and worms by various people, my research on software forensics with Steve Weeber and Ivan Krsul, and evidentiary audit trail issues explored by Peter Sommer at the London School of Economics were some of the earliest academic works in this area. The signs were clearly present then that forensic technologies would need to be developed in the coming years—technologies that have resulted in the emergence and consolidation of a new and important specialist field, a field that encompasses both technology and the law. There are professional societies, training programs, accreditation programs and qualifications dedicated to computer forensics. Computer forensics is routinely employed by law enforcement, by government and by commercial organizations in-house.

The adoption of personal (desktop) computers by domestic users and by industry in the 1980s and early 1990s (and more recently the widespread use of laptop computers, PDA's and cell phones since the 1990s) has resulted in an enormous volume of persistent electronic material that may, in the relevant circumstances, constitute electronic evidence of criminal or suspicious activity. Such stored material—files, log records, documents, residual information, and information hidden in normally inaccessible areas of secondary storage—is all valid input for computer forensic analysis. The 1990s also saw enormously increased network connectivity and increased ease of access to the Internet via the WWW. This has led to an explosion in the volume of e-mail and other communications traffic, and correspondingly in the volume of trace information or persistent electronic evidence of the occurrence of such communication. The Internet and the Web present forensic investigators with an entirely new perspective on computer forensics, namely, the application of computer forensics to the investigation of computer networks. In a sense, networks are simply other—albeit, large and complex—repositories of electronic evidence. The projected increase in wireless and portable computing will further add to the scale and complexity of the problems.

Increased connectivity and use of the WWW has also led to the large-scale adoption of distributed computing—a paradigm that includes heavy-weight government and commercial applications employing large distributed databases accessed through client-server applications to provide consumers with access to data, for example, their bank accounts and medical records. Society relies on the security of such distributed applications, and the security of the underlying Internet and Web, for its proper functioning. Unfortunately, the rush to market and the shortage of experts has led to many infrastructure components being deployed full of glaring errors and subject to compromise. As a result, network and computer attacks and intrusions that target this trust have become a prime concern for government, law enforcement and industry, as well as a growing sector of academia.

The investigation of such attacks or suspected attacks (termed "intrusion forensics" in this book) has become a key area of interest. The earliest widely publicized large-scale attack on the Internet was the Morris Internet Worm, which took place in 1988 and that I analyzed and described at the time. (It appears that my analysis was the first detailed forensic report of a such an attack.) The Worm incident demonstrated how vulnerable the Internet was and indicated the need for improved system and network security. Unfortunately, for a number of reasons including cost, increased connectivity and time-to-market pressures, our overall infrastructure security may be worse today than it was in 1988. Our systems today are still vulnerable and still need improved security. The Carnegie Mellon University CERT Coordination Center reported an increase by a factor of five in incidents handled from 1999 to 2001, from approximately 10,000 in 1999 to over 50,000 in 2001, and an increase by a factor of six in the number of vulnerabilities reported, from approximately 400 in 1999 to over 2,400 in 2001. With this increase, there has been a greater need to understand the causes and effects of intrusions, on-line crimes, and network-based attacks. The critical importance of the areas of computer forensics, network forensics and intrusion forensics is growing, and will be of great importance in the years to come.

Recent events and recent legislation, both national and international, mean that this book is especially timely. The September 11, 2001 terrorist attacks have led directly to the passage of legislation around the world that is focused on providing national authorities with streamlined access to communications information that may be relevant in the investigation of suspected terrorist activity. (It is important to note that the increased access can also be used to suppress political or religious activity and invade privacy; we must all ensure these changes are not so sweeping as to be harmful to society in the long run.)

In a recent address to the First Digital Forensic Research Workshop held at the Rome Research Site of the Air Force Research Laboratory, I noted that for the future, we needed to address more than simply the technical aspects:

Academic research in support of government, as well as commercial efforts to enhance our analytical capabilities, often emphasizes technological results. Although this is important, it is not representative of a full-spectrum approach to solving the problems ahead. For the future, research must address challenges in the procedural, social, and legal realms as well if we hope to craft solutions that begin to fully "heal" rather than constantly "treat" our digital ills. This full-spectrum approach employs the following aspects:

> ▸ *Technical:* "Keeping up" is a major dilemma. Digital technology continues to change rapidly. Terabyte disks and decreasing time to market are but two symptoms that cause investigators difficulty in applying currently available analytical tools. Add to this the unknown trust level of tools in development, and the lack of experience and training so prevalent today, and the major problems become very clear.

> ▸ *Procedural:* Currently, digital forensic analysts must collect everything, which in the digital world leads to examination and scrutiny of volumes of data heretofore unheard of in support of investigations. Analytical procedures and protocols are not standardized nor do practitioners and researchers use standard terminology.

> ▸ *Social:* Individual privacy and the collection and analysis needs of investigators continue to collide. Uncertainty about the accuracy and efficacy of today's techniques causes data to be saved for very long time periods, which utilizes resources that may be applied toward real problem solving rather than storage.

> ▸ *Legal:* We can create the most advanced technology possible, but if it does not comply with the law, it is moot.

Whatever the context presented by the relevant national jurisdiction(s), the task of the computer and intrusion forensics investigator will become more critical in the future and is bound to become more complex. Having standard references and resources for these personnel is an important step in the maturation of the field. This book presents a careful and comprehensive treatment of the areas of computer forensics and intrusion forensics, thus helping fill some of that need: I expect it to be a significantly useful addition to the literature of the practice of computing. As such, I am grateful for the opportunity to introduce the book to you.

Eugene H. Spafford
February 2003

Eugene H. Spafford is a professor of Computer Sciences at Purdue University, a professor of philosophy (courtesy appointment), and director of the Center for Education Research Information Assurance and Security (CERIAS). CERIAS is a campuswide multidisciplinary center with a broadly focused mission to explore issues related to protecting information and information resources. Spafford has written extensively about information security, software engineering, and professional ethics. He has published over 100 articles and reports on his research, has written or contributed to over a dozen books, and he serves on the editorial boards of most major infosec-related journals.

Dr. Spafford is a fellow of the ACM, AAAS, and IEEE and is a charter recipient of the Computer Society's Golden Core Award. In 2000, he was named as a CISSP. He was the 2000 recipient of the NIST/NCSC National Computer Systems Security Award, generally regarded as the field's most significant honor in information security research. In 2001, he was named as one of the recipients of the Charles B. Murphy Awards and named as a fellow of the Purdue Teaching Academy, the university's two highest awards for outstanding undergraduate teaching. In 2001, he was elected to the ISSA hall of fame, and he was awarded the William Hugh Murray medal of the NCISSE for his contributions to research and education in infosec.

Among his many activities, Spafford is cochair of the ACM's U.S. Public Policy Committee and of its Advisory Committee on Computer Security and Privacy, is a member of the board of directors of the Computing Research Association, and is a member of the U.S. Air Force Scientific Advisory Board.

More information may be found at http://www.cerias.purdue.edu/homes/spaf

In his spare time, Spafford wonders why he has no spare time.

Preface

Computer forensics and intrusion forensics are rapidly becoming mainstream activities in an increasingly online society due to the ubiquity of computers and computer networks. We make daily use of computers either for communication or for personal or work transactions. From our desktops and laptops we access Web servers, e-mail servers, and network servers whether we know them or not; we also access business and government services, and then—unknowingly—we access a whole range of computers that are hidden at the heart of the embedded systems we use at home, at work and at play. While many new forms of illegal or anti-social behavior have opened up as a consequence of this ubiquity, it has simultaneously also served to provide vastly increased opportunities for locating electronic evidence of that behavior.

In our wired society, the infra-structure and wealth of nations and industries rely upon and are managed by a complex fabric of computer systems that are accessible by the ubiquitous user, but which are of uncertain quality when it comes to protecting the confidentiality, integrity, and availability of the information they store, process, and communicate. Government and industry have as a result focused attention on protecting our computer systems against illegal use and against intrusive activity in order to safeguard this fabric of our society. Computer and intrusion forensics are concerned with the investigation of crimes that have electronic evidence, and with the investigation of computer crime in both its manifestations—computer assisted crime and crimes against computers.

This book is the result of an association which reaches back to the 11th Annual FIRST Conference held in June 1999 at Brisbane, Australia. Together with a colleague, Alan Tickle, we were involved in organizing and presenting what turned out to be a very popular computer forensic workshop—the

Workshop on Computer Security Incident Handling and Response. Soon afterwards we decided that we should continue the collaboration. It has taken a while for the ideas to bear fruition and in the meantime there have been many excellent books published on the related topics of computer forensics, network forensics, and incident response, all with their own perspective. Those we know of and have access to are referred to in the body of this book. Our perspective as implied by the title is two-fold. First, we focus—in Chapters 1 to 4—on the nature and history of computer forensics, and upon current practice in 'traditional' computer forensics that deals largely with media acquisition and analysis:

- *Chapter 1:* Computer Crime, Computer Forensics, and Computer Security

- *Chapter 2:* Current Practice

- *Chapter 3:* Computer Forensics in Law Enforcement and National Security

- *Chapter 4:* Computer Forensics in Forensic Accounting

The second focus (Chapter 5 to 7) of this book is on intrusion investigation and intrusion forensics, on the inter-relationship between intrusion detection and intrusion forensics, and upon future developments:

- *Chapter 5:* Case Studies

- *Chapter 6:* Intrusion Detection and Intrusion Forensics

- *Chapter 7:* Research Directions and Future Developments

We hope that, you, our reader will find this book informative and useful. Your feedback will be welcome, we hope that this book is free of errors but if not—and it would be optimistic to expect that—please let us know.

Finally, we would like to note our special thanks to Gene Spafford for writing the Foreword to this book. We the authors are privileged that he has done so. There is no better person to introduce the book and we urge you to start at the beginning, with the Foreword.

Acknowledgments

The field of computer forensics has come a long way in a short time, barely 15 years. The pioneers and pioneering products, that helped fashion the field are, as a result in many cases still in the industry, a fortunate and an unusual outcome. The field owes an enormous debt of gratitude, as do the authors of this book, to the pioneers and product developers who hail from across academia, law enforcement and national security agencies, and the industry.

We have been fortunate to have colleagues and graduate students interested in the area of computer and intrusion forensics who have assisted us with developing or checking material in the book. We would like to thank and acknowledge the contributions of Detective Bill Wyffels (Eden Prairie Police Department), Gary Johnson (Minnesota Department of Human Services), Bob Friel (U.S. Department of Veterans Affairs Office of the Inspector General), Detective Scott Stillman (Washington County Sheriffs Department), Matt Parsons (U.S. Naval Criminal Investigative Service), Steve Romig (Ohio State University), Neena Ballard (Wells Fargo), Dr. Alan Tickle (Faculty of Information Technology, Queensland University of Technology), and Nathan Carey (Faculty of Information Technology, Queensland University of Technology). We would also like to acknowledge the constructive comments of our reviewer for the improvements that have resulted. We are grateful to all these people for their contributions. Needless to say, any errors remaining are ours.

Finally, we wish to thank our publisher, Artech House, for their guidance and, in particular, for their forbearance when schedules were difficult to meet. Special thanks and acknowledgments are due to Ruth Harris, Tim Pitts, and Tiina Ruonamaa.

Disclaimer

Any mention of commercial or other products within this book is for information only; it does not imply recommendation or endorsement by the authors or their employers nor does it imply that the products mentioned are best suited or even suitable for the purpose. Before installing or using any such products in an operational environment, they should be independently evaluated for their suitability in terms of functionality and intrusiveness.

The book contains legal discussion. This should, however, not be taken as legal advice and cannot take the place of legal advice. Anyone dealing with situations of the sort discussed in the book and which have legal implications should seek expert legal advice.

CHAPTER 1

Contents

1.1 Introduction

1.2 Human behavior in the electronic age

1.3 The nature of computer crime

1.4 Establishing a case in computer forensics

1.5 Legal considerations

1.6 Computer security and its relationship to computer forensics

1.7 Overview of the following chapters

References

Computer Crime, Computer Forensics, and Computer Security

Computers are a poor man's weapon.
Richard Clarke, Special Advisor to the U.S. President on Cyberspace Security.

In some ways, you can say that what the Internet is enabling is not just networking of computers, but networking of people, with all that implies. As the network becomes more ubiquitous, it becomes clearer and clearer that who it connects is as important as what it connects.
Tim O'Reilly, ''The Network Really Is the Computer.''

1.1 Introduction

Computers undeniably make a large part of human activity faster, safer, and more interesting. They create new modes of work and play. They continually generate new ideas and offer many social benefits, yet at the same time they present increased opportunities for social harm. The same technologies powering the information revolution are now driving the evolution of *computer forensics:* the study of how people use computers to inflict mischief, hurt, and even destruction.

People say that the information revolution is comparable with the industrial revolution, as important as the advent of print media, perhaps even as significant as the invention of

1

writing. The harm that can be inflicted through information technology invites a less dignified comparison. We can make analogies, for instance, with the mass uptake of private automobiles during the last century. By this we mean that although cars, roads, and driving may have changed life for the better, modern crimes like hijacking or car theft have become accessible to a mass population, even though most drivers would never contemplate such acts. Old crimes, such as kidnapping or bank robbery, can be executed more easily and in novel ways. Drivers can exploit new opportunities to behave badly, committing misdemeanors virtually unknown before the twentieth century, such as unlicensed driving or road rage. The point of this analogy is that an essential, freely accessible, and widely used Internet can be adapted for every conceivable purpose, no matter how many laws are passed to regulate it.

In 1979, the U.S. Defense Advanced Research Projects Agency (DARPA) developed the ARPANET network, the parent of the modern Internet. The ARPANET consisted initially of a comparatively small set of networks communicating via Network Control Protocol (NCP) that was to become the now ubiquitous Transmission Control Protocol and Internet Protocol (TCP/IP) suite. At that stage, its main clientele consisted of an élite scientific and research population. Its popular but primarily text-based services, including applications such as e-mail, File Transfer Protocol (FTP) and Telnet, still demanded nontrivial computer skills at the time when its public offspring was launched in 1981. As the Internet expanded, so did the opportunities for its misuse, the result of a host of security flaws. For instance, e-mail was easy to spoof, passwords were transmitted in clear and connections could be hijacked. Nevertheless, most users had no real interest in security failings until the 1988 Internet Worm case, which provided a glimpse of how damaging these defects could be.

From then onwards, Internet security has never been off the agenda. Introduced in the early 1990s, the Hypertext Transfer Protocol (HTTP), Hypertext Markup Language (HTML) and various Web browsers have made the Internet progressively more user friendly and accessible. On the Web, it was no longer necessary to understand how different applications worked in order to use them. Yet with such a huge information source available to them, novice users could relatively easily become expert enough to exploit vulnerabilities in networks and applications. One important reason contributing to Internet reliability is that the same software is run on many different nodes and communicates via the same protocols, so that for a user with criminal inclinations, there are multiple targets, vulnerabilities and opportunities.

The title of this book, *Computer and Intrusion Forensics*, refers to its two main themes:

1. *Computer forensics,* which relates to the investigation of situations where there is computer-based (digital) or electronic evidence of a crime or suspicious behavior, but the crime or behavior may be of any type, quite possibly not otherwise involving computers.

2. *Intrusion forensics,* which relates to the investigation of attacks or suspicious behavior directed against computers per se.

In both cases, information technology facilitates both the commission and the investigation of the act in question, and in that sense we see that intrusion forensics is a specific area of computer forensics, applied to computer intrusion activities. This chapter sets out to explain the shared background of computer forensics and intrusion forensics, and to establish the concepts common to both. The Internet provides not only a major arena for new types of crime, including computer intrusions, but also as discussed in Chapter 6 a means of potentially tracking criminal activity. In any case, not all computer-related offences (an umbrella term by which we mean offences with associated digital evidence such as e-mail records—offences which do not otherwise involve a computer—as well as offences targeted directly against computers) are executed via the Internet, and many perpetrators are neither remote nor unknown. Prosecuting a computer-related offence may involve no more than investigating an isolated laptop or desktop machine. It is increasingly obvious that the public Internet has become the vehicle for an escalating variety of infringements, but many other offences take place on private networks and via special-purpose protocols.

An important point to note is that while computer forensics often speaks in legal terms like *evidence*, *seizure*, and *investigation*, not all computer-related misdeeds are criminal, and not all investigations result in court proceedings. We will introduce broad definitions for computer forensics and intrusion forensics which include these less formal investigations, while subsequent chapters will discuss the spectrum of computer forensic and intrusion forensic techniques appropriate in various criminal and noncriminal scenarios.

This chapter briefly reviews the social setting that makes the exercise of computer forensics a priority in law enforcement (LE), government, business, and private life. Global connectivity is the principal cause of an unprecedented increase in crimes that leave digital traces, whether incidentally or whether perpetrated through or against a computer. We outline a spectrum of ways in which people perpetrate familiar crimes or invent new ones. This chapter then highlights that while computer forensics and intrusion forensics are rapidly gaining ground as valid subdisciplines of traditional forensics, there are both similarities and important differences between computer forensics and other forensic procedures. These differences are particularly significant with regard to evidence collection and analysis methods.

This chapter also outlines first the interest groups and then the legal framework within which the computer forensic discipline has developed and is developing. Both computer forensic analysis and intrusion forensic analysis have a symbiotic relationship with computer security practices, and utilize many of the same techniques. In some ways, the two activities are mutually supportive, while in other respects their objectives conflict: best security practice prefers to prevent untoward incidents rather than to apportion blame afterwards. Finally, we review relevant network and security concepts, before introducing topics to be covered in subsequent chapters.

1.2 Human behavior in the electronic age

There are various estimates of the number of people now connected to the Internet, all of which acknowledge an enormous rise in on-line activity. A typical example [1] shows more than a 10-fold rise in connectivity from 1996 to 2002, ranging from 70 million to nearly 750 million people. What are all these people actually doing? The shortest answer is that they are busy doing what comes naturally to them: interacting.

During the Internet's rapid expansion in the 1990s, individuals, businesses, and other organizations immediately took advantage of something technologists had long predicted: that computer networks are a personal and social as well as a technological and economic resource. For these newcomers, a network interface was taken for granted as a kind of accomplice in the household or workplace. Exploiting it has become an extension of normal human behavior and what people are doing is as good and as bad as in the pre-Internet days. Now, however, they are doing on the Internet: they are not only enthusiastically talking, listening, buying, selling, teaching, learning, playing, and creating but also lying, cheating, stealing, eavesdropping, exploiting, destroying, and even in extreme cases actually planning or executing a murder. That such extreme cases can and will occur was widely publicized and discussed following the September 11, 2001 attacks in the United States. A crime, the public now realizes, can be initiated, planned, and partly executed in cyberspace.

What information technology has achieved by connecting people and computers in one large network is the first significantly global social system. From its beginning, the Internet exhibited self-organizing behavior as any other social system does, but much more rapidly. Public spaces—newsgroups, chat rooms, file resources—developed first a good behavior code (netiquette), then a monitoring system (moderation), and then a set of punishments (exclusion). In the same way, privately owned spaces on the Internet tried to protect themselves by plugging vulnerabilities and installing safeguards. The security policies they evolved aimed to control how the entire system, including its users, should behave. From this point of view, all components of the worldwide system including its end users are expected to behave both cooperatively, to achieve common objectives, and correctly to avoid violating the rules.

Good behavior is notoriously difficult to reconcile with competitive objectives. For example, a commercial Web site (if its administrators are conscientious) encloses its core processes with several layers of rules. Although the site's primary objective is to support a business, not everything the system is capable of doing is productive, and not everything productive is legal, let alone socially desirable. Laws, regulations, and ethics are sometimes in conflict with business aims: It might, for instance, be cheaper in cost-benefit terms to abandon user authentication or audit trails, but it may also be illegal for a business to do so. Typically, workplace rules also constrain employees (e.g., from excessive private Web surfing, from browsing sensitive information not covered by privacy laws, or from using inappropriate language in e-mail). Such normative rules are increasingly found in application interfaces, typical examples being Web site censors, or word processor vocabulary and style monitors.

An idealized picture of an ethical system is represented in Figure 1.1; of all possible system actions, comparatively few will be desirable, legal, and ethical, but no known system architecture supports such a view of operations. Instead, computer systems fragment their rules and regulations across networks, implementing them through such diverse forms as user authentication, intrusion detection systems, encryption and access control, with the result that traces of any offence are also fragmented. A network user now has the potential to cause an undesirable event anywhere in the connected world, and can deliberately or not offend on a global scale, leaving an equally far-flung trail.

The terms *computer forensics* and *intrusion forensics* refer to the skills needed for establishing responsibility for an event, possibly a criminal offence, by reassembling these traces into a convincing case. But the case may have to be convincing in the eyes of the law, and not merely in

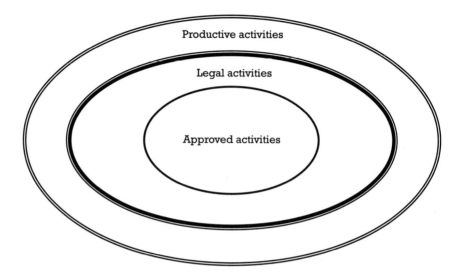

Figure 1.1   A sociolegal view of computer system activity.

the personal view of a system administrator, auditor, or accountant. In particular, to satisfy a court of law, an investigation needs to be legally well founded as well as convincing in the everyday sense. The term *forensics* as applied in information technology confronts civil society with a whole new array of problems in conceptualization. How is a crime actually proved with computer-related evidence? How is criminal responsibility allocated? What would be the elements of a valid defense? Can a computer be an accessory? Worst of all, could the computer actually cause an apparent crime, and could it then be made to appear that some innocent person is responsible?

1.3   The nature of computer crime

Computer forensics involves the investigation of computer-based evidence, and this necessarily requires that investigators understand the role played by computer technology. This cannot be done without some understanding of computer technology. As noted, many investigations need not end in a criminal case (e.g., those related to civil action or internal disciplinary procedures) but they still need to be performed if responsibility is to be justly assigned. The scope of an investigation includes detecting planned acts and acts in progress, as well as acts in the past, so the investigators (whether humans or their system surrogates, such as intrusion detection systems) can also play a role in crime scenarios. This section looks at the fluid nature of the term *computer crime* in this context.

[handwritten margin note: A feeling of unhappiness or disappointment]

[handwritten margin note: to commit a crime]

Computers have inspired new types of misconduct, such as hacking and denial of service. Since these acts demand some computer expertise from a perpetrator, they retain a certain glamour in some circles, which regard them as heroic rather than criminal. Perhaps more dismaying for law enforcement is the rate at which ordinary, inexpert people find new opportunities for older crimes like credit card fraud, embezzlement, and even blackmail. In the electronic age, people behave as unlawfully as ever, but ever more imaginatively:

> Unlawful activity is not unique to the Internet—but the Internet has a way of magnifying both the good and the bad in our society...What we need to do is find new answers to old crimes. (U.S. Vice President Al Gore, 1999).

Vice President Gore's remarks reflect a sense of public unease about loss of control. There is ample evidence that computer-related crime rates rise in step with the rate of connectivity [2], as the general public has not failed to perceive. Up to 75% of respondents in a U.S. November 2001 survey by the Tumbleweed Communications Corporation [3] thought that they were at risk from using the Internet, agreeing that they were worried about the misuse of personal information both by government and by persons unknown. Less than 20% of respondents trusted the ability of the U.S. government to prevent computer-based attacks on their agencies.

Although these survey figures probably reflect a heightened level of public anxiety following the September 11, 2001, World Trade Center attacks, the results are consistent with preexisting perceptions of personal vulnerability in relation to information on privacy and security. This sense of unease is not difficult to source. Ten years ago, no newspaper published an information technology section of more than a few pages. Computer hacking incidents and service failures of any kind were rarely reported in main news. Now, the IT section in a newspaper can run to 20 or more pages, with many items personally relevant to the average user. A single edition of a single newspaper's IT pull-out section, for example, includes the following articles that could directly or indirectly have forensic implications (*The Australian*, January 23, 2002):

1. Investigators find that the scrubbed computers of a failed mega-corporation still contain a large amount of retrievable data.

2. Second hand computers being sold off at auction are found to contain confidential company and personal records.

3. Internet service providers (ISPs) proposing to collect Internet users' phone numbers to identify spammers find their e-mail servers

being blocked overseas because of the increasing amounts of dross e-mail passing through.

4. A domain name regulator is reducing its holdings of personal information in order to comply with privacy regulation.

5. The American Civil Liberties Union voices its opposition to a plan for a unified national database system of driver identification.

6. A U.K. supercomputer suffers over a million dollars' damage when thieves steal printed circuit boards worth $200,000 each.

Meanwhile, in the main news pages, figures such as the following appear routinely (from the U.S. Office of Public Information, values in USD):

1. In 2001, software to the value of $5.5 billion was stolen via Internet-based piracy.

2. Over $1 billion in income has been lost by phone companies through use of stolen or faked phone credit card numbers on the Internet.

3. Over $3 billion has been lost by credit card issuers through use of faked or stolen credit cards.

4. Some 2 million laptops were stolen in 2001.

5. Estimates claim that computer crime may cost as much as $50 billion per year.

6. Fewer than 10% of computer crimes are reported.

7. Fewer than 2% of these reported crimes result in a conviction.

8. Hackers committed an estimated 5.7 million intrusions in 2001 alone.

Computers will probably be involved in crimes that no one has ever imagined. New kinds of computer-related or assisted crimes emerge constantly, even if only new in the sense that information technology is now able to facilitate and record them. There is, however, a generally accepted classification of computer crime:

- The computer (by which we mean the information resident on the computer, code as well as data) is the target of the crime, with an intention of damaging its integrity, confidentiality, and/or availability.

✓ ▸ The computer is a repository for information used or generated in the commission of a crime.

✓ ▸ The computer is used as a tool in committing a crime.

These categories are not mutually exclusive, as a report from the U.S. President's Working Group on Unlawful Conduct on the Internet explains [4]:

Computers as targets   One obvious way in which a computer can be involved in an unlawful conduct is when the confidentiality, integrity, or availability of a computer's information or services is attacked. This form of crime targets a computer system, generally to acquire information stored on that computer system, to control the target system without authorization or payment (theft of service), or to alter the integrity of data or interfere with the availability of the computer or server. Many of these violations involve gaining unauthorized access to the target system (i.e., hacking into it).

Computers as storage devices   A second way in which computers can be used to further unlawful activity involves the use of a computer or a computer device as a passive storage medium. As noted above, drug dealers might use computers to store information regarding their sales and customers. Another example is a hacker who uses a computer to store stolen password lists, credit card or calling card numbers, proprietary corporate information, pornographic image files, or "warez" (pirated commercial software).

Computers as communications tools   Another way that a computer can be used in a cybercrime is as a communication tool. Many of the crimes falling within this category are simply traditional crimes that are committed on-line. Indeed, many of the examples in this report deal with unlawful conduct that exists in the physical, off-line world—the illegal sale of prescription drugs, controlled substances, alcohol and guns, fraud, gambling, and child pornography. These examples are, of course, only illustrative; on-line facilities may be used in the furtherance of a broad range of traditional unlawful activity. E-mail and chat sessions, for example, can be used to plan or coordinate almost any type of unlawful act, or even to communicate threats or extortion demands to victims.

The term *computer crime* has a precise sense deriving from its use in laws framed specifically to prohibit confidentiality, integrity, and availability attacks. In this usage, it approximately corresponds to public perceptions such as those aired in the previously cited November 2001 survey:

computer crime there refers specifically to activities targeting computers in order to misuse them, to disrupt the systems they support, or to steal, falsify or destroy the information they store. Broad as it is, this definition fulfils only the first category quoted earlier—"computers as targets." Casey [5], for instance, views computer crime as a special case of the comprehensive term *cybercrime*, where the latter applies to all three categories—in fact, to any crime leaving computer evidence. If computer crime is to be confined to infractions targeting computers, then there is a need for a term such as *computer-assisted crime* or *computer-related crime* to embrace the other two categories. In these, the act causes no harm to the computer but instead enrols it as an accessory (i.e., as a tool or data repository in the above sense).

Nevertheless, it is not uncommon for computer crime to refer to a broader spectrum of acts than just those targeting computers. The term is often applied to all three categories of crime, and we shall adopt this comprehensive usage throughout the book except where it is otherwise noted. Accordingly, in this frame of reference, the following convictions under U.S. federal law (a sample, from the year 2001) are all computer crimes, illustrating the multiplicity of computer-related acts we can address when using the term in its more comprehensive sense [6]:

1. A demoted employee before leaving the company instals a date-triggered code time bomb, which later deactivates hand-held computers used by the sales force.

2. Someone advertizes for goods on eBay, the Internet auction site, but on receiving payment never supplies the goods.

3. Another one advertizes collectible items via eBay; these prove to be fakes.

4. An ex-employee sends a threatening e-mail.

5. An employee in a law firm steals a trial plan in order to sell it to an opposing counsel.

6. A disgruntled student sends a threatening e-mail, leading to closure of his school.

7. A Web site advertizes fake identification documents.

8. Employees of a hardware/software agency sold bonafide copyright products and pocketed the proceeds.

9. Numerous people sold illegal satellite TV decryption cards.

10. A ring of software pirates used a Web site to distribute pirated software.

11. A software company employee is indicted for altering a copyright program to overcome file reading limitations.

12. Someone auctions software via eBay, claiming it is a legal copy, but in fact supplies a pirate copy.

13. Two entrepreneurs pirate genuine software and make CD-ROM copies; sell these through a Web site; and use e-mail sent through an employer's account to contact potential purchasers.

14. A hacker accesses 65 U.S. court computers and downloads large quantities of private information.

15. Another hacker accesses bank records, steals banking and personal details, and uses these for extorting the account owner.

16. Via hacking, others steal credit card numbers for personal use (credit card theft is a variety of identity theft).

17. A hacking ring establishes its headquarters on unused space in an unsuspecting company's server; this stolen space is used to exchange hacking tools and information.

It is clear from the above that there is no such thing as a typical computer criminal with a typical criminal method. Perpetrators of the above include males, females, nationals, foreigners, juveniles and mature adults. Their motives ranged from revenge through greed, mischief and curiosity to simple pragmatic convenience. Some perpetrators applied extensive planning and computer expertise; others just used universally available software. Some criminals targeted computer components or the information stored by these components. In other crimes, people or organizations were targeted by means of a computer. Some of these human targets must have colluded in the act, knowing it was illicit. In other cases, the crime had no particular person as a target: perpetrators did no more damage than helping themselves to superfluous file store or processing cycles.

Computer forensics and intrusion forensics are used to investigate cases like these, crimes now so common that forensic approaches have evolved in response. The International Organization on Computer Evidence (IOCE) notes the nature of the investigatory frameworks for some of the more common subtypes of computer crimes which include on-line auction fraud, extortion, harassment, and stalking as well as hacking and computer piracy.

For example, in the case of an extortion investigation, an investigator would begin by looking at the following: "...date and time stamps, e-mail, history log, Internet activity logs, temporary Internet files, and user names" [7]. In contrast, a computer intrusion case suggests both more computer expertise and more computer-based planning on the perpetrator's part. Hence, the investigator will include a greater variety of sources: "...address books, configuration files, e-mail, executable programs, Internet activity logs, IP address and user name, IRC chat logs, source code, text files...sniffer logs, existence of hacking tools...network logs, recovering deleted information, locating hidden directories..." [7].

While we have already distinguished broadly between crimes which target computers or computer systems, and computer-assisted or related crime where the computer itself is not adversely affected but is an accessory to the act, the above highlights the clear differences between investigating a computer-assisted crime like extortion, and catching an intruder or hacker. The distinction arises not only because the hacking investigation needs more and qualitatively different evidence, but also because acts targeting computers (even if only potentially targeting them) require a faster response than post hoc analysis. Consequently, we use the term *intrusion* as a special sense of computer as target: intrusions are intentional events involving attempts to compromise the state of computers, networks, or the data present, either short- or long-term, on these devices. Such attempts need different investigatory techniques because, in effect, the investigation ideally would take place before the crime occurs. Chapter 6 presents a detailed discussion of intrusion investigation techniques. For the present, we note that intrusions are a special kind of computer crime, and that intrusion forensics is correspondingly a specialization of computer forensics.

1.4   Establishing a case in computer forensics

Section 1.3 distinguished between crime assisted by computers and crime specifically targeting computers in order to establish the difference between computer forensics and intrusion forensics. Both, however, rely upon computer-based evidence that must meet the formal evidentiary requirements of the courts if it is to be admissible in a court of law. Here, we explore the special characteristics of computer-based evidence, and its place within the forensic tradition. We can then introduce adequate definitions for both computer forensics and intrusion forensics.

Computer forensics and intrusion forensics, in both the broad sense (using any computer evidence) and narrow sense (focusing on court-admissible evidence only) are made up of activities quite different from those of traditional forensics, with its foundation in the physical sciences. In computer forensics, there is no unified body of theory. Its raw material is not a natural or manufactured product, nor are its tools and techniques discoveries. Both the evidence itself and the tests applied to it are artifacts developed not in research laboratories but in a commercial market-place. Instead of independent, standardized tests conducted in sanitized conditions, computer forensics aims to assign responsibility for an event by triangulating separate streams of evidence, each furnishing a part of the scenario. It is the computer data stream itself that forms the evidence, rather than any conclusions about what a test result means. Hence, the tasks of identifying, collecting, safeguarding, and documenting computer (or digital) evidence also include preserving test tools and justifying their operation in court. The same obligation of care operates when investigations do not aim to take court admissibility into account. In these cases, a plausible explanation rather than proof of guilt may satisfy the investigators.

Concepts about digital evidence have been developed in a bottom-up fashion. Until recently, few lawyers or law enforcement officers had qualifications in information technology and thus there has been limited success in relating existing law to a new language that speaks of intrusions, downloads, masquerading, information integrity, or update. The problem court officers faced was that the familiar language of evidence had evolved for discoursing about physical traces—paper records, blood spatters, footprints, or wounds. Evidence in computer crime cases had no such physical manifestation. In consequence, no general agreement has yet emerged on admissibility and weight of computer-based evidence, although some progress has been made, as Chapters 2 and 3 will discuss.

Admissibility of evidence is treated differently across different jurisdictions, and there is growing pressure for a global legal framework to deal with transborder computer crime, as Section 1.5 shows. Computer-based evidence never publicly challenged or recorded, such as that collected for an internal employee disciplinary case, does not need to meet admissibility requirements. It is not intended for production in court, but its reliability is no less important for that. The same is true when we consider the role computer evidence plays in information warfare (see Chapter 6) and other applications of preventative surveillance.

In Section 1.4.1 we overview the genesis of computer forensics and its emergence as a professional discipline, a topic treated in detail in Chapter 3.

1.4.1　Computer forensic analysis within the forensic tradition

Although computer forensics is a comparatively new field, it is developing within a tradition that is well established. In classic forensics, the practice of "freezing the scene" to collect potential crime traces is more than 100 years old. Advances in portable camera technology allowed Paris police clerk Alphonse Bertillon to introduce in 1879 a methodical way of documenting the scene by photographing, for example, bodies, items, footprints, bloodstains in situ with relative measurements of location, position, and size [8]. Bertillon is thus the first known forensic photographer, but this is not his only contribution. *Bertillonage*, his system of identifying individuals over 200 separate body measurements, was in use till 1910 and was only rendered obsolete by the discovery that fingerprints were unique:

> His was something of a radical notion in criminal investigation at the time:
> that science and logic should be used to investigate and solve crime. [9]

Among those influenced by Bertillon's scientific approach was his follower Edmond Locard, who articulated one of the forensic science's key rules, known as Locard's Exchange Principle. The principle states that when two items or persons come into contact, there will be an exchange of physical traces. Something is brought, and something is taken away, so that suspects can be tied to a crime scene by detecting these traces. Although forensic analysis has developed enormously since Bertillon and Locard, the three ideas they introduced—crime scene documentation, identification, and trace analysis—were a major advance in criminal justice. Unless there is evidence, no hypothesis is of any use and it is as if there had been no crime. Unless a perpetrator can be validly identified, and placed at the crime scene via unadulterated evidence, the case cannot be justly solved. These principles are also foremost in computer forensics.

Forensics is not by itself a science (*"forensic:* of, used in, courts of law"— *Concise Oxford Dictionary*). The term can describe any science, but more commonly applies to technologies of a science, rather than to the science itself. A forensic scientist will be an expert in, for example, gunshot wounds, organic poisons, or carpet fibers rather than in chemistry or surgery, as an FAQ from http://www.forensics.org explains:

> Forensic means to apply a discipline, any discipline, to the law. It is the job
> of forensics to inform the court. So, you can be a computer scientist, and if
> you apply computer science to inform the court, you are a forensic
> computer scientist. There are forensic specialities [...]: questioned

documents expert, profiler, medical examiner and coroner, anthropologist, blood spatter expert, DNA technician, ballistics expert, dentist, computer expert, civil engineer, auto crash investigator, entomologist, fingerprint expert, crime scene reconstruction expert.... .

Forensic specialties therefore can become obsolete along with their technologies. But in any case, other skills besides up-to-date expertise in a current technology are needed. A key skill in forensic computer science is the challenge that lies in "informing the court": not only knowing how the event might have happened, but also assembling event traces into acceptable legal evidence in a form that tells a complete and convincing story, without distorting any of it. This requires specialized expertise and training in a range of computing and noncomputing skills—legal knowledge, evidence management, data storage and retrieval, and not least, courtroom presentation.

While later chapters, especially Chapter 3, will return to the topic of law and the nature of legal evidence, it should be noted here that formal computer forensic methods are still in development, as is their status in court evidence. For example, the Daubert standard applicable in the U.S. courts [10] specifies that admissible expert evidence must satisfy strict criteria. Given that a witness can establish his/her personal standing in the discipline, for example via experience, publication and teaching, any expert evidence also needs to pass these tests:

‣ Any method and technique used to form the expert's opinion must have been tested empirically (i.e., able to be confirmed or refuted independently in repeat experiments, by other experimenters, and with different data);

‣ Methodology and techniques should have been subjected to peer review and publication, and should be accepted in the corresponding scientific community;

‣ There should be known error rates for methodology and techniques.

What has to be made clear in court is the operational detail, that is, how the observed result was achieved. The Daubert criteria focus on test techniques supported by scientific theory. For computer forensics, this is a central difficulty: there are no generally accepted tests per se, and to explain methods and theory is the equivalent of explaining how computers work. Every test individually reflects the interaction of the event and the entire system, and no two event sequences are exactly alike.

This last observation supports the argument that digital evidence presentation needs its own special standard, one that does not rely on Daubert-type criteria. Such a standard will have wide applications. Governments, businesses, and individuals require high quality digital evidence in many contexts, as much to pursue legitimate objectives as to frustrate illicit ones.

Figure 1.2 shows the complex influences creating layers of restrictions on employers, employees, and other users. The arrows denote responsibility pathways under legal and/or company restrictions of various kinds (i.e., where a potential for violating restrictions can occur). Digital evidence analysis can be applicable in any of these pathways. For example, users abuse their rights, organizational policies ignore legal requirements, or security enforcement inadequately captures security policy. Even organizational policy can be illegally framed, or framed in such a way that it contravenes overtly expressed organizational culture, but it might be that this state of things could only be proved through evidence retrieved from computers (e.g., e-mail evidence). Although not all these violations will result in court action, all may require a high standard of digital evidence to be resolved, and all could be candidates for computer forensics investigations.

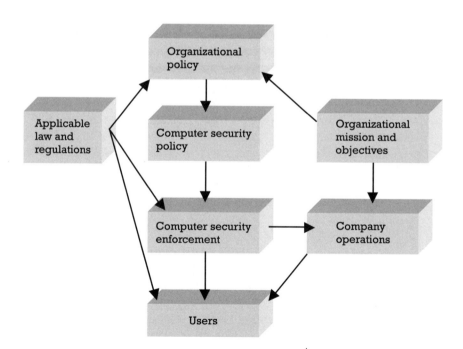

Figure 1.2 The organizational setting for digital evidence potential.

What Figure 1.2 omits is the emerging international framework for computer forensic investigations (see Chapter 3 and Section 1.5 for an overview) which, while it will promote faster investigations and better quality digital evidence, also potentially exposes users to multiple jurisdictions. An act that constitutes a computer crime in one country or culture may be acceptable in another. An event can be actionable in one country but not in another, so that international history is regularly being made as the first on-line defamation cases come to court. An example case [11] exploited national defamation law differences by winning the right to sue a U.S. on-line publisher in Australia, rather than in the United States where, it was claimed, defamatory material had originally been uploaded. The advantage to the complainant was that under U.S. law the case would have been less likely to succeed.

Evidence extracted from computer storage has been used in courtrooms since the 1970s, but in its earliest phase the computer was regarded as no more than a device for storing and reproducing paper records, which constituted the real evidence. Printed versions of accounting records were accepted as the equivalent of hand-kept or typed business records. Opportunities for computer fraud were limited to creative accounting, destruction or theft of equipment and such exploits as siphoning away cent division remainders. Computer evidence presented a challenge even in these limited conditions, as in some jurisdictions the workings of the system that produced it had to be explained in detail to the court. For example, under the U.K. Police and Criminal Act (PACE), Section 69 of which governed admissibility and weight of computer evidence, introducing computer evidence in a court case was not straightforward. The computer had to be certified as "operating properly" in the same sense as a device like a lamp or radar speed detector [12].

Forensic computing emerged in the mid-1980s, firstly because of the increasingly common cases of stolen or counterfeit hardware and software, a consequence of the escalating personal computer market; and secondly, because masquerading outsiders could now access mainframes remotely and anonymously. Viruses began to proliferate and mutate via local-area networks (LANs) and wide-area networks (WANs). Businessmen and the government began to show a greater interest in formalizing their computer security policies, and implementing these via suitable countermeasures. Many of these detection or prevention mechanisms produced, almost as a side effect, the raw material for computer forensics: computer-based evidence. The term *computer forensics* and the standardization of associated evidence-handling procedures began to gain acceptance during the late 1980s.

From Table 1.1, it can be seen that computer forensics as a standardized discipline arrives comparatively late in computer systems evolution. Only in

Table 1.1 Forensic Computing's Historical Context

Time	Technology	Computer Crime	Computer Forensics
1950	Transistors	None	
1960	Commercial applications	Local fraud	
1970	Silicon	*Insider* crime	
	10-baud lines	*Outsider* crime	
	Databases	Hacking	
	ARPANET		
1980	Personal computers	Violating security standards	
	Telnet	Stolen hardware	Local crime units
	LAN, WAN	Copyright violations	National crime units
		Viruses	
1990	Internet goes public	Online fraud	
		Web pornography	National task forces
	The Web	Cyberstalking	
		Web site hacking	
		Information warfare	Global task forces
		Identity theft	
		E-mail abuse	
2000		Corporate fraud	Training and
		Global terrorism	certification in computer forensics

the past few years, as Section 1.5 shows (and Chapter 3 discusses in more detail), have national and international organizations taken on the task of creating global frameworks for computer crime prevention, detection, and punishment. The following list of stakeholders, though incomplete, shows how rapidly potential applications for computer forensics and intrusion forensics are appearing:

1. *National security:* Initiatives such as the Clinton administration's National Infrastructure Project highlighted national dependence on information technology, and put the prospect of information warfare on every nation's agenda. Since the attacks on September 11, 2001, a sharper national security focus has emerged: as well as investigating past Internet-based attacks on information, a critical priority lies in discovering computer-based clues about planned real attacks.

2. *Customs and excise:* Customs agencies deal with potentially criminal importations. Examples include counterfeit software and hardware, or prohibited obscene materials in soft copy. When suspected pornography in digital form (e.g., an image buried in a computer

game) is seized, it is a nontrivial task to determine whether the images embedded in software actually contravene the law.

3. *Lawyers:* Counsel for both prosecution and defense can find themselves working with criminal cases where evidence is wholly or mainly computer-based.

4. *Civil courts:* These courts need to use computerized business or personal records in cases such as bankruptcy, divorce proceedings, or workplace harassment.

5. *Police:* Law enforcement agents will retain computer forensics specialists for advice on the extent of evidence to collect during a raid, and to analyze seized evidence during the investigation that follows.

6. *Businesses:* While they often prefer not to publicize internal offences, businesses will use forensic services to assemble evidence of breaches such as embezzlement, industrial espionage, stealing confidential information, and racial or sexual harassment.

7. *Insurance firms:* These firms can use computer-based evidence to establish complicity and fraud in accident or workers' compensation claims. Examples include e-mail evidence, phone records, or financial records.

8. *Corporate crime:* Such crime as the Enron 2002 bankruptcy case involves acts by the business entity as a whole rather than by individual employees. Such investigations look for evidence of deliberate policy implementation as well as of specific events. For example, according to report [13] the accountants and auditors for Enron not only used e-mail to communicate but also subsequently deleted these e-mails. Both the retrieved e-mail fragments and the evidence of intentional deletion would be of interest to investigators.

9. *International (transnational) crime:* Investigation of these crimes demands computer forensic analysis on a global scale. Drug cartels and other organized conventional crime entities increasingly resemble mega-corporations in their scale, complexity, business methods, and dependence on information technology.

10. *At the personal level:* It is now nearly impossible to find anyone who produces no computer-based trail and has no stake in its use as evidence. Such a person would have no bank account, no phone, and no personal computer; would pay no tax, receive no official income or state benefits. He/she would never own a car, travel by

plane, use a credit card, legally own a gun, buy a house, take out insurance, receive medical care, or work in any but the most basic industry.

Computer-based evidence now can be found almost everywhere, and almost everyone has a stake in its existence, even if not in its analysis. Computer forensics has a wide scope that needs an equally broad definition. For this book's purpose, we need something less procedurally oriented than this:

> Computer forensics involves the preservation, identification, extraction and documentation of computer evidence stored in the form of magnetically encoded information. [14]

but more generally applicable than this:

> Inforensics (Information Forensics) is defined as the application of forensic techniques to investigate crimes, involving either directly or indirectly, information and computer technology and information storage media. [15]

Computer forensics, we have established, can now be applied to investigate or prevent acts of enormous national and global importance. Increasingly, computer security is becoming national security, as Chapter 3 shows. Security policy at a national level is part of national defense policy and includes information warfare strategies, where variants of computer forensic techniques apply (Chapter 6 will discuss this interrelationship). Typically, a computer attack threatening the national information infrastructure is an asymmetric attack: a small enemy injures a powerful opponent through surprise and stealth. This introduces a real-time aspect to forensics. It is vital to know who and where your enemy is, and especially what the enemy is planning. Defensive information warfare is the process by which opposing sides use computer forensics to try to find out what the other side is planning or has planned, in order to thwart the plan or at least to mitigate its effects. Offensive information warfare launches computer attacks of its own on the enemy's information infrastructure. Both offensive and defensive warfare are tools that can be activated against any enemy, not only those threatening national interests. Hence, from a pro-active point of view, computer forensics is an activity carried out after, during, and before the crime occurs.

This is a very broad view of computer forensics, but it is the one this book needs to invoke. Our own definition takes into account the issues touched on in this section and expanded in later chapters, all of which deserve to be included:

Computer forensics is the identification, preservation, and analysis of information stored, transmitted, or produced by a computer system or computer network in order to reason about the validity of hypotheses which attempt to explain the circumstances or cause of an activity under investigation, in a manner intended to meet evidentiary requirements. There is a generally applicable and broader definition, which omits the evidentiary requirements. [16]

Intrusion forensics can now be perceived as one specialization of broad computer forensics:

Intrusion forensics is the recovery and analysis of information from a computer system or computer network suspected of having been compromised or accessed in an unauthorized fashion, information which includes host-based data and will typically also include communications traffic and payload data, with analysis also of information very possibly from other sources, for example, call records, personal digital assistant (PDA) flash memory contents, and business organizational structure, in order to allow investigators to reason about the validity of hypotheses attempting to explain the circumstances and cause of the activity under investigation, and possibly provide evidence to support litigation either criminal or civil.

See Chapter 6 for a full discussion of this and other intrusion forensics terminologies.

1.4.2 The nature of digital evidence

Evidence is what distinguishes a hypothesis from a groundless <u>assertion</u>. Evidence can confirm or disprove a hypothesis, so evidence reliability and integrity is the key to its admissibility and weight in a court of law. There are several special characteristics of digital or computer evidence, and of the computer systems and proprietary and public networks involved, that make evidence interpretation especially challenging:

1. *Too many potential suspects:* With traditional offences, the offending act or event is usually manifested—there is a corpse, a theft or at least a complaint to work with. Usually, as well, there is a starter list of potential suspects: Who knew the victim? Who had physical access to the scene? Who had a motive? The Internet, due to its user friendliness and anonymity, can offer some 700 million suspects according to the connectivity figures mentioned earlier.

2. *Identifying the crime:* In computer crime and in computer-related or evidenced crime, the nature of the event is often less obvious and immediate. For example, when a hacker steals confidential information, victims may not find out what has been stolen unless informed by the system administrators, who in turn may not notice until long after the hacker has gone. Identity theft, described as "the fastest-growing financial crime in America and perhaps the fastest-growing crime of any kind in our society" [17], may take years to be exposed.

3. *Too much potential evidence:* In computer forensics, particularly in its inclusive meaning (i.e., not only "computers as target") it may be necessary first to hypothesize tentatively, then progressively refine this hypothesis as the investigation proceeds. Hence, as Chapter 2 discusses in more detail, it can be difficult to decide how much at the scene is actually the evidence. Just as investigators may not know initially exactly what the crime is, they may be unsure of a virtual crime scene's limits. "Freezing the scene" in a home office may only reveal part of a global scenario.

4. *The evidence is easily contaminated:* Traditionally, evidence at the scene is sent for independent forensic laboratory testing while investigators pursue their enquiries elsewhere until the results come back. But in computer forensics, all investigatory aspects—naming the crime, identifying the perpetrator, following the evidence trail, and constructing the modus operandi—use the same digital analysis techniques. Hence, computer forensic handling is especially vulnerable to errors. Just as blood samples or fingerprints can be contaminated at the scene, digital evidence can be damaged during collection unless strict procedures are followed. Rebooting a system, for example, immediately changes the system state and destroys possible traces.

5. *Contaminating some evidence may ruin it all:* Each item of physical evidence is only a single component of the case, often an independent one, and the prosecution might still succeed without it. In contrast, digital evidence is highly interconnected. Proving a hypothesis— what, how, and who—means recreating the scenario step by step in a time-line where invalidating a single step means the loss of the entire case. To achieve this, an investigator needs to preserve the integrity of evidence contained in seized devices and material. Integrity refers not only to physical (bit-level) copies of

information but also to structures such as indexes and directories used to turn the data into information, and to the software used to view it.

For these reasons, law enforcement agencies stress correct "search and seizure" procedures, as discussed in Chapter 2. As with physical evidence, the investigation will have to demonstrate that the digitally recorded evidence was correctly collected, that all relevant traces were collected, and that it has remained uncontaminated ever since (i.e., the *chain of custody* has been maintained).

It is clear from the list of U.S. criminal convictions given earlier that in future fewer and fewer digital crime investigations will begin and end with seizure of a single computer in one location. Further, the scope of evidence collection is likely to go beyond a single state or national jurisdiction. With this in mind, the following proposed set of *International Principles for Computer Evidence* was developed in 1999 on a G-8 initiative:

> Upon seizing digital evidence, actions taken should not change that evidence;
> When it is necessary for a person to access original digital evidence, that person should be forensically competent;
> All activity relating to the seizure, access, storage or transfer of digital evidence must be fully documented, preserved, and available for review;
> An individual is responsible for all actions taken with respect to digital evidence while the digital evidence is in their possession... [18]

This extract is one example of international pressure to harmonize computer forensic procedures and laws in the face of accelerating global computer crime (see Section 1.5 for a summary of the national and global legal frameworks).

1.4.3   Retrieval and analysis of digital evidence

While seizure of digital evidence needs carefully controlled procedures to avoid contamination, it is often unclear exactly what needs seizing. Further, examining seized evidence uses a variety of analysis techniques depending on the suspected offence, the applications used in it and the suspected modus operandi. This section introduces the range of tool types used in computer forensics and explains briefly why they are important. A detailed discussion of specific tools appears in later chapters.

The generic computer consists of one or more central processing units for executing code, a memory module for dynamic working storage, a backing store or disk, and input and output devices (Figure 1.3). During execution, the state information describing what the computer is and has been doing, is distributed across several components:

- The *central processing unit* is executing its current operation on data in dynamic storage; its working registers show what the operation is doing, and with what.

- *Dynamic storage* contains fragments of the operating system, applications currently being executed, and pages of temporary data—about to be retrieved from or written to the backing store.

- The *backing store* (such as fixed or removable disks, tapes, and CDs) contains *persistent* data—files retained from one session to the next.

- *Network components* record, for example, data packets sent, phone connections, router tables accessed.

As later chapters explain more fully, the proliferation of laptops, special purpose devices, and embedded systems means that the basic computer function described above can be represented in a variety of technologies. However, computer forensics is interested not so much in function itself as in the form in which past functional information has been stored, because this will influence methods of evidence collection. Hence, a forensically useful

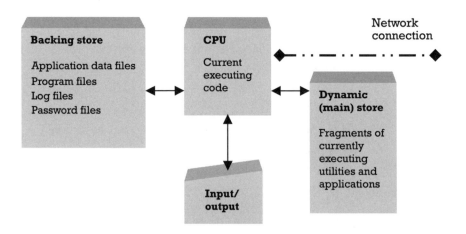

Figure 1.3 Sources of evidence on a basic desktop computer.

way of classifying information sources, according to Bates [19], is by *information store* type:

> ▶ *Temporary* form, powered by an external source without which the information ceases to exist: typically RAM;

> ▶ *Volatile* form, with an internal power source such as a battery (again, the information is lost if the battery is removed, as in CMOS or the RAM in a battery-powered laptop);

> ▶ *Semipermanent* form, as with persistent but changeable data: diskettes, hard disks, flash memory;

> ▶ *Permanent* form, such as ROM (Bates notes that CD-ROM and other WORM technology no longer qualifies as permanent, since data can be altered).

All these "forms" can provide forensically relevant clues—their presence or absence may be as much of a clue as actual content, as with log files which ought to exist but have disappeared. Consequently, an investigator needs to know how to handle all forms appropriately. Most forensic digital information content will be located in semipermanent form. Analysis of an entire network adds complexity of several orders of magnitude.

Accordingly, any computer-related criminal investigation is faced initially with the need to reduce a potentially large information search space to manageable proportions. The information space can be very large indeed, since it comprises both the initial information set of direct source evidence (e.g., files) and also any other metainformation or metadata generated during the investigation (e.g., file timestamps, file author, and file permissions). During the initial phase of an investigation, the total amount of information processed by the investigator will significantly exceed the imaged case evidence. However, as the case progresses and more forensic tools are used, the amount of information investigated will rapidly be reduced to manageable levels. The information reduction process usually needs iterative filtering in order to produce a shortlist of suspicious activities, so that the oscillating pattern of Figure 1.4 can converge towards a solution. During this phase, it is important not to filter out any relevant data nor, if possible, generate too many irrelevant data.

Each iteration of the forensic filtering process (generally performed by one or more applications of a forensic tool) generates some additional case data but, at the same time, reduces the overall information used in the investigation. The reduced set of information can subsequently be used to perform more in-depth analysis in the next iteration. The final set of results

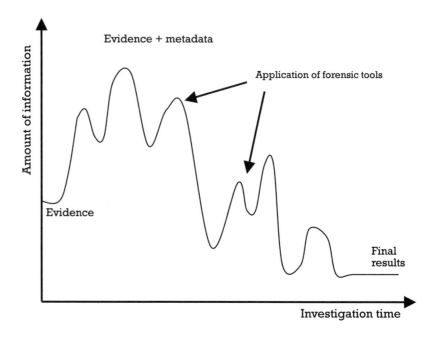

Figure 1.4 Amount of information produced during an investigation.

will be the one required for the successful (or not successful) prosecution of the perpetrator of the crime. The shape of the information–investigation curve will ultimately depend on the investigator, as will the choice of the sequence of forensic tools, and the type and quality of data generated at each step. The operational model for computer and intrusion forensic investigations is summarized in Figure 1.5.

Later chapters will discuss specific tools in detail, but a typical forensic computing toolkit could include the following range for different investigatory phases:

> ' Tools for imaging and information-capture at the scene, or subsequently in the laboratory;

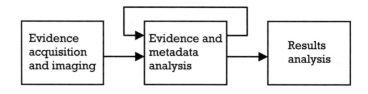

Figure 1.5 Operational model for computer and intrusion forensics.

- Tools for post-seizure analysis;

- A more general class of analysis tools supporting *hypothesis development*, including application-dependent tools;

- Finally, *investigation management tools* for producing a clear, complete reconstruction, demonstrating not only that the original evidence is clean, but also that the analysis procedure was properly and securely carried out.

If the investigation of an offence leads to a court case, the forensic specialist needs to be in a position to explain how and why tools like these are used. In any case, he/she must be prepared to vouch for and possibly to demonstrate the integrity of all these aspects of investigation:

1. *Collection:* to describe processes by which the evidence was gathered, showing that the collection process does not alter it;

2. *Chain of evidence:* to show that the evidence remained uncontaminated after it was gathered, and during analysis;

3. *Authentication:* to show the evidence is unaltered in any way from its state on the original computer, typically with file signatures;

4. *Recovery:* to explain how deleted files and file fragments are recovered, what the system logs, swap files and temporary files contain, and how the perpetrator's actions can be inferred from these;

5. *Verifiability:* to confirm that these inferences are standard, and can be confirmed by an independent third party's analysis.

Hence, a forensic specialist needs not only a detailed knowledge of operating systems, file formats, popular applications and computer security but also needs to know how to maintain the validity of the evidence, and in particular how retrieval and analysis tools work. However, the evidence in a single desktop is less useful without corroborating evidence from other sources. The scope of the investigation will normally extend to mainframes and service providers, where the amount of obtainable evidence depends on other distant computers.

1.4.4 Sources of digital evidence

At the scene, investigators must decide how much to take away from any part of a network involved in an offence, but how do they know what to seize? A basic desktop machine can now have many gigabytes of disk storage, far too much to print. If they do not take enough, they jeopardize the prosecution case; if they take too much, not only may they frustrate their own analysis but they may also exceed the terms of their own search warrant. An example of this dilemma emerges in the following extract from the U.S. Department of Justice search manual [20]:

> ...computer hardware, software, and electronic files may be important to a criminal investigation in two distinct ways: (1) the objects themselves may be contraband, evidence, instrumentalities, or fruits of crime and/or (2) the objects may be used as storage devices that contain contraband, evidence, instrumentalities, or fruits of crime in the form of electronic data. Rule 41 of the Federal Rules of Criminal Procedure permits the government to search for and seize computer hardware, software, and electronic files that are evidence of crime, contraband, instrumentalities of crime, and/or fruits of crime.

Hence, software, hardware, or information may be seized if they are

- Contraband, something which should not be in anyone's possession (e.g., an illegally manufactured TV decoder);

- Fruits of crime (i.e., something that should not be in the suspect's possession, such as stolen copyright software);

- An instrumentality that may be used in committing a crime, such as a password-cracking program;

- Evidence of crime: If the computer's role in an offence is not certain, almost anything digital could be evidence—yet how else to establish the role except by seizing and analyzing it? Not the least of computer forensics skills is a talent for excluding many unlikely hypotheses on the spot.

While later chapters present more details on specific crimes and types of related evidence, there are several readily apparent sources at the average scene. Fixed or removable hard disks and the semipermanent storage referred earlier contain current files and file data, deleted file fragments, and application and operating system temporary files that reveal potential clues such as chat room participation, Web-surfing, or e-mail in and out. Analysis of the latter can recreate a partial recent history. Diskettes at the scene, matched with patterns of hard disk activity, indicate what the suspect was doing, and can even indicate attempts to conceal it. Printed records can be useful as well, although often only as confirming evidence, since the average desktop laser output shows little to identify its product. Depending on operating system or network settings, there will be more or less evidence in system logs showing processes started and stopped, dates and times; and audit trails recording actual or attempted file accesses and updates. Generally, standalone household desktops hardly use logs and audit trails, and even those used are easily modified. However, once the desktop is connected to the Internet, the event involves an ISP and a telephone connection, where other, less corruptible records of logons, file downloads, and line usage can be obtained. Unassailable cases cannot be built from a single digital evidence source.

Clearly, a broad interpretation of [20] would yield vast quantities of data, most of it useless, and would involve shutting down or crippling operations in many network nodes. This is why forensic specialists rely on specialist toolkits to perform on-scene information filtering.

1.5 Legal considerations

While subsequent chapters will discuss in detail legal aspects of computer forensic practice, this section overviews recent trends in the law's attitude to digital evidence. There are two distinct foci in legal concerns. Of these, the more formal relates to "search and seizure" by law enforcement officers as mentioned in Section 1.4.4 and discussed in detail in Chapter 2. This focuses on the question of what investigators can legally seize, without jeopardizing a prosecution case either by missing crucial evidence or by violating the civil rights of an individual. Here, we are speaking of cases where a specific crime is suspected, and procedures are dictated by the immediate jurisdiction(s) of the investigation.

Until recently, applicable law in most western countries reflected a piecemeal approach. Computers can play a role in almost any crime, and as the incidence of prosecutions based on computer or electronic evidence began to rise, gaps in existing law became apparent. Legislative terminology describing wiretaps, voice recordings or paper documents is not easily adapted to computer crime. Digital child pornography is a case in point: how to apply the law framed for the seizure of printed material? What constitutes possession, when suspects can claim they had not intended to download the graphic? Should an ISP be held responsible for helping pedophilia? Publicity about such Internet-based cases raised community support for computer evidence law reform in many countries, for the reason that unlike hacking or spoofing, the reform's intentions were well understood and popular, an issue politicians could articulate without the need to understand technical details. Likewise, with data privacy: the average person can readily grasp the idea of personal information being collected by stealth, or used for unacknowledged purposes.

What is not so readily accepted by the public is the paradox that, in the electronic age, protecting privacy and solving computer-related crimes are almost contradictory activities. Law enforcement agencies now need global access to computer-based records for activities like banking, travel, e-mail, and phone. From the individual's point of view, there are dangers to civil rights in such techniques as electronic surveillance, techniques for mining mass personal electronic data, and cultural profiling. Law enforcement and national security agencies have a different agenda, being less directly concerned with civil rights. From their perspective, the point is to streamline computer crime and terrorism investigations through police and government cooperation, and to harmonize computer crime laws by international agreement. National law and international agreements should therefore reflect a special characteristic of digital evidence—that voice, text pictures, and sound are now all stored in the same form, transmitted globally in seconds, and may also be encrypted to baffle detection. A particular concern is to have ISPs act as proxy evidence collectors, and to encourage or if necessary enforce their cooperation on a global scale.

For example, the U.K. Regulation of Investigatory Powers Act [21] as well as extending the power of companies to tap internal phone calls and e-mail, can require ISPs to track data traffic. With encrypted messages, refusal to supply the decryption key is punishable. A significant international agreement now exists in the Council of Europe treaty on cybercrime [22] signed in 2001 by 32 countries including the United States and the United Kingdom. The treaty binds signatories to enact mutually supportive computer crime laws using similar definitions and procedures:

> The Convention is the first international treaty on crimes committed via the Internet and other computer networks, dealing particularly with infringements of copyright, computer-related fraud, child pornography, and violations of network security. It also contains a series of powers and procedures, such as the search of computer networks and interception . . . Its main objective, set out in the preamble, is to pursue a common criminal policy aimed at the protection of society against cybercrime, especially by adopting appropriate legislation and fostering international cooperation. (from the *Summary*)

The public profile of computer forensics rose dramatically after the events of September 11 in the United States, as it was reported that hijackers had used e-mail to communicate plans, and that large quantities of incriminating files had been discovered on seized computers, or on machines abandoned in the Afghanistan war zone. Typical of the computer forensic-enhancing powers consequently enacted in different countries are the U.S. Patriot Law and the U.K. Anti-Terrorism, Crime, and Security Bill. These and similar developments in law and computer forensics demand continued serious debate by an informed public.

1.6   Computer security and its relationship to computer forensics

While computer forensics as an occupation is a recent invention, it has several related precursors in information technology. These parent disciplines include not only the theory and practice of hardware, software, and programming but also notably computer security, which has matured alongside the information technology it aims to protect, not only against attacks but also against errors and accidents.

There are two important observations that come to mind regarding this relationship between computer forensics and computer security. First, remote networking and on-line business had, after 1991, made it increasingly difficult to establish the scope of a security policy. Earlier, it was possible to draw a boundary around operations, to distinguish between *insiders* who were permitted to use the system and *outsiders* who were not, and to focus on building the boundary securely. This distinction has become meaningless as businesses and government rushed to establish an Internet presence. Now, focus has shifted to role enforcement, where permitted roles are described in terms of a hierarchy of access rights. Second, risk assessors have always warned that it is exactly these insiders acting in inadequately monitored roles who will probably commit the greater part of computer crime—up to 80% in some estimates.

The objective of computer security in its widest sense is to preserve a system as it is meant to be—specifically, as its security policy dictates. Computer forensics—or at least intrusion forensics—sets out to explain how the policy came to be violated, a process that may reveal fatal gaps in the policy itself. There are several reasons for this security–forensics discrepancy:

> ▸ Historically, computer security has been seen as the costly antithesis of good performance. Business objectives such as transaction speed and easy access traditionally conflicted with time-consuming countermeasures like logging or encryption/decryption.

> ▸ In the global framework that this book addresses, "security policy," as Section 1.1 points out, is really several layers of policy. When planning policy, a security manager needs to consider not only the owners' protection rules but also the requirements of national or even international laws. In a climate of close shareholder scrutiny, it is not so easy now for businesses to deal with breaches internally.

> ▸ As technology changes fast, systems are never technologically stable for long, often going through periods of frenzied expansion and overnight obsolescence, a particularly marked phenomenon during the recent e-commerce and dot-com surge. Since computer security enacts in technology a set of nontechnological concepts and rules, it can never be demonstrated for certain whether a policy is being enacted all the time, everywhere.

This section outlines the foundations of computer (network) security which are also relevant to computer and intrusion forensics. The section summarizes the basics of network communications security, and overviews mainstream computer security policy development and implementation, emphasizing the aspects most useful in computer and intrusion forensics.

1.6.1   Basic communications on the Internet

The system in Figure 1.3 describes only the simplest of desktop computers, yet any number of these devices takes part in the network of networks formed by the Internet. To understand how network security is relevant to computer forensics, we need first to understand how networks behave. A network is no more than a set of interlinked computers. Home users typically connect to the Internet remotely, while workplace users can get access via their own LANs or WANs.

1.6.1.1   Remote connection

A user wishing to access a remote computer will typically connect via a workstation or personal computer to that remote host through a phone line and dial-up modem. To do this legitimately, the user needs an account with an ISP. The ISP can be an independent service, profitable or nonprofitable; or the service can be offered by the user's employer or a special interest group. The important point here for computer forensics is that, once the connection is made there are two evidence streams instead of one.

1.6.1.2   Local-area network

A LAN connects several or many small computers located fairly close to each other, typically in the same building, with the objective of sharing files and other local services such as printing, a diary, or a business database. It is common to have one LAN connected to another, which may be connected in turn to still others. Desktop computers on a LAN usually have private local filestore and software as well as network access. The typical LAN configuration shown in Figure 1.6 contains several *server* computers for which the desktops are *clients,* issuing requests for the server's attention (e.g., for printing).

Within a LAN, the nodes communicate via multiple *protocols,* common languages ensuring that the packets of information making up a message are delivered to the right destination. A LAN may be freestanding, or may be connected on a backbone to other LANs in the same business. If one of the services to which a LAN or the backbone connects is an Internet service, this connects the individual desktops to the network of networks that is the Internet, and to Internet services such as e-mail, the Web, newsgroups, and chat rooms. These are the major applications that have attracted so many new users.

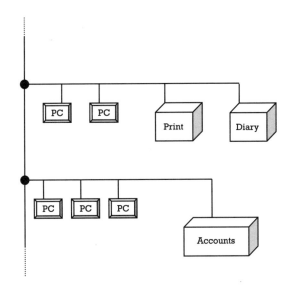

Figure 1.6   A LAN.

1.6.1.3 Communicating on the Internet

Communicating on the Internet is layered as follows:

- The application layer for delivering messages in an application-specific protocol, the most common being HTTP (Web), SMTP and POP (mail), NNTP (news), FTP (file transfer), and Telnet (terminal connections);

- The transport layer, where TCP (usually) controls the disassembly into packets of a complete message, and vice versa on the return journey;

- The network layer, where IP turns message packets into datagrams (and vice versa), translating IP addresses into the appropriate physical address for transmission in the underlying network;

- The data link and physical layers, which control how packets are sent and received by signals on the communication channel, using protocols such as Ethernet.

As there are many possible routes between client and server, layered message handling promotes reliable message delivery, failsoft network layout (through redundancy), and modular expansion. However, a key problem for computer security and now for computer forensics is that it multiplies the points at which electronic evidence of misuse can be located.

Consequently, an investigation may have to find evidence from many nodes: source desktops, service providers (including the providers' authentication servers, routers, and application servers), domain name servers (DNSs, multiple worldwide servers mapping hostnames, such as http://www.first.qpt.edu.au to IP addresses, such as 24.53.13.12, and vice versa). These can physically be located in several different countries, and evidence either may not exist or has not been collected in a sufficiently detailed form. For example, several known attacks can be made at IP level because it does not provide a reliable method of authenticating that a packet comes from its putative source. These include IP spoofing—persuading a host to accept a packet with a false but acceptable IP address—and the IP session hijack, where the attacker takes over an existing legitimate session by persuading the user that the host has dropped out. Neither of these attacks could be solved using IP level evidence alone.

1.6.2 Computer security and computer forensics

The mission of computer security is to guarantee that computer-related assets (e.g., hardware, software, files, services) remain in an appropriate state of assurance, as demonstrated by

- *Confidentiality:* that records, business intelligence, proprietary software, system state data, and other digital assets are not revealed to the wrong user or to an agent masquerading as an approved user;

- *Integrity:* that digital assets remain in the state expected, representing a true record of actual events;

- *Availability:* that system usage is available when needed, providing approved access to digital assets;

- *Authenticity:* that all events taking place are the intentional, correct result of actions sourced to approved users or their agents.

The concept of authenticity is related to that of integrity. Message authenticity requires that the sender is the purported sender, that the message was not intentionally or accidentally corrupted, and that message metadata such as date and time are also a true record.

Security attributes can be compromised in many ways, by accident or design. Policies are composed of many different rules aimed at preserving different facets of each attribute; for example:

- The response time for ATM balance enquiries should be no more than 10 seconds (availability).

- Employees should check for viruses daily, and should not download alien software from the Internet (integrity).

- A student's course results may only be viewed by the student or by course administrators (confidentiality).

- Only bonafide customers may log on remotely (authenticity).

As an earlier section argued, these rules are a mixture of safety, business, legal and ethical rules, and are designed for people, not networks. The rule base tends to accrete over time, as externally imposed restrictions and internal objectives change. Hence, it is difficult enough to verify policy completeness and consistency at the policy level, let alone at implementation level after rules and rule enforcement have been translated into software and hardware measures.

Each of the above rules could be implemented via several counter-measures, and each countermeasure can counter several threats. While redundant protection may seem like a good idea, the system's owners have cost-benefit in view and usually want the least expensive countermeasure set that satisfies system security performance objectives. This means that there will always be limits to the total amount of system assurance, and the unassured areas are at risk. Owners can also choose among different protection modes:

- Preventive countermeasures, such as strong access control or centralized virus scanning, that aim at zero loss;

- Mitigating countermeasures intended to reduce the loss to an acceptable level (e.g., virus scanning at the individual desktop level);

- Transferring countermeasures that move the loss to a third party (e.g., by insurance);

- Recovery countermeasures restore the system to a previous correct state after the event.

To further complicate things, many countermeasures serve more than one role: as well as protecting, they may support other nonsecurity functions such as performance monitoring and recovery. Audit trails for database accesses, for example, can double as recovery countermeasures when the database is corrupted. Session logs can be used both for authentication and for charging.

There is a degree of overlap between the raw material of computer security and that of computer forensics, but the two functions have different and sometimes opposing aims. Many security measures would, if fully implemented, support computer forensic function: event logs, database access logs, exception logs, attempted physical or computer accesses are only a few. Yet normal practice is to apply only minimal logging levels because of performance overheads. Logs are also routinely rolled over and deleted to save on persistent storage, so that continuity for forensic data mining is lost.

Some countermeasures such as file signatures also find application both in computer forensics and in intrusion forensics. System log files secured by these means can be produced as evidence that the log was tamper-proof before seizure. Countermeasures like card-controlled access, user passwords, or limits on the number of login attempts all leave digital evidence for later analysis. However, there are situations where recommended security practices work against the interests of computer forensics. Many other countermeasure types are based on cleaning up the system environment to prevent users from using unauthorized resources or information, to remove the consequences of unauthorized activity as soon as possible, or simply to facilitate system housekeeping. Evidence-destroying countermeasures like these include disk defragmentation, virus cleaning, storage flushing, or even waste paper shredding (see [23] for a full discussion of security counter-measures and their relation to computer crime).

In summary, the type of computer security risk management and threat evaluation commonly prescribed provides effective protection if properly carried out, but because its main focus is prevention rather than investiga-tion, crucial forensic data is likely to be incomplete or missing. Com-puter security's cost-benefit approach relies on evaluating the information resources underpinning an organization's mission, and then constructing a cost-appropriate safety net. A computer forensics investigation, aiming only to prove whether something happened or not, has no interest in the size of the organization's loss. Hence, while security measures such as logs may pro-vide the best available system evidence, they are often likely to be inade-quate for forensic purposes.

1.7   Overview of the following chapters

In this chapter, we introduced a fundamental theme for this book, one that assumes a broad understanding of the term *computer forensics* and so takes in all investigations of criminal, offensive, or simply unwanted behavior, whether past or planned, where these leave trails of computer-based evidence. We also introduced a special computer forensics subdiscipline that is treated at length in Chapter 6. The chapter surveyed the ever-increasing range of relevant offences and highlighted a need for more and better computer and intrusion forensics. We briefly reviewed computer forensics' stakeholders. The chapter outlined basic concepts of computer evidence, networks, and computer security underpinning computer and intrusion forensic analysis. Technical detail on all these topics has been kept to a minimum and will be expanded where necessary in later chapters.

In Chapter 2, we review current practice in computer forensics. Here, the focus is on the technical detail of digital evidence—where it is found, its retrieval and processing, its presentation, and its legal framework. The chapter discusses computer-related evidence in detail from a law enforce-ment point of view: problems associated with establishing the *chain of evidence*, taking unchallengeable copies of evidence through disk imaging, some common tools for analysis and hypothesis testing, and recovering deleted or damaged data. The chapter ends with a review of major national and international developments that have influenced the development of computer forensics, and will continue to influence it in the future.

Chapter 3 begins with an introduction to the origin and development of computer forensics practice, with specific reference to its application in law enforcement and, by extension, in national security. This chapter is a detailed account of the role of the computer forensic examiner in law enforcement cases, showing how the technical information covered in Chapters 2 and 3 is applied in real situations, and supplying much practical advice on the conduct of a case. Typical elements of computer forensics toolkits, such as free and proprietary software, are reviewed along with a range of other resources. It describes how law enforcers acquire formal computer forensics certification, and discusses a range of cases under various jurisdictions, ending with an outline of the computer forensic role in national security and infrastructure protection.

The focus of Chapter 4 is the forensics of financial fraud, where auditors and forensic accountants play a key role. These cases can be especially lengthy, as their techniques require an auditor's skills and patience in tracing complex chains of transactions. As large amounts of data usually need to be sifted, the analyst needs systematic methods over and above the court-directed procedure of the typical hacking or net pornography case. As well as the range of techniques surveyed in Chapter 2, fraud detection has its own set of tools and techniques. The chapter surveys a range of internal and external fraud types, and discusses fraud detection tools, in particular data mining, statistical concepts, link or relationship analysis, and time-lining, before describing a staged model for fraud detection.

Chapter 5 discusses several in-depth case studies to show how the tools and techniques of computer forensics (including intrusion forensics) are applied in practice. It presents detailed accounts of several crimes or attempted crimes, including an instance of organized crime, an intrusion case, the Melissa virus attack and the bombing of the New York World Trade Center in 1993.

Chapter 6 is concerned with network systems, intrusion detection, and intrusion forensics, focusing specifically on illegal activity targeting a computer system, rather than use of a computer to carry out a noncomputer crime. Hence, it focuses on the central support methods of intrusion forensics—log and audit trail analysis supplemented by computer security expertise. The resulting information might not qualify as court evidence but is vital in establishing and testing hypotheses while an investigation is in progress.

Finally, Chapter 7 reviews emerging new directions for computer forensics. One particular interesting development relates to establishing standards (and thus enhanced legal status) for the forensics tools of the future. As previously noted, an investigation faces major obstacles, not the least of which is the sheer amount of data to be examined for clues. The chapter describes a range of emerging techniques for forming and pursuing hypotheses. These include text, software and image authorship, advanced pattern detection, and methods for retrieving evidence that has been hidden through encryption or steganography.

References

[1]   InterGov survey, http://www.intergov.org/public_information/general_information, visited Jan. 2002.

[2]   *U.S. Department of Justice Financial Year 2001 Budget Requests*, http://www.usdoj.gov/jmd/2000-budget/fy2001.htm, visited Jan. 2002.

[3]   Information Technology Association of America (ITAA) and Tumbleweed Communications Corporation, "Americans show Net security concern," http://www.itaa.org/infosec/121101survey.htm, visited June 2001.

[4]   *The Electronic Frontier: The Challenge of Unlawful Conduct Involving the Use of the Internet* (March 2000), http://www.usdoj.gov/criminal/cybercrime, visited April 2002.

[5]   Casey, E., *Digital Evidence and Computer Crime: Forensic Science, Computers and the Internet*, San Diego, CA: Academic Press, 2000.

[6]   "Computer Crime and Intellectual Property Section of the Criminal Division of the U.S. Department of Justice," http://www.cybercrime.gov/, visited June 2002.

[7]   "Elements for the Testing of Internet Investigators," *IOCE 2000 Conference*, Rosny-sous-Bois, France, Dec. 2000. See also: http://nfstc.org/ioce/Draft Files/Internet Evidence Elements for testing.doc, visited Sept. 2002.

[8]   Evans, C., *The Casebook of Forensic Detection*, NY: Wiley, 1998.

[9]   Chisum, W. J., and B. Turvey, "Evidence Dynamics: Locard's Exchange Principle and Crime Reconstruction," *Journal of Behavioral Profiling*, Vol. 1, No. 1, Jan. 2000.

[10] Brodsky, S. L., *The Expert Expert Witness: More Maxims and Guidelines for Testifying in Court*, Washington, DC: American Psychological Association, 1999.

[11] http://www.adlawbyrequest.com/international/NetLibel010702.shtml, visited Jan. 2002.

[12] Sommer, P., "Digital Footprints: Accessing Computer Evidence," *Submission to the Criminal Courts Review*, March 2000, http://www1.bcs.org.uk/DocsRepository/00500/505/footprints.htm.

[13] "Enron Bankruptcy Case Highlights E-Mail's Lasting Trail," *Computerworld*, Jan. 21, 2002.

[14] Lunn, D., "Computer Forensics: An Introduction," http://rr.sans.org/incident/forensics.php, visited Feb. 2002.

[15] TNO Physics and Electronics Laboratory, "Inforensics: Information Forensics, Reconstruction and Recovery," http://www.tno.nl/instit/fel/intern/wkisec17.html, visited July 2002.

[16] *Computer Forensics Past Present and Future*, Technical Report for DSTO, ISRC QUT 1999.

[17] Hoar, S. B., "Identity Theft: The Crime of the New Millennium," *United States Attorneys' USA Bulletin*, Vol. 49, No. 2, March 2001.

[18] Scientific Working Group on Digital Evidence (IOCE), "International Principles for Computer Evidence," *Forensic Science Communications* Vol. 2 No. 2, April 2000, http://www.ioce.org/2000, visited May 2002.

[19] Bates, J., "Fundamentals of Computer Forensics," *International Journal of Forensic Computing*, Jan./Feb. 1997, http://www.forensic-computing.com/archives/fundamentals.html, visited March 2002.

[20] *Searching and Seizing Computers and Obtaining Electronic Evidence in Criminal Investigations*, http://www.usdoj.gov/criminal/cybercrime/searching.html, visited Jan. 2002.

[21] *Explanatory Notes to Regulation of Investigatory Powers Act* (2000), http://www.hmso.gov.uk/acts/, visited Feb. 2002.

[22] Council of Europe Convention on Cybercrime, Budapest 23.11.2001, http://conventions.coe.int/Treaty/EN/projets/FinalCybercrime.htm, visited April 2002.

[23] Icov, D., K. Seger, and W. VonStorch, *Computer Crime: A Crimefighter's Handbook*, Sebastopol, CA: O'Reilly & Associates, 1995.

Current Practice

Contents

2.1 Introduction

2.2 Electronic evidence

2.3 Forensic tools

2.4 Emerging procedures and standards

2.5 Computer crime legislation and computer forensics

2.6 Networks and intrusion forensics

References

2.1 Introduction

This chapter provides a perspective on current practice in computer forensics and the influences that shape it, and will continue to shape it into the future—the technology and the regulatory context, both national and international, in which this plays out. We focus largely though not exclusively on current and emerging practice relating to standalone computer systems. The forensics of network systems and of network intrusions is treated separately in Chapter 6. Chapter 3 addresses some of the topics of this chapter in more detail in the context of law enforcement and national security.

Section 2.2 focuses on electronic evidence, what it is and how it may be analyzed. The section also focuses on disk and file imaging and subsequent analysis, and includes discussion of the forensic examination of two other essentials of modern life: the mobile phone and the PDA. We then examine the nature of modern forensic toolsets in Section 2.3, and in order to provide the reader with a point of reference, present a detailed description of three such toolsets: the very popular EnCase from Guidance Systems, ILook Investigator, and Computer Forensics Investigative Tool (CFIT)—a sophisticated experimental system currently under development, from the Defence Science and Technology Organization (DSTO) of Australia. Section 2.4 identifies current and emerging procedures and standards in the field of computer forensics and discusses some of the most important national and international guidelines and resources relating to digital evidence and how to deal with it.

Section 2.5 then examines the national and international legal and regulatory framework within which computer forensics operates. It does not attempt to provide a comprehensive navigation chart of the law; rather it highlights a number of recent developments with legislative and procedural implications for computer crime and computer forensic investigation:

1. The Council of Europe Convention on Cybercrime of 2001 [1].

2. The nature of the U.S. Carnivore network surveillance system, and its use by the FBI [2].

3. The Regulation of Investigatory Powers Act (RIPA) in the United Kingdom [3].

4. The U.S. "Provide Appropriate Tools Required to Intercept and Obstruct Terrorism" (Patriot) Act [4].

5. Data retention legislation passed recently by the United Kingdom [5] and the EU [6].

2.2 Electronic evidence

The complexity of computer forensic cases and the computer evidence upon which they rely has increased significantly over the years with the increased sophistication of standalone computer systems and increased use of the Internet. The Internet provides both a framework for increasingly sophisticated applications that are themselves vulnerable to computer crime (e.g., e-commerce applications) and a communication channel by means of which crime that is otherwise not computer-related can be planned, managed, or facilitated.

Consequently, while computer forensics has in the past tended to rely upon computer evidence that consisted essentially of standalone disks or disk files, more recent cases have relied increasingly upon electronic evidence gathered from a variety of sources. For example, the seminal article on admissibility of computer evidence in hacking cases by Peter Sommer [7] points out, that in addition to hard disk information belonging to the "Datastream Cowboy," there were five other streams of digital evidence relevant to the case. These included telephone logs, ISP logs, and network logs. Nonetheless, evidence obtained from the individual magnetic disks of individual personal computers continues to be important and often crucial and it is evidence of this sort and the procedures that have evolved to seize, manage, and analyze such evidence that we address in this chapter.

As a result of the continued use of magnetic media by the vast majority of both mainframe and personal computers, there is a well established and growing service industry in the areas of data recovery and electronic evidence discovery from magnetic media. The former is focused upon the restoration of information from deliberately or accidentally damaged media, the latter upon the recovery of information from storage media for investigative purposes, either for forensic purposes or for in-house purposes that may or may not lead to litigation. The services usually extend to other mass storage media, such as magnetic tape and optical technologies. The vast bulk of both investigative and restoration cases involves magnetic disk media. This reflects the reality that magnetic disk continues to be the medium of choice despite inroads for specialist purposes by magnetic tape and optical media. Technology trend data indicates that this will continue due to cost factors and anticipated progress with respect to bit densities [8].

Finding sought after information from a disk may be as simple as locating application files on a disk by name and then verifying their contents. Alternatively, such information may need to be extracted fragment-by-fragment from system or unused areas on the disk and then carefully pieced together. The term *ambient data* is used to refer to such normally inaccessible areas; we return to this term later in the chapter, and examine a number of related terms also.

Searching for and extracting information from a disk, whether from a bona fide file or from ambient data must occur in a manner that will satisfy the courts as regards integrity of the evidence so obtained. It relies therefore upon careful noninvasive imaging of the original data disk to a faithful *bit-by-bit* duplicate, which can then be searched at leisure without the possibility of jeopardizing the original. Techniques and procedures used to accomplish this are examined below. The fundamental requirement in order to meet the *chain of evidence* requirements of the courts is the assurance of the integrity of the file imaging or disk imaging procedures employed. In those cases which do not end up in court, the need for assurance of the integrity of the imaging procedures is less acute but still clearly highly desirable.

Imaging and investigative procedures for devices such as mobile telephones and PDAs, are not as well established as they are for laptops, desktop computers, and computer disk storage devices. Yet, both mobile phones and PDAs have become a standard item of business and domestic life and both are likely to contain considerable personal information and information about a person's activities, which can be of significant value in investigating crime. There are some differences between the handling of these devices and the handling of computers and disks which arise out of their different technology and makeup. In particular, there is much less standardization amongst devices of this sort and because they are more recent, there has been less experience of how to deal with them. Nonetheless, as a result of their rapidly increasing market penetration, LE has already been dealing with mobile telephones as a matter of routine, and to some extent also with PDAs. van der Knijff points out [9] that mobile phones and PDAs are specific instances of embedded devices, which are, as a whole, becoming pervasive in the home, in the automobile, and at work, and that all such devices and their stored information present great potential for the forensic investigator.

2.2.1 Secure boot, *write blockers* and forensic platforms

Before discussing some of the topics foreshadowed immediately above, we turn to an important corollary that follows from the *chain of evidence* requirements of the courts. This relates to the booting of a system prior to its investigation and imaging. System boot in such situations must be undertaken in a secure and controlled manner which is noninvasive, that is, in a manner which precludes any modification to the content of disks and/or files being examined and imaged. It should also preclude any modification to the metainformation or descriptive information that describes either the system or the disks and/or files (e.g., time last written to). To ensure this, forensic investigators need to circumvent the default boot process (the normal boot process which boots from the system hard drive), and to boot the system using their own specialized boot diskette or CD that is configured to boot the system in a controlled manner from the diskette or CD, typically to DOS or Linux. This will require redirecting the boot source to a floppy or CD drive at system startup. The need for such a controlled boot process to guarantee that nothing will be written to the evidence disk(s) prior to their investigation becomes apparent when one considers the variety of implicit checks and operations carried out at boot and initialization time by some operating systems, in particular the more powerful Microsoft operating systems. These operations can for instance include registry updates in the case of Microsoft Windows 2000/NT or file decompression that will result in updated file timestamps. Such operations are invasive or intrusive and compromise the integrity of any related information being examined or imaged.

Once the system is booted, assurance of continued protection of original disks and files from being written to by the system relies on *write blocker* technology. A *write blocker* can be implemented in either hardware or software (see NIST's "Hard Disk Write Block Tool Specification" [10] for an account of write blocking software). The function of a *write blocker* is, in any case, to make absolutely certain that any unknown or unexpected disk or file write—a write unknown to and unexpected by the investigator—is blocked. This ensures that no writes to the disk or file that is about to be imaged can take place before the imaging. As a general rule, safeguards implemented in hardware are more reliable than those implemented in software, and *write blockers* are no exception. Software *write blockers* based upon interception of interrupt calls are commonly used; they can conceivably, however, be circumvented and there is a recent trend towards hardware *write blockers*. Imaging of seized disks and files should only take place with the appropriate write blocking in place.

In any event, in addition to the use of secure boot and write blocking, an investigator needs to use trusted software comprising a secured command line interface or shell and a forensically sound copying program, both executed from removable media and thus trusted, in order to ensure the integrity of the file or disk imaging.

Digital Intelligence Inc. [11] provides a range of disk *write blockers* while Guidance Software has released FastBloc [12], which is a hard drive duplication device that allows investigators to duplicate disks noninvasively in Microsoft Windows environments. FastBloc may be used in conjunction with Guidance Software's EnCase system, in which case the acquired data can then be managed as part of the EnCase methodology.

Many recently developed computer forensic tools, developed specifically with forensics in mind, have been targeted at Microsoft Windows systems. Published procedures for forensic examination too have tended to focus on Microsoft platforms, with some notable exceptions listed below. The reason for this apparent bias is simply that as a result of its long history, UNIX already has an established plethora of tools which achieve what is needed—there has been much less of a need to develop new tools specifically for forensic investigation when using UNIX systems. For instance, the September 2000 issue of *SC Magazine* reviewed a number of forensic imaging tools, including Linux *dd* 6.1 (Red Hat Inc.) and gave it their top rating [13]. Furthermore, there are several highly regarded UNIX/Linux forensics toolsets, which have been in wide usage for some time including: the Linux Forensic Toolkit (LFT) from NASA, the Coroner's Toolkit by Farmer and Venema [14], and ForensiX from Fred Cohen & Associates [15]. While the last is available only to LE, the related White

Glove/PLAC distribution [16] is available to the general public. Three accounts of forensic procedures relating specifically to the UNIX platform are as follows:

1. *Basic Steps in Forensic Analysis of UNIX Systems* by University of Washington's Dave Dittrich [17];

2. The chapter *UNIX Systems Analysis* by Seglem, Luque and Murphy in Casey's *Handbook of Computer Crime Investigation* [18];

3. The series of articles describing The Coroner's Toolkit in *Dr Dobbs Journal* by Dan Farmer and Wietse Venema [14].

Linux has achieved a special position as a forensic platform in recent times on account of its rich utility set inherited from UNIX and due to the large number of different file systems that it understands, including nfs, ntfs, and vfat. Having obtained a disk partition image using the UNIX/Linux utility *dd*, an investigator using a Linux platform can then analyze that partition image by mounting it *read-only* in loopback mode, which provides the specific file system support needed to analyze it, be it a Microsoft file system, or a UNIX-variant, or some other file system. This provides the investigator with the powerful facility to analyze not just files within the file system or partition but also the so-called ambient (supposedly unused) disk space on the disk or partition while mounting the partition *read-only* provides the write blocking safeguard discussed earlier. Other powerful UNIX/Linux forensic capabilities include generalized string searching using *grep* and other utilities, and file integrity checking using *md5sum* and related utilities.

2.2.2 Disk file organization

The main function of the file management services provided by an operating system is to store the information content of a file on disk in such a manner that it can be found quickly when needed. In essence, irrespective of the type of computer operating system, whether it is a UNIX-style operating system or a Microsoft Windows operating system, the way in which the information content of a file is arranged on disk is remarkably similar.

Information is invariably stored in 512 byte sectors in concentric circular tracks of decreasing radius. The outermost track includes the first disk sector, that is, the sector with an identifying number or sector address of zero and the remaining sectors on that track then have sector addresses ranging in a sequence from 1 upwards. The number of sectors per track varies with the particular disk technology being used. The last disk sector of the innermost disk track on a recording surface (a disk may consist of one or more *platters*, each with two recording surfaces) is then the last sector on that recording surface and has a sector address of $n-1$ where n is the storage capacity (in sectors) of one surface of the disk (Figure 2.1). Sector numbers on successive surfaces are then logically numbered from n, $2n$, $3n$, ... upwards. Finding information on a disk is then, in principle, a case of the file management software keeping a record (referred to later as the *filemap record*) of each filename and the sector addresses of the sectors holding the information content of that file; we refer to the collection of these records as the *filemap*. Note that the sectors constituting the content of any particular file may be fragmented (i.e., scattered all over the disk), they do not necessarily follow on from one another. It is precisely for this reason that defrag utilities are so useful and popular. The more fragmented or scattered are the files on a disk, the less efficient they are to access, so a defrag utility will attempt to reorganize the sectors of each file as much as possible to be contiguous on the disk, hence making sequential processing of a particular file that much more efficient. Of course, for each particular file, the filemap lists the sector addresses in *file logical order,* that is, the first sector address in the filemap record for a particular file is the sector address of the first sector of the file, the last sector address in the record is the sector address of the last sector of the file. It follows that the list of sector addresses for any one file will

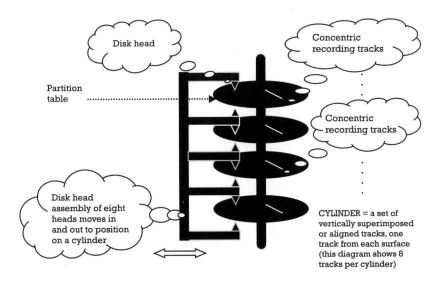

Figure 2.1 Schematic of disk organization.

generally not be in arithmetic order; using a defrag utility will update the filemap accordingly.

For readers interested in more detail regarding disk addressing, we point out that the simple sector addressing scheme that we have outlined above is somewhat more complicated in order to cope with concepts of disk cylinders and heads, as well sectors, and the different *disk geometries* of different disks. This leads in practice to an addressing scheme which is called CHS addressing—CHS being an acronym for cylinder, head, sector. A related complication which can conceivably affect an investigation is that the physical CHS geometry of the disk is generally hidden from the application by disk address translation logic which is part of the system BIOS. For an informative and detailed account of disk geometry and disk access at the hardware level, the reader is referred to Sammes and Jenkinson [19]. James Holley presents the case for developing a standard methodology for assessing computer forensic software tools and provides an interesting account of how different tools portray a different view of the logical CHS structure of the same physical disk [20].

In summary, the file management software of the operating system maps filenames to disk (sector) addresses using information stored in what we call a filemap. It does so in a way which allows for the fact that file content will typically require many disk sectors and which allows for the fact that these disk sectors will not necessarily be at contiguous locations of the disk. This noncontiguity or fragmentation occurs as the natural outcome of dealing efficiently with the fact that files expand and contract dynamically.

Sectors are relatively small; they are typically formatted to be 512 bytes in size (i.e., they can accommodate 512 bytes of information). As a result, when allocating or deallocating disk space on file expansion or contraction, it turns out to be more efficient to allocate/deallocate in larger units which are called clusters. A cluster consists of a fixed number of physically contiguous sectors. Cluster size is configuration dependent and has a significant effect both on performance (file access times) and on disk store utilization. A large cluster size favors performance; a small cluster size favors disk utilization. On average, given randomly distributed file sizes, only half of the last cluster of a file is used for bonafide data, the rest of the cluster is wasted. However, *end of file* over-run problems are avoided as the system information for a file includes the actual size of the file excluding the waste at the end of the last cluster and the operating system observes that actual size by not reading beyond it. Typical cluster sizes are 8 and 16 sectors.

The filemap is itself stored on the disk, at a known fixed sector address on the disk, usually near the start of the disk. Furthermore, a physical disk may be subdivided into a number of separate *logical disks* called partitions, each of which can then accommodate a distinct file system and each of which has its own filemap situated at the start of that partition. For instance, a single physical disk may comprise two partitions, one formatted for a File Allocation Table (FAT) file system, the other for a New Technology File System (NTFS). This allows the user to use both file systems from a single platform. (FAT and NTFS are discussed further later.) Information is stored usually in the first sector(s) of a physical disk that lists the number, type, and size of the partitions on the disk. A physical disk may have just a single partition, which is commonly the case, or it may have a number of partitions. Note that disk partitions may be configured so as not to cover the entire disk thereby providing a potential hiding place for data between partitions.

In UNIX and UNIX-like systems, the role of the filemap is fulfilled by the *index node list* (i-list). Microsoft Windows systems support a number of different file management systems including both FAT and NTFS. In the case of FAT file systems, it is the FAT that fulfils the function of the filemap while in the case of NTFS file systems, it is the Master File Table (MFT) which does so. As an aside, it is worth noting that the directories or folders maintained by a user in order to keep related files together are essentially just data files, which contain the names of those related files. Therefore, directories or folders are treated for the most part just like any other regular application file or data file, and the information contained in a directory or folder simply comprises the names of application data files and subdirectories or subfolders.

2.2.3 Disk and file imaging and analysis

2.2.3.1 Disk imaging and physical analysis

In cases where the investigation involves actual seizure of a computer, disk imaging takes place once the computer has been seized according to warrant and is properly in custody. Once the imaging is complete, further analysis can then take place on the duplicate(s) with an assurance that the integrity of the original evidence is not compromised.

In some situations, imaging of disks or files will take place at the site of the investigation, for instance, if the case involves an individual within an organization whose employers sanction the imaging of that person's desktop. In such situations, the question arises as to whether the entire disk is to be physically imaged (disk imaging) or whether instead the investigator should identify the required files and copy just those for later analysis in the laboratory. In the past, and still today in many cases, the entire disk is imaged in order to allow subsequent analysis at both physical and logical levels. This is known also as a *bit-by-bit* or *sector-by-sector* or, increasingly, a *bit-stream* image or duplicate. There is also partition aligned imaging which allows each separate partition to be duplicated to a separate image; this is particularly useful if the intention is to use Linux to mount such a partition image (i.e., the file to which it has been imaged) in loopback mode as mentioned previously to access the file system it constitutes.

Increases in disk size and in particular the increased use of RAID disk technology, with its ability to stripe a file logically across multiple physical disk drives, have seen a recent change in that imaging an entire disk can in such circumstances be impracticable as one requires an identical RAID system for the image. As a result, there is a trend to duplicate files at the site of the investigation rather than the entire disk. This is sometimes referred to as *file-by-file imaging* or *logical imaging*. While in some circumstances inevitable, this carries with it the disadvantage that subsequent analysis at the physical level (of the entire disk image) is not possible.

Disk image analysis at the physical level allows an analysis of the disk image, *sector-by-sector*, and by definition provides for analysis of each disk partition in the image and possibly of nonpartition areas also. Analysis at the physical level is important in situations in which there is the likelihood of forensically valuable information appearing on the disk in ambient data areas. In this case, a detailed physical analysis of the disk image needs to be carried out, either at the site of the investigation or later on off-site. Information that is residual on or hidden in ambient data areas will otherwise be overlooked.

Before leaving the topic of disk imaging, and given that there are a number of variations on the theme, it is useful to note the related terms or phrases used to describe these variations. The terms are not orthogonal:

1. *Sector-by-sector:* The image disk comprises the same sectors in the same sequence as does the original disk.

2. *Disk mirroring:* Sammes and Jenkinson [19] argue that this term should be avoided as it implies a target disk which is a physical replica of the original disk from which it has been duplicated which is typically not the case. The point here is that the physical CHS geometries of disks vary, some have more cylinders and fewer heads (recording surfaces), others have fewer cylinders and more heads. For this and other hardware-related reasons, disk input or output involves what is known as *CHS translation* and as a result, while one can make a faithful *sector-by-sector* copy of one disk to a target disk

with a different CHS geometry, it will not be a mirror image of the source disk with its different physical CHS geometry. Sammes and Jenkinson [19] provide a very thorough explanation of this and many other hardware-related aspects of the forensics of personal computers.

3. *Physical imaging:* This is a commonly used term to signify a *sector-by-sector* duplicate which is what is needed and acceptable in court.

4. *Bit-by-bit imaging:* This is an ambiguous term, it can be used as a synonym for disk mirroring, or it can mean a *sector-by-sector* duplicate.

5. *Partition imaging:* This is *sector-by-sector* imaging of a single partition.

6. *Bit-stream duplicate,* and *qualified bit-stream duplicate:* These terms [21] have been defined by the NIST Computer Forensics Tool Testing (CFTT) project in developing their standard requirements and terminology for disk imaging tools. The first, *bit-stream duplicate* is defined as "a bit-for-bit digital copy of a digital original document, file, partition, graphic image, entire disk, or similar object." The second, *qualified bit-stream duplicate* is *bit-stream duplicate* but allowing for identified portions of the *bit-stream* which differ.

2.2.3.2 File imaging and authentication

to consider it for long time

If disk imaging of an entire disk is not contemplated, either because seizure of the entire computer is for some reason not possible, perhaps not permissible by warrant or because of the intractable size of the disk, then it is up to the investigator to identify the file or files that need to be examined, and to image or duplicate just those files for subsequent analysis. Whether the file imaging is carried out at the site of the investigation or in a police laboratory, in either case—as with entire disk images—the subsequent analysis must take place on a duplicate, not the original. Furthermore, in either case, whether it is a file image or a disk image that is to be analyzed, both the original and the copy or copies of the original must be authenticated. That is, there must be assurance of the continued integrity of the original and it must be demonstrable that the copies are exact, bonafide, copies of the original, so that any conclusions drawn from analysis of the copies are valid. This involves what amounts to computing a kind of checksum [see one-way hash function (OWHF) later] on the original file or disk at the time the copies are made, and an assurance that the copying process is noninvasive, that is, that it leaves the original in pristine condition, unchanged in any way including in respect of the modification, access and creation (MAC) date and timestamps which are discussed in more detail later in this chapter.

The imaging or duplication operation will, if possible, use a set of procedures and utilities for making an image of the disk or file(s) without relying upon invoking the system's native operating system program. This is typically accomplished by using the secure boot and write blocking measures discussed earlier and trusted imaging software. As alluded to earlier, disk imaging may in some circumstances be impossible, for instance in the case of servers running RAID arrays. In this case, the best that can be done is to do *file-by-file imaging.*

Use of a *read-only* medium onto which to copy an image of the original disk or file(s), such as write-once CDs (recordable CD—CD-Rs) or DVD-Rs, provides an assurance that the image retains its integrity throughout successive steps of analysis. It is not uncommon in the interests of performance for forensic teams to use very large SCSI disks (e.g., 182 GB) for the images, whether they are files or disks or disk partitions. These then may be write protected using *write blocker* technology, or physically setting a jumper on the SCSI drive or alternatively under Linux/UNIX the partition/drive can be mounted *read-only.* Archive copies of the images will also be made, typically to CD-R/DVD-R.

Most of the commonly used forensic toolsets provide a capability for file image authentication and disk image authentication. This capability typically makes use of what is known as a cryptographic OWHF [22]. OWHF make use of block cipher (cryptographic) algorithms to compute a fingerprint or digest (typically a value represented by a 128-bit string or longer) of a file or disk. This value is sensitive to the change of even a single bit in the original file or disk and so provides an authenticator for that file or disk. It is necessary to fingerprint the seized files or disk at the earliest possible point (e.g., before or at imaging time), so that later on it is possible to compute the OWHF for the copy or copies and by comparison demonstrate that the copy or copies are faithful. This assists in demonstrating integrity of the evidence and that the *chain of evidence* has been maintained but in turn relies upon the assurance that the original OWHF value(s) were calculated on the uncompromised original data. The most commonly used OWHF technologies are SHA-1 [23] and MD5 [22].

We note that another application of OWHF which occurs in a related area is that of integrity checkers for operational file systems. The best known such integrity tool is Tripwire originally from Purdue University and now marketed by Tripwire Inc. [24]. Such integrity checkers detect changes in designated files or directories and notify the administrator when this occurs.

For example, SMART Watch from WetStone Technologies Inc. [25] provides real-time notification of unexpected file changes to alert the administrator of unexpected system behavior. Hash databases (repositories of computed OWHF values for commonly deployed files) are discussed in Section 2.3, and also in Chapter 3. Section 2.3 also examines the hash capabilities of EnCase and ILook Investigator.

2.2.3.3 Hidden areas

Physical analysis will identify forensically useful information which is not part of any file listed in a directory or in the filemap, information which has persisted accidentally as the residue from previously deleted files, or which occurs as a result of a deliberate attempt to hide information in an unexpected disk location. Deliberate hiding of data can be accomplished in a variety of ways so that its presence may not be apparent from a simple listing of all the directories in each of the disk partitions. One means of hiding data on disk as noted earlier is by manipulating disk configuration information, such as partition tables. See for example PartitionMagic [26], a popular partition management tool that allows users to configure their partition tables. A potential variation on this is noted by Sanderson [27]—this involves exploitation of a characteristic of devices compatible with the ANSI AT Attachment Interface specification (ATA-4 and beyond), which allows disks to be configured with system areas that are then not visible to applications [28]. Data can also be hidden by using a steganographic file system as in [29]. Other simpler ways of hiding data on a disk include the use of files and directories with nonprintable characters for names and in a UNIX file system to mount another disk over the directory that contains the hidden files. Reference [20] gives a detailed breakdown of the many ways in which information can be hidden on a disk.

The term *ambient data* refers to those areas on disk that are not accessible at the logical or application level. The term actually encompasses a number of separate areas on disk where forensically useful information may reside and from which it may be recoverable. One of the most important is the so-called file slack space, which refers to the space *left over* in the last cluster allocated to a file. Residual information appearing in file slack space is not accessible using standard file processing utilities that are designed to prevent a user from reading past the *end of file* in order to avoid the processing of meaningless data. Nonetheless, data residing in file slack space is potentially of forensic value. For example, if the cluster size is 8 sectors, a file of size 11.5 sectors will result in 4.5 sectors of file slack space which may contain data useful to the investigator. Figure 2.2 shows this concept using the file

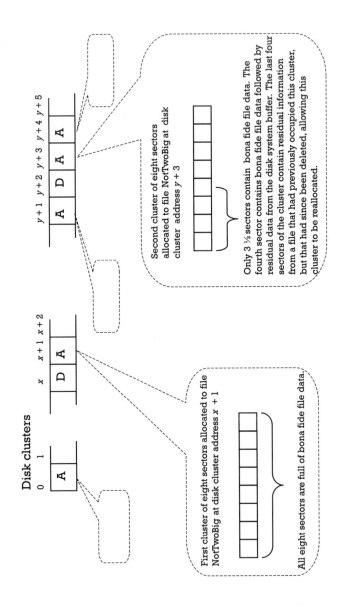

Figure 2.2   Ambient data areas on disk.

NotTwoBig. There are in fact two kinds of slack space at the end of a bonafide file: the unused space within the last sector of the file (assuming the file size is not an exact multiple of sector size), and the wholly unused sectors (if any) in the file's last cluster. If files are written to disk from a RAM system buffer one sector at a time, the last sector of the file (the 12th sector of the file in Figure 2.2) will comprise some bona fide file content and some residual information from the RAM system buffer while the four entirely unused sectors may represent residual information from a file (now deleted), which had previously occupied the cluster. Both of these may be forensically useful. In the case of the latter, it may simply reflect residual data or it may reflect a deliberate attempt to hide data. An implementation of a MS-DOS program to store encrypted data in file slack space is provided by Johnson [30].

Unallocated disk sectors too may contain unintended residual data or be used to hide data deliberately although to conceal data is somewhat more problematic to implement than in the case of file slack space as the operating system may well overwrite such hidden data when allocating clusters for new files. Deleting a file removes the filemap reference to the file and marks the no longer needed clusters as *deallocated*. However, this leaves the original data in the clusters where it was previously resident and unless the user makes special provisions, the system does not actually overwrite that data until and unless it allocates those clusters at some future time to a new file. So until that happens, the information is no longer easily accessible but it is still there. As a result, and unless those clusters have been reallocated in the meantime, the original information may be recovered during physical analysis by the use of specialized software tools that search the disk at a sector level rather than relying upon the filemap. One can reconstitute a previously deleted file by searching deallocated clusters (e.g., looking for text string matches) and piecing together the pieces of the jigsaw, of course not all the pieces may necessarily exist which makes it more difficult. Some operating systems will leave obsolete possibly fragmented filemap records *lying around*, and if located these will assist the task enormously.

System swap-files (known as page-files in Microsoft Windows NT onwards) too will contain residual data, which is of potential value. Swap-files will include residual information, such as from previously opened files and print spooling, and the more sophisticated forensic tools allow an investigator to check such areas.

Latent data and fragile data are two related terms of interest. Latent data reflects the fact that disk information even after being overwritten one or more times can be recovered using specialized techniques including magnetic force microscopy (MFM). This has obvious implications for forensic investigations and data recovery, and has in fact spawned a whole mini-industry devoted to the development of the so-called safe file deletion software—safe in the sense of not recoverable. Section 2.2.4 deals further with this topic. Fragile data is intended to emphasize the ease with which digital data (e.g., disk or file images), can be altered and the extent to which it is thus vulnerable to claims in court of having been mishandled, which is exactly why integrity and *chain of evidence* considerations are particularly vital in the context of electronic evidence.

There are a host of highly respected and widely used computer forensics toolsets or systems which carry out disk analysis at the physical level and which inter alia recover unallocated space for analysis. Some comprise both hardware and software, for instance, DIBS [31] comes as a configured workstation. Most are software tools, such as EnCase from Guidance Software [32] (discussed in Section 2.3), the Law Enforcement Computer Evidence Suite from New Technologies Inc. (NTI) [33], The Coroner's Toolkit (TCT) [14] developed by Dan Farmer and Wietse Venema, The Forensic Toolkit™ (FTK™) from AcessData [34], DataTrail FacTracker from Ontrack Data International [35], and GenX from Vogon International Ltd. [36].

2.2.3.4 Logical analysis (*file-by-file analysis*)

Logical analysis, sometimes known as *file-by-file analysis*, analyzes disk files at the application level, something that is far more convenient and efficient—and in some ways more effective—than physical analysis. Logical analysis investigates the contents of a file using the application that produced the file or an application-specific tool designed to read files produced by the application. This is the natural way of accessing and inspecting a file, and provides two benefits:

1. It overcomes the principal shortcoming of physical-only analysis, which by its nature will overlook search strings split across two logically consecutive sectors of a file if they happen to be not physically consecutive on disk.

2. It provides a high-level or semantic view of the file contents. For instance, inspecting a file directory is much simpler using a Microsoft Windows "dir" or UNIX "ls" command than using a simple text editor. A similar and more powerful example occurs in the case of inspecting the Microsoft Windows NT/2000/XP Registry, which is much facilitated by use of the program *regedit*.

The files may be identified by an investigator by filename or extension type or both with appropriate listing and visualization utilities employed, following which the investigation will employ key word or key phrase searches. File extensions can, however, be deliberately corrupted in order to attempt to confuse an investigation based simply on file extensions. For example, a JPEG image file may be given a *.doc* extension in order to attempt to hide the existence of an image. The commonly available forensic packages deal with situations in which extensions have been deliberately corrupted, and report on files that do not match their extension, for example, a .gif picture file stored as an .xls spreadsheet program. In UNIX systems one can use the "file" command to check the *magic number* at the start of the file to attempt to determine the file type, while in Microsoft systems file identification is normally based upon hexadecimal file header information. Of course, magic numbers could also be corrupt.

Logical analysis may lead to the discovery of encrypted files, which brings its own set of challenges. This is addressed further in Chapter 7.

2.2.4 File deletion, media sanitization

File deletion and media sanitization present the forensic investigator with the same problem and the same irony that encryption does—if done well, the job of the forensic investigator is then that much harder; on the other hand, ineffective file deletion will provide an investigator with information that may be forensically useful, just as in the case of data found in the file slack space. This section addresses the reasons why file deletion is not necessarily a simple matter and how it is that deleted data may yet be recoverable.

Ineffective file deletion can occur for a variety of reasons:

- In Microsoft Windows systems, to guard against unintentional deletion, deleting a file merely removes a reference to the file from the local folder or directory and moves the reference to a recycle or trash folder where it remains available.

- Deleting all references to a file, including from both the local folder and the recycle or trash folder, will typically remove all references to the file at the directory or filemap level and will free up the no longer used clusters. However, this will not overwrite the contents of the freed up clusters; hence, these remnants will be available to the investigator at the physical analysis level unless those clusters are in the meantime reallocated to a new file and thus overwritten.

▶ Copies of a file or fragments of a file may continue to exist in system areas such as the swap area or in temporary storage areas on disk if the application had made temporary copies for application-specific purposes.

It is for the last two reasons that a recommended method of effective file deletion is

▶ Delete the file (from all folders);

▶ Defrag the disk;

▶ Load some new large files such as JPEG images or similar (to fill the partition).

Thus overwriting (probably) the clusters of the original file. Nonetheless as we shall see later, this may still be insufficient, although doing the above several times increases the probability of achieving a safe deletion.

The fact is that even after the 0s and 1s stored in a file cluster have been overwritten, say by *zapping* the cluster with all 1s, it is still possible to reliably retrieve the previous, overwritten, data using specialized hardware. There are two excellent publications which describe why it is that such old or overwritten *shadow* data persists and how techniques such as MFM can be used to detect it. NTI's Curt Bryson and Michael Anderson [37] provide an excellent and intuitive account of how the writing of bit streams to a disk is analogous to the paint spraying of a line along the middle of the road from a moving vehicle. Repeated spraying of the same line causes a wandering line due to slight variations in steering, car bounce affects the intensity and spread of the paint hitting the road, and the porosity of the road surface effectively provides for different layers of road surface to reflect successive generations of data. The paper by Peter Gutmann [38] presents the theoretical basis for the number of overwrites and the nature of the overwrite patterns needed to foil such retrieval techniques for disk technology prevalent at the time of his writing in 1996. The reader is referred also to Gross [39] which includes further references to work in MFM. While techniques such as MFM may retrieve data that is potentially useful forensically, the fact is that the necessary equipment is expensive and exists only in the laboratories of the various national security laboratories and their related agencies. One speculates that it is used relatively infrequently. The usefulness of such techniques in law enforcement (LE) is limited as the retrieved data can be vouched for only on the basis of probabilities, due to interference from multiple shadows, and is thus likely to be unacceptable in court. Bryson and Anderson report that it is possible programmatically to retrieve low level information from disk drives using standard hardware and thus retrieve some *shadow* data information sufficient for experimental purposes if not necessarily for a confident identification of the previously recorded data.

The industry recommendation regarding deletion of data is simple: data to be deleted should be *safely deleted* using validated file delete or *file shredder* utilities, such as Norton's Wipeinfo, which uses multiple overwrites in accordance with recommended practice. Where absolute security is required, then complete incineration of the medium is necessary. Degaussing of the medium is prescribed in some circumstances by government agency regulations but the medium is afterwards to all intents and purposes unusable, so that complete destruction appears to be simplest unless regulated otherwise.

The previous considerations regarding safe file deletion naturally apply equally to e-mail as was seen recently with the widely publicized Enron case. In this context, the Houston Chronicle reports [40] of a case in which Lotus Notes e-mail messages that had been deleted up to 8 months earlier were subsequently retrieved. E-mail as noted in Section 2.3 is a particularly valuable resource in an investigation as it provides not just content but also date and time, and sender and receiver information.

2.2.5   Mobile telephones, PDAs

A recent report regarding the forensic operations of PriceWaterhouseCoopers in New Zealand [41] highlights the increased incidence of cases in which mobile telephones and PDAs play a part and which require forensic investigation of such devices. The report mentions Zert [9], a tool developed by the Netherlands Forensic Institute and Paraben's *PDA Seizure* for the PALM Pilot [42], two of the few tools built specifically for investigating PDAs and mobile phones. Zert, which consists of both hardware and software components can image mobile phones and PDAs, including retrieving the passwords of PDAs, but is reported to be unavailable to the general public. The Netherlands Forensic Institute has also developed related software-only tools, such as *Cards4Labs* and *TULP* which are used for reading smart cards and mobile phones, respectively. All these tools are described in more detail by van der Knijff in [9]. Paraben's PDA Seizure for the PALM Pilot is available commercially but nothing has yet been published regarding experience with using it. PDAs are now approaching the desktops of a few years ago in storage capacity, and one is tempted to call them minicomputers (they are sometimes called hand-held PCs) except that the phrase has already been used to describe the minicomputers that came after mainframes in the 1970s. As a result, there is enormous potential value in seizing and analyzing the PDAs of individuals in a case and the development of forensic tools targeted specifically at PDAs will no doubt escalate.

The prerequisites to an investigation of such devices, as with any forensic examination, is that the proper circumstances exist which allow the seizure and search of the phone or PDA and that requisite court authorization has been obtained. The fundamental constraint on the investigator then is preservation, that is, to ensure the continued existence of all potential evidence, that is, to ensure that power to the device is not interrupted so that any information in volatile storage is not lost. At the earliest stage, it is therefore, important to identify the nature of the power requirements for the device and to avoid removing any batteries, at least until all information resident in volatile memory has been extracted. Clearly, in the case of devices that come with chargers, the charger too should be seized.

In the case of mobile phones, once volatile memory has been preserved, then the Subscriber Identity Module (SIM) smart card at the heart of the phone can be examined using smart card analysis tools such as *Card4Tools*. (However, some phones, do not use SIM cards and the manufacturer may need to be contacted to allow extraction of the data.) We note that the familiar cycle of preserve, image, and analyze is once again observed, as it is in the computer forensics of larger systems. van der Knijff [9] provides a comprehensive and detailed account of the examination of mobile phones, in particular the smart cards without which most mobile phones are useless while Gibbs and Clark [43] describe the architecture of mobile or cellular phone networks, treating them as a special case of circuit switched wireless networks. Gibbs and Clark identify four sources of forensic evidence available in investigations involving such networks:

Area 1—Equipment (if any) connected to the mobile device;
Area 2—The mobile device itself;
Area 3—The wireless network in which the mobile device functions;
Area 4—The subsequent network (if any) that the caller accesses.

It is the second and third of these that are of direct interest here, the mobile itself and the wireless network of Mobile Switching Centers (MSCs) at the heart of a cellular phone network. These two can provide access to the following information:

1. Numbers called, numbers that called, time stamps for each call, duration of call, caller/callee location;

2. Address/phone book;

3. Voicemail messages;

4. Sent and received Short Message Service (SMS) messages;

5. Possibly a diary or calendar.

Some of this information exists at the various MSC receiver and switching centers and is available via the mobile service provider from MSC logs subject to proper authorization. Much of the information will, however, be directly available from a seized phone. The de facto standard mobile phone technology used throughout most of the world, with the notable exception of the United States, is global system for mobile communications (GSM), and Gibbs and Clark provide a detailed account of the technology. In essence, a working GSM mobile phone consists of two components (apart from power source): the mobile phone handset and a smart card called the SIM card. This card is inserted into the handset and is exchangeable between handsets although there is an increasing trend by providers to link handset and SIM card by storing a SIM card specific code in the handset. The phone will not work without the SIM card. The handset has stored in it a unique international mobile equipment identity (IMEI) number which serves to identify the handset. Other important subscriber information, the subscriber phone number and other unique subscriber information, are stored on the SIM card. So too is information entered by the operator such as phone/address book information, and sent and received SMS messages. This SIM card information can be recovered using standard smart card reader hardware coupled with the appropriate software such as *Cards4Labs* discussed by van der Knijff who also provides a detailed account of how to approach the investigation of a PIN-locked mobile and describes *TULP*, a program which can read certain GSM mobile phones directly via data cables or an infrared communications port.

The booklet *Best Practices for Seizing Electronic Evidence* [44] produced jointly by the U.S. Secret Service and the International Association of Chief of Police, and referred to in Section 2.4.2 lists other storage devices—other than computers—that may need to be seized and analyzed and provides general guidelines as how to do so. The devices listed are as follows: wireless telephones, electronic paging devices, facsimile machines, caller ID devices, and smart cards.

2.2.6 Discovery of electronic evidence

Forensic tools can be divided up roughly into three categories: imaging, analysis, and visualization or in the language of Chapter 4. The last two categories, analysis and visualization, tend to overlap and there are certainly tools required at various stages of an examination which do not fall neatly into one or other of these categories. Earlier sections focused upon imaging and analysis, here we list briefly some of the important additional capabilities that need to be provided, capabilities such as link analysis which relates data from separate files or sources, and provides an effective visualization of that information. These tools rely in turn upon time-lining tools and sophisticated search engines with fuzzy logic capability (e.g., NTI's IPFilter program, which can identify patterns of text associated with prior Internet activities).

Link analysis explores and visualizes the key nodes and structures within a data network (i.e., a collection of related data). It is an important tool for exploring relationships in data when investigating complex cases such as fraud that involve large volumes of data such as e-mail or audit data. Link analysis examines a large number of potentially dissimilar records of data and establishes links among those records based on data fields with identical or related values using artificial intelligence (AI) techniques such as heuristic methods to find the links between the records [45]. This bottom-up approach to constructing networks is quite different to techniques that rely on statistical methods. A good introduction to the concept of link analysis can be found at [46]. One of the best known link analysis tools used in computer forensics is the Analyst's Notebook from i2 Inc. [47]. Analyst's Notebook is a link analysis and data visualization product that has been used in criminal and fraud investigations worldwide. It consists of two main tools, one for link analysis and one for case management. The latter also provides a time-line analysis capability, a capability whose importance cannot be overestimated. Time-lining is a recurring theme in this chapter (Section 2.4.1) and Chapters 3, 4, and 6. Both EnCase [32] and CFIT [48] examined in Section 2.3 support time-lining. The case studies listed on the i2 site include New Scotland Yard and the Gloucester Police as two users of the Analyst's Notebook [49]; in addition, the FBI has recently signed a $2 million contract with i2 while the U.S. Postal Inspection Service is also a user of this tool. Netmap is a link analysis tool widely used by LE in the United States [50] while Watson from Xanalys [51] is also widely used for link analysis and data visualization in both LE and in the finance sector. The latter was successfully used recently by the Durham Police (United Kingdom) to analyze over 4,000 e-mail messages as part of a child pornography investigation, leading to a heavier conviction against the offender.

Data mining tools too are becoming increasingly important for identifying previously unknown associations. These employ a range of Knowledge Discovery and Data Mining (KDD) techniques such as pattern recognition, neural nets, and rule induction for investigation and analysis of large volumes of data in transaction situations. Data mining tools are used increasingly, for example, in cases involving possible financial fraud. Two well-known products that are used routinely for this purpose are Clementine from SPSS Inc. [52] and IBM's Intelligent Miner for Data/Text [53]. Data mining and link analysis for forensic purposes in the accounting and finance sectors are discussed further in Chapters 4 and 7, respectively.

2.3 Forensic tools

The growth of the data recovery and electronic evidence discovery industries referred to in the previous section has been accompanied by similar strong growth in the number of computer forensic tools available and in use. More importantly, there has been a trend towards sophisticated tools or integrated packages that perform a greater range of forensic functions. For instance, the Computer Forensic Investigative Toolkit (CFIT®) software, described later provides facilities for analysis of data streams (such as disk drives, network data, disks, and telecommunications call records), the ability to add and integrate a variety of specialized interactive forensic tools into a common, easy-to-use visual framework, and the ability to capture the history of an investigation in a simple visual manner. It is the last which is perhaps the most noteworthy development of recent times in the functionality of some of the commonly used tools namely, the integration of various aspects of a forensic investigation into a *case-based* portfolio.

We can, in general, identify three categories of forensic functionality: imaging, analysis, and visualization. These categories can naturally be subdivided further and as noted earlier an increasing number of tools integrate these functionalities within one toolkit or workbench:

1. *Imaging:*

   a. Imaging volatile memory (including on PDAs and mobile phones);

   b. Disk and file imaging;

   c. *write blockers*;

   d. Integrity code generators and checkers.

2. *Analysis:*

   a. Ambient data recovery and the searching of raw disk data for text strings, by sector (typically including unused areas);

   b. Data and file recovery;

   c. Disk and file system integrity checking tools;

> **d.** File conversion (i.e., conversion of proprietary files into text files or vice versa, or between proprietary formats, to facilitate further processing);
>
> **e.** Data filtering by date last modified and other file properties such as file or application type such as e-mail, graphics, word processing, spreadsheets, or presentation files;
>
> **f.** Search tools, sophisticated search engines with fuzzy logic capability;
>
> **g.** Data mining tools.

3. *Visualization:*

> **a.** Time-lining;
>
> **b.** Link analysis tools.

E-mail is an application type that deserves special consideration here because it is by nature more than just a record of data. It is also a record of communication which identifies not only the content of the communication but also its originator, its recipient and time and date information and so presents the forensic investigator with a potential cornucopia of information which needs to be given special attention. (We note that the CFIT® software will in future versions include software to provide a capability for authorship attribution of e-mail.) For a case involving purpose built tools to accomplish preprocessing and conversion of e-mail to facilitate forensic investigation, see [54]. The same site presents a penetrating and amusing account of the futility of using so-called self-deleting e-mail as a means of limiting the legal liability which may otherwise arise through discovery of incriminating e-mail. It points to Michael R. Overly's *E-policy—How To Develop Computer, E-Mail, and Internet Guidelines To Protect Your Company and Its Assets* for organizations wishing to limit their liability exposure. We explore some current research directions with regard to e-mail, in particular, concerning authorship attribution of e-mail, in Chapter 7.

SC Magazine has recently reviewed a number of forensic tools, first in September 2000 [13]:

> ‣ Byte Back (Tech Assist Inc.), Drive Image Pro 3.0 (PowerQuest Corporation), EnCase 2.08 (Guidance Software Inc.), Linux *dd* 6.1 (Red Hat Inc.), Norton Ghost 2000 Personal Edition (Symantec Corporation), SafeBack 2.0 (New Technologies Inc.), SnapBack DatArrest 4.12 (Columbia Data Products Inc.);

and then in April 2001 [55]:

‣ Byte Back (Tech Assist Inc.), DriveSpy (Digital Intelligence Inc.), EnCase (Guidance Software Inc.), Forensic Toolkit (AccessData Corporation), MaresWare Suite (Mares and Company).

Automated Computer Examination System (ACES) [56] is worthy of special mention. It has been developed by the FBI to provide a forensic tool for LE, which inter alia supports the identification of known files (e.g., executables), and thus their exclusion from further investigation and considerably facilitating the work of the investigator. Common estimates are that a typical personal computer or workstation will contain of the order of tens of thousands of standard files, which can safely be excluded from analysis. Further development of ACES functionality appears to have been subsumed into the NIST projects referred later. Another tool used by U.S. government agencies, including the Department of Treasury and the IRS, and also by the Australian Federal Police, is ILook Investigator [57] examined in detail later. FBI CART has recently announced that it has suspended further training in and development of ACES in favor of ILook Investigator. A number of tools including both ILook Investigator and EnCase support the import and use of hash sets from the hashkeeper database of the U.S. DOJ National Drug Intelligence Center [58]. Another highly regarded tool with a built-in Known File Filter (KFF) capability is Forensic Toolkit from AccessData Corporation [34], which also provides Password Recovery Toolkit, one of the leading password recovery packages. A recent survey [59] of 151 U.S. LE agencies and other federal organizations (including the FBI, OIG/NASA, NIPC) found that some 69% of investigations use Encase, 55% use Safeback, and 27% use ILook. Noteworthy was that 41% of the agencies/organizations surveyed were dissatisfied with the tools at their disposal.

NIST's Information Technology Laboratory is currently working on two computer forensics research projects, which are relevant here, the National Software Reference Library (NSRL) and the CFTT [60]. The former, the NSRL project, makes use of integrity checking technology (Secure Hash Algorithm SHA-1) to characterize files that are commonplace by calculating a fingerprint for each file so that during an investigation a file purporting to be one of these standard files can be fingerprinted in like fashion and if its fingerprint matches that in the NSRL database (the Reference Data Set or RDS), then it may safely be excluded from further forensic examination. For added assurance and flexibility, the RDS also makes optional use of other OWHF such as MD5 (Message Digest 5). The latter project, the CFTT project, is intended to develop specifications and test methods for assessing computer forensic tools in order to identify their level of performance.

We present an overview of three different forensic tools currently available: EnCase, ILook, and CFIT. This is an overview of some of the functionality and features of each tool and, as such, is not intended to be a comprehensive or detailed analysis. For more details concerning the tools we suggest the reader consult directly with the software developer or look at some previous works [61].

There is not now nor is there ever likely to be one forensic tool that does it all, a single tool that does everything an investigator may require. A significant number of forensic investigations will require some form of analysis that no standard tool currently provides (e.g., correlation of events). In these cases, the analysis may make use of specialized tools and even one-off tools as was the case in the previously cited case which involved e-mail analysis via purpose built tools [54]. A common occurrence is the uncovering of corrupt, unusual or rare media and file systems in which case once again, custom tools may have to be developed.

Three areas present a continuing challenge that must be addressed by the tools of the future:

1. The increasing volumes of data with which the analyst is faced as a result of the increased bandwidth of most Internet connections;

2. The need to provide software that supports and encourages collaborative working by multiple examiners who may be geographically separated and possibly from different jurisdictions;

3. The need to be able to accommodate new forensic tools to interoperate with existing tools and systems in order to be able to correlate forensic data from a wide variety of logs and records (i.e., the need for extensibility in order to support at least to some degree the notion of a *generic forensic* capability in the face of changing technology).

Most software requires constant development and support in order to adapt to changing software and hardware environments. Forensics software development is particularly affected by such a changing environment. The problem is further compounded by the rapid change in the types of computer or computer-based crimes being investigated and the complexity of investigations (e.g., multiple actors distributed over multiple computers, in different geographical locations). Forensic software needs to keep abreast of these changes in order to be of use in a wide spectrum of investigations.

2.3.1 EnCase

EnCase (Encase3, June 2001), distributed by Guidance Software [62], is a computer forensics software product used by many LE and information security professionals. Since its ongoing development from 1998, it is one of the few fully integrated Microsoft Windows–based products for forensics investigations. EnCase is a direct descendant of the Expert Witness software previously distributed by ASR Data (pre-1998) and early versions of EnCase were very similar to Expert Witness.

The EnCase integrated environment means that the EnCase software acquires the evidence as a verifiable, proprietary *bit-stream image* (called an Evidence File, EF), mounts the image EF as a *read-only* virtual drive, and reconstructs the file system structure utilizing the logical data in the image. This integrated procedure eliminates the time-consuming sequence of steps normally associated with traditional command-line-based imaging and ensures all the evidence and meta-evidence (such as timestamps) remains forensically unaltered. The acquired EF is available as a loss-less compressed image, and includes cyclic redundancy checks and a MD5 hash value to ensure data integrity. EnCase can image different forms of media, such as SCSI/IDE drives and Zip/Jaz drives as well as RAID disk sets. The investigator can also bypass the acquisition of an EF by *prescanning* an evidence drive using a parallel port or 10-BaseT network cable between the investigator's computer and the target computer and invoking the *remote preview* feature. This makes it easy for an investigator to quickly undertake a perfunctory forensic analysis of the drive without incurring the overheads of an EF creation. Previewing is useful when a preliminary look at the evidence storage media is warranted by time constraints, such as during on-site inspections. Unfortunately, in the review preview mode, the investigator is unable to save any of his/her findings, such as search results as all of these will be lost once the computers are disconnected.

Once the EF has been created, the investigator can then apply one of the several integrated multitasked tools within a common graphical user interface to analyze the file system. File systems, such as Microsoft Windows FAT and NTFS and UNIX can be reconstructed. The user interface displays several *Encase Views* such as the Case View, Bookmarks View, and Keywords View together with associate supporting views, such as Table View, Gallery View, Timeline View, and the Case Report.

The Case View displays all the evidence included in a case for analysis in a convenient tree of folder structure as found in a Microsoft Windows Explorer view. It can also display *recovered folders* for an EF folder, that is, subfolders and files found in the unallocated disk clusters that have not been overwritten, as well as perform a signature analysis of every file in the EF. Signature analysis is useful for (1) identifying any discrepancies between a file's extension and the file's header, and (2) building hash sets for file filtering. Hash sets are used in the context of search operations for eliminating well-known files, such as operating systems files, or for including selected files and bringing these to the investigator's attention, such as porn files or noncompliant software. E-mail attachments are ripped, zip files are automatically unzipped and compound documents (such as Microsoft Word documents) can be recovered. An associate Table View displays all of the subtree of folders and files of a Case View tree node, together with the file type (e.g., deleted file, unallocated space, deleted, and overwritten file) and a set of attributes (e.g., file name, file extension, and file timestamps) that can be sorted. File contents can always be viewed in text or hex format in the bottom pane of the EnCase GUI. The associated file clusters are displayed in the Disk Surface View, together with their disk geometry location values.

The Keywords View enables the investigator to build a set of search terms that can be placed in a set of *keyword folders* (a keyword folder is a type of user dictionary). Keyword search can be case sensitive, *grep*-based (i.e., regular expression), or Unicode. Images can also be searched and displayed in a thumbnail picture viewer called the Gallery View.

The Bookmark View displays the bookmarks, such as EF, text fragments and images, that the investigator has previously bookmarked. The Bookmark View also displays keyword search results as the search hits automatically become bookmarks. Bookmarks are a convenient way of identifying, for example, particular clues and files and writing comments in each bookmark entry. Selected bookmarks can then be incorporated in the case report.

The Timeline View is a basic graphical display of the time attributes of EF. It provides a quick way to identify patterns of file activity in time. File time attributes include creation, modification, deletion times, and these are displayed in a calendar-like display at different levels of granularity (e.g., seconds and minutes). A potential problem with any timestamp interpretation in EnCase is that all time information is based on the system clock of the investigator's computer. The investigator has to change his/her clock to coincide with the subject's machine clock to ensure that evidence sourced from a different time zone than the investigator's machine will be interpreted incorrectly. The problem is compounded when, for example, evidence comes from different sets of time zones.

An useful feature of EnCase is the inclusion of a scripting language, Escript Macro Language, which allows the more adventurous investigator to construct his/her own custom forensic tools and filters for execution within

EnCase. This requires some knowledge of object-oriented programming, as it is based on the C++ language paradigm.

EnCase is a comprehensive (based on file system and media type) and integrated forensic tool that allows investigators to do some useful and basic forensic analyses. Its user interface is simple and easy-to-use and provides some useful functionality (such as the instant decoding of nontext data for meaningful interpretation and integrated reporting). The interface, however, could potentially become more cluttered as more forensic tools are included in future.

2.3.2 ILook Investigator

The ILook Investigator, or simply ILook, forensic software is developed and owned by Elliot Spencer and the Criminal Investigation Division of the United States Internal Revenue Service (U.S. IRS), U.S. Treasury Department. It can be downloaded from their Web site (http://www.ilook-forensics.org), though it is only available to "LE personnel, forensic personnel working for LE agencies with a statutory role, national security, and military police agency staff." To use the software, a password is needed that can be obtained by registering with the authority. It is claimed on the Web site to be used by "thousands of LE labs and investigators around the world" [57], including some Australian agencies, and has been adopted by the U.S. IRS and FBI as a forensic analysis platform. ILook is designed to allow an investigator to access the partition file system(s) imaged during the evidence gathering process and undertake an extensive forensic analysis. Currently ILook (version 7, July 2002) is only supported if installed on a Microsoft® Windows® NT or Microsoft Windows 2000 or XP operating system. Investigators using ILook need a relatively high degree of technical knowledge to drive it effectively.

ILook can be used to image any attached media device, however, it relies on an alternative write blocking mechanism. In addition to its own imaging tool, ILook can identify and reconstruct raw bit stream images, ISO, and CIF CD images, VMware virtual disks as well as image files generated by other forensic software (e.g., Encase and Safeback). Investigators can investigate the image map by traversing the image to examine the partition structures and can probe the image for specific meta-structures such as boot records and partition tables that could be used to recover a (broken) file system. ILook can reconstruct Microsoft's FAT, VFAT, NTFS, Macintosh's HFS and HFS+, Linux's Ext2FS and Ext3FS, Novell's NWFS, and CDFS file systems.

The ILook software (see Figure 2.3) provides a Microsoft Windows–like Explorer interface that consists of various window frames; for example:

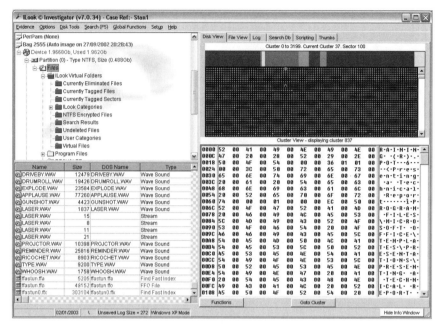

Figure 2.3 ILook Investigator. (*Source*: Elliot Spencer, 2002, ILook Manual, reprinted with permission.)

1. An EvidenceWindow frame allows the investigator to view and navigate the partitions and file system structure of a suspect disk. It also displays a set of additional *Virtual Folders* that contain pointers to undeleted file streams, files that have been eliminated from the investigation, files or unallocated sectors that have been tagged for a specific purpose, files identified in previous searches, and files with user-defined specifications or categories (e.g., deconstructed files).

2. A FileWindow frame lists all the files and file properties stored in the selected folders in the EvidenceWindow.

3. An InfoWindow frame gives the investigator access to groups of information related to the objects selected in the EvidenceWindow and FileWindow frames. These groups of information are arranged as a set of tab window panes and include

   a. A Disk View pane that displays the disk partition layout together with a Norton-like two-dimensional partition cluster map and cluster content;

   b. A File View tab pane that displays file contents in their intended manner or in raw text and/or hex view;

    c. A tab pane displaying the audit log of an investigator's activity during a session (e.g., date/time and actions undertaken);

    d. A search results window tab pane;

    e. A tab pane incorporating an editor and execution engine for processing data in specific ways and for undertaking repetitive tasks using a BASIC-like scripting language;

    f. A tab pane for displaying thumbnail graphical images.

Investigators can undertake string term searches with the help of one of three search engines—a standard search engine for a small number of search terms (with Boolean combinations), a bulk search engine for a large number (up to 1,000 search terms stored in a file) of simultaneous searches, and an indexed search engine for fast repetitive searches (requiring the investigator to generate an index of the case data prior to invoking any indexed searches). Searches can be undertaken on all the data associated with a case (e.g., files, slack, and free space), as well as compressed archive files and file signatures (file magic numbers stored in the first few bytes of a file). Magic numbers can also be used for salvaging (or "carving" in ILook parlance) files in free space (e.g., deleted files that cannot be undeleted using a Norton-like recovery method). ILook also allows the investigator to search for files based on specific attributes (e.g., name, date, and MACtimes). Date/time-based searches use a basic calendar as the basis for date/timestamp selection and viewing. A simple frequency analysis of the file MACtimes is also displayed with the calendar representation. (NB: ILook allows the investigator to manipulate date/times on a partition-wide basis.) An interesting search facility is the *search bot,* an autonomous search engine that runs in the background thereby enabling the investigator to continue his/her examination at the same time as the search is being performed.

As indicated in the introduction to this section, an useful file filtering facility offered by ILook is based on the hash analysis of file content. Each file can be identified by a unique message digest (one-way hash), which is used to either include the file, or exclude it, from the investigation by performing a match of the file content hash. Hashes can be generated in two ways namely, internally from files selected by the investigator, or from known files such as operating systems files. In the first case, the investigator can generate and export his or her own hash data set using standard CRC32, MD5, and SHAN ($N = 1$ or 2) formats, depending on the level of the false positive rates required. Specifying a small number of false positives necessitates the use of hash algorithms with larger message digest sizes (such as SHA). In the second case, ILook will perform the match hash analysis of known files (including cryptography and steganography programs. Steganography is discussed in detail in Section 7.7). These hash sets are available as standard *hashing toolsets,* such as U.S. DOJ NDIC's Hashkeeper [58]; see also the NIST, NSRL'S Reference Data Set [60], and Chapter 3 for some further discussion of these databases. The investigator can choose to perform either positive hash analysis (searching for files that do match) or negative hash analysis (eliminating matching files, thereby reducing the number of files that require further investigation).

File deconstruction, that is, the interpretation of a limited set of compound file formats (such as Microsoft Outlook Express files, netscape cache files, AOL mailboxes), and extraction of data therein can also be performed using the ILook forensic software. Once data extraction has been undertaken by the ILook "deconstruction engine," the contents of the extracted data structures can then be investigated. Content analysis can subsequently be undertaken using one of the three search engines mentioned earlier.

2.3.3   CFIT®

The CFIT® is an integrated computer forensics tool developed by the DSTO, Department of Defence, Australia [48]. CFIT® provides efficient and flexible automated forensic methods for analyzing the content of data streams such as disk drives, network data, disks, and telecommunications call-data, thereby enabling investigators to discard data that are peripheral to their investigation. CFIT® provides a forensic problem-solving environment that integrates tools in a visual framework for investigating the unauthorized use of computer and network facilities. The main advantages of CFIT® are (1) the ability to integrate multiple interactive forensic tools into a common, easy-to-use visual framework; (2) the facility for adding new specialized forensic tools to the framework; and (3) the ability to capture the history of an investigation in a simple visual manner.

The basic investigative environment in CFIT® is the *case*, in which investigators can work individually or as a team to solve one or more criminal cases. Networked multiple investigators can investigate a case at the same time using CFIT®. The CFIT® platform includes case management, forensic data stream access and manipulation, data visualization, and forensic processing. CFIT® incorporates a two-dimensional visual language environment, called *Picasso*, for graphically expressing a forensic case on a visual framework or workbench. Forensic tools that analyze the case data can be *dragged and dropped* onto the workbench, interconnected, and executed. Investigators use the interactive visual workbench to undertake an investigation and share their results with other investigators working on the same case.

Forensic tools included in CFIT® include a hard disk analyzer, file system analyzer (currently ext2 and FAT), log extractor, ontological search engine, unallocated space extractor, time event resolver, and time-lining tool. Investigators can interconnect these tools using flows within *Picasso*, though the interconnections are not always universal since some tools cannot interconnect with other tools due to semantically incompatible data types. CFIT® also ensures the consistency in the interpretation of time differences from computers running different operating systems (which may interpret time in different ways), located in different countries and possibly covering multiple time zones. It does this by associating each piece of case evidence, or metadata generated by forensic tools, with a time reference defined by the investigator. These time references are then automatically mapped into the common UTC timeframe (see Section 6.4 for further information).

An example computer forensic tool available in CFIT® is the *Ferret Discovery Engine*—a tool for textual concept ontology generation, navigation, and searching. It can be used for searching files or documents for particular concepts and identifying those documents that might have a forensic significance. It is particularly useful for searching text-based files, though it can also be used for searching text in nonprintable files such as binaries (executable files) and even network packets. *Ferret* allows the investigator to

 ▸ Discover suspicious text byte streams, such as files/documents from one or more file systems;

 ▸ Establish the inherent relationships between the streams based on a set of concepts.

An *ontology* is a domain of discourse where one or more keywords (or *terms* in the *Ferret* terminology) is organized as a domain-specific graph-based concept structure that best describes knowledge or information about a given domain. A concept is a set of one or more terms and their sets of relationships with other concepts (we define *relationship* later). In *Ferret*, a concept is initially restricted to containing a single term (e.g., money, transfer, and, account) and its set of relationships with other concepts.

The most basic function available in *Ferret* is to perform searches for a set of terms on some input data streams defined by upstream forensic tools in CFIT®. This can be most useful when searching unallocated space and hidden space on disks. Investigators can select a set of search options (such as stop terms, allowable errors in each term, and case-sensitivity) as well as being able to subselect data streams and run multiple concurrent searches. Figure 2.4 shows the results of a search operation for a single term on eight input data streams (in this case, the streams are Linux log files). The terms *kernel* and *apollo* have been found in four of the data streams and can be viewed in the *messages log* file in the lower panel.

The term nodes in the graph-based concept structure are usually related to each other by one or more relationships (represented as the arcs of the graph), such as generality, specificity, synonymy, and meronomy. Semantic relationships may also arise in the context of the text language model employed. The language model captures and characterizes the regularities in the natural language used in the text stream. For example, short- and long-distance textual information such as *N-grams* and *triggers* describe the underlying associative relationships used in a text document. An *N*-gram [63] is a sequence of contiguous words in a text stream with its significance

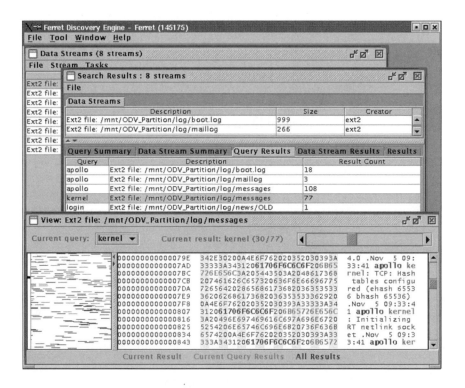

Figure 2.4 Search results using *Ferret*.

in that text stream defined by the conditional probability of one word given the preceding sequence of words. Consequently, *N*-grams capture well short-term and local dependencies in the text stream. *N*-grams can, unfortunately, capture nonsensical text frames that are unrelated to their linguistic role. A *trigger* is a pair of terms that cooccur, usually within a fixed word window size, in the text stream. *Triggers* are effectively long-distance bi-grams (2-grams) capable of extracting relationships from a large-window document history [64]. *Triggers* have been shown to be effective in capturing semantic information over small-to-medium text stream window sizes (distances up to 5) [65]. *Ferret* uses *triggers* to extract semantic relationships within text documents.

The ability to describe semantic relationships using a concept graph allows the investigator to visualize the concept domain of the case under investigation much more succinctly. The graph combines both language semantic relationships as well as data-driven semantic relationships (e.g., *triggers*). This allows the investigator to navigate the concept domain and possibly discover new relationships. Figure 2.5 displays the concept graph

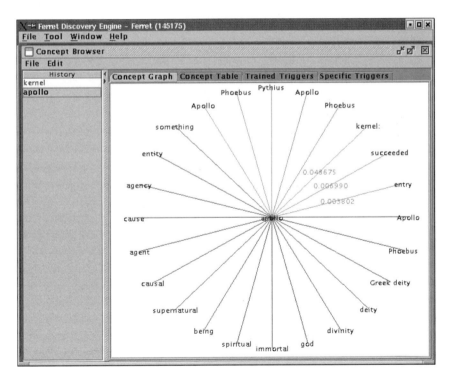

Figure 2.5   Concept graph using *Ferret*.

within the *Ferret* concept browser window, showing the different concepts derived from the input data streams, as selected in Figure 2.4, related to the central concept *apollo*. Some of these concepts are semantically related by generality and specificity (e.g., *agency,* and *supernatural*), and to others derived from data stream *triggers* (e.g., *kernel, entry,* and *succeeded*) shown together with their associated mutual information index values (a measure of the distance or strength between the two terms).

In summary, CFIT® is an easy-to-use forensic investigation environment that provides the ability to integrate multiple interactive forensic tools into a common, visual framework. It is an ongoing development and requires the addition of more file system support (e.g., NTFS is currently being included) and the inclusion of an improved reporting facility.

2.4   Emerging procedures and standards

Procedures and standards relating directly or indirectly to computer forensics have developed quite quickly over the past decade. Progress has occurred on three fronts:

1. Procedures relating to the seizure and analysis of electronic or digital evidence;

2. Standards relating to a consistent understanding within and across national boundaries of forensic procedures, and as to what constitutes computer crime;

3. Issues relating to matters of legal jurisdiction, both national and international.

Section 2.4.1 addresses the first point. It presents a discussion of the standard procedures that have evolved with regard to the seizure and examination of personal computers and workstations and the like. In many cases, this is exactly a case of searching disk files and disks for directly incriminating information (e.g., strings of incriminating text, such as the supposedly unknown phone number or address of a stalking victim). Having said that, if the defense is likely to dispute ownership or authorship of the incriminating information—on whatever basis—then it is incumbent upon the investigator to provide proof to the contrary. It is here that *meta* or *system*-information, as opposed to the directly incriminating information, becomes crucial. Such information may be as simple as a record in an event log indicating the identity of the author (e.g., a computer account name) and the time and date of the writing of that information. Section 2.4.1 addresses various kinds of such metainformation that are part and parcel of an investigation and the related concept of time-lining which establishes a chronological history of events based upon metainformation and possibly external information also.

The second point, which relates to standards and a common understanding within and across national boundaries of forensic procedures and what constitutes computer crime, is addressed partly in Section 2.4.2 and partly in Section 2.5.

The third point, relating to legal jurisdiction, both national and international, while referred to incidentally in Section 2.5 is not considered in detail in this book.

2.4.1   Seizure and analysis of electronic evidence

Accepted wisdom regarding standard procedure to be adopted by LE for the seizure and examination of personal computers and workstations and the like is based upon the following sequence of steps:

1. *Preparation:* A careful review of the scope of materials covered by the court order and preparation of a plan of the materials likely to be present and seized.

2. *At the site of the investigation:*

    a. Take notes of everything at the scene (cabling, the lot), photograph everything especially the screen for a system that is switched on.

    b. Document all actions.

3. *Shut down:* This step needs careful consideration and depends on circumstances and the competence of the investigator—there are variations depending on whether a disk is seized only (unusual) or an entire computer, whether or not the computer is networked, and depending upon the platform being used (see [66] for a reference to U.S. Department of Energy shut down guidelines for different platforms such as MS-DOS, UNIX/Linux, Mac, and Microsoft Windows 3.x/95/98/NT):

    a. If the computer is switched off, the accepted wisdom is that a computer which is switched off should be left that way, this immediately allows disk removal in cases where only the disk(s) are to be seized.

    b. If the computer is switched on:

      (1) The previous step has recorded the display on the screen.

      (2) If there is a display on the screen, note the time displayed; in any case enter whatever command is needed to display the time depending on circumstances and the competence of the investigator (the time is useful in order to synchronize, for example, file access times with real time); display the time maintained in the system CMOS.

      (3) There are circumstances in which an experienced investigator will enter commands to display the processes being executed, files that are currently open and the currently open network connections. This step may be important if it is suspected that network connections are relevant or are being used to store what may be relevant evidence.

        For what to do with a networked computer system, the booklet produced jointly by the U.S. Secret Service and the International Association of Chief of Police says simply [44]: "Networked or business computers—Consult a Computer Specialist for further assistance—Pulling the plug could: Severely damage the system, Disrupt legitimate business, Create officer and department liability." For further discussion of what to consider when dealing with a networked computer, see Chapter 6.

        In the absence of expert advice, which could be from a member of the team or from a reliable employee not implicated in the investigation, the system should be shut down as follows:

      (a) For a Microsoft Windows platform, disconnect the power cord from the back of the computer.

      (b) For a UNIX box, do a safe shutdown.

    c. If permissible by court order (or after seeking a further order), take other relevant material and at least note its presence—this may be information or material unconnected to or with the computer (e.g., fax machine, boxes of floppies, and tapes).

    d. Document all actions.

4. *Seizure:*

    a. Carry out appropriate labeling and packaging.

    b.   Document all actions.

5.   *Imaging* (at the laboratory, or sometimes at the site of the investigation):

    a.   Boot to a known, trusted operating system (see Section 2.2); alternatively, execute a secured command line interface or shell from removable media.

    b.   Ensure that the appropriate write-blocking software or hardware measures are in place (see Section 2.2).

    c.   Image and authenticate disks or files to be duplicated (see Section 2.2).

    d.   Document all actions.

6.   *Physical analysis* (at the laboratory, but possibly also as a preliminary step to logical analysis, note, however, that this step may in some circumstances be omitted entirely):

    a.   *Sector-by-sector* analysis of disk image (ambient data), to identify hidden data or accidental residues or suspicious disk structure.

    b.   Document all actions.

7.   *Logical analysis* (at the laboratory, or sometimes at the site of the investigation):

    a.   Boot to an operating system that supports the file system of the seized disk image.

    b.   *File-by-file analysis* for keywords, phrases and keeping a record of file metainformation in particular all time information (e.g., time created, time last accessed, time last written to).

    c.   Document all actions.

Guidance Software's *EnCase Legal Journal, Second Edition* (2001) [67] is a very detailed and informative account of computer forensic seizure and search practices albeit naturally focused on its widely used Encase forensic software. That publication also provides a careful examination of legal issues and precedents in the U.S. context, relating in particular to the admissibility and authentication of electronic evidence and issues relating to conformance with warrants, a topic we return to in Section 2.4.1.1. (admissibility is addressed also in Section 3.3.2).

2.4.1.1 Seizure and warrants

One of the legal complications that can seriously undermine the successful prosecution of an investigation relates to the particularity of the search warrant used to authorize seizure of a suspect computer or disk. Guidance Software reports on this issue [67] in the context of the use of their EnCase software. That report gives a detailed discussion of recent U.S. cases, especially *United States v. Carey,*[1] which turned on the extent to which a search warrant pitched in general terms at seizure of a desktop or laptop computer can allow an investigator to go fishing for any and all files resident on the disk or computer and indeed the extent to which search warrants need to be more specific in terms of the particular file type or file content allowed to be investigated. The report concludes with the observation that post-*Carey* practice and expectations are that searching an entire drive or cartridge is permissible as long as the search is restricted to the terms of the warrant. Furthermore, the report notes that this includes the recovery of deleted files as determined in *United States v. Upham*[2] in which context "the court held that the recovery of deleted files pursuant to a search warrant . . . was valid and did not exceed the scope of the warrant." If material outside the terms of the issued warrant is encountered, then the investigator needs to apply for a supplemental warrant specifying the broadened scope in order to search explicitly for materials of the (new) sort.

Bartlett v. Weir Ors in the Federal Court of Australia: Tasmania (1994) [68] likewise turned on this issue. The offence with which the applicant was sought to be convicted was "Imposition and operation of a bank account in a false name under s 29B Crimes Act 1914 (Cth) or s 24(2) Cash Transaction Reports Act 1988 (Cth)." The search warrant authorized the seizure of

> . . . things being: 3.5 and 5.25 in. computer disks containing information in relation to the payment or receipt of monies involving the Unemployed Workers' Union, including the Community Resource centre; A–Z Desktop Publishing Bureau and Wholefoods store, computers of any brand name, including visual display units, keyboards, control units, printers, modems, diaries, receipt books, receipts, telex, cash books, . . .which there are reasonable grounds for believing that the same will afford evidence as to the commission of an offence against Section 29B *Crimes Act* 1914, or Section 24(2) *Cash Transaction Reports Act* 1988, both laws of the

---

1. 172 F.3d 1268 (10th Cir. 1999)
2. 168 F.3d 532 (1999)

Commonwealth namely the imposition and Operation of a Bank Account in a false name.

The disks seized included approximately 400 floppy disks containing several thousand files including a box of floppy disks seized from Wright's room. Wright at that point informed the Australian Federal Police (AFP) officers that the disks contained Wright's personal material, which were not connected with the Unemployed Workers' Union (UWU). One such disk after being examined on one of the UWU computers was returned to Wright's possession. Some of the floppy disks seized were neither clearly labeled nor indexed. The crucial issue turned out to be that the warrant did not permit the seizure of all floppy disks found at the premises. It rather permitted the seizure of floppy disks containing information in relation to the payment or receipt of monies involving the above-named entities and in the view of the court no attempt had been made to ascertain such contents. Given the availability of a computer expert, the AFP officers made no attempt to ascertain the information stored on the disks, with the exception of the inspection of Wright's one returned disk, which was only conducted upon Wright's express request. As a result, it was concluded by the court, given the AFP officers' inability to distinguish the contents of files without examination they were not justified in simply removing all floppy disks found with the intention to ascertain their contents at a later time. In removing the floppy disks in this fashion, the AFP officers failed to comply with the fundamental obligation imposed upon persons executing search warrants. The outcome of the case was that the evidence obtained pursuant to the search warrants was held inadmissible due to the invalid execution of such warrants.

The clear lesson deriving from this case as with *United States v. Carey* is that the party seizing the evidence must ensure that the items seized fall within the terms of the warrant. A point of interest regarding how jurisdictional differences can have an impact is that had the *Carey* case occurred in Australia, the seizure of the additional material would have been allowed by warrant and admitted in court given that Australian warrants for seizure are less restrictive.

2.4.1.2 Time attributes of files, metainformation and event logs

Metainformation is the term used for the information that describes the properties or attributes of an information object such as a file or e-mail message. The prefix *meta* serves to emphasize that metainformation is not the actual information itself (i.e., it is not the actual content of the file or e-mail) but rather a description of some attribute of the information such as when it was generated or how it is managed or who it is addressed to. The most easily available such metainformation when it comes to disk files is the time last accessed for the file. That information can easily be displayed using a "listdir" command (i.e., a ListDirectory command) or its equivalent. That command will list the name of each of the files within a folder or directory and for each file it will also display file metainformation such as date and time last accessed.

The term *MACtimes* [69] has been used for the following three time attributes of a file:

1. LastWriteTime (M for modification);

2. LastAccessTime;

3. CreationTime.

This information can be vital in a forensic investigation in establishing when a file with relevant content was generated or accessed, and in correlating this information with other time-lining information (time-lining is discussed in Section 2.4.1.3). While these times, accessible via the UNIX "ls" command and various proprietary tools on Microsoft Windows platforms, provide a potential mine of information, Farmer [69] notes that such information needs to be gathered and treated with care as systems can interpret time last accessed in different ways. In particular, one needs to know how accurate the timestamps are, both in terms of how accurate the system clock for the host system is and also that the timestamps have not been altered. In UNIX, for instance, there is the touch command which allows the time last accessed attribute of a file to be updated to current time, something which can be useful in certain circumstances in a software development environment but if not taken into account in an investigation will mislead.

Farmer also notes that systems unfortunately do not typically keep a log of all such times, that is, do not keep a log of all the times when a file was accessed or modified; while adding to the minutiae which an investigator would need to manage, this would in many situations be an enormous boon. As it happens, some such information will be available if the investigator is fortunate enough to have access to file archives—in this case, the investigator will have access to file MACtimes relevant at the time that a file archive was produced, and furthermore to the contents of files archived at that time. This may be useful in its own right (in the case of deleted files), serve as extra time-lining information to aid an investigation (with regard to the evolution of a file over time), or serve simply as an useful and separate check of imaged files. Event logs are themselves archived from time to time by most systems, and once again both the MACtimes and contents of such archived logs may be available and if so may be useful in their own right or as an independent and useful check of imaged log files.

While the UNIX "ls" command and the various Microsoft Windows utilities will provide MACtimes as well as file size and possibly other information also such as access rights, they will typically not identify the last person to access a file. Nonetheless, the access rights associated with a file will at least indicate which users are allowed to access a particular file thereby providing some extra information about who could have accessed it. To get more detailed information than this requires information from log files that are maintained by the operating system or by some applications such as Web browsers.

UNIX-based operating systems have a sophisticated logging capability and can be configured to record a wide range of events. For instance, one can typically configure UNIX systems to record events of the following sort in files in the directory */var/adm* or */var/log*:

1. */var/adm/messages*—system messages;

2. */var/adm/lastlog*—the most recent login time for each user in the system;

3. */var/adm/utmp* and */var/adm/wtmp*—information such as the terminal line, login time, logins and logouts since reboot;

4. */var/adm/acct*—the system accounting file which if enabled records accounting information (username, command, CPU time used, timestamp of the process, status);

5. */var/adm/sulog*—records everyone who has sued on the system;

6. */var/log/syslog*—miscellaneous events notified through the system's *syslogd* facility.

These event records can then be processed and correlated using a variety of utilities, and indeed intrusion detection systems (IDS), a topic we return to in Chapter 6, rely precisely upon such logs for their success, at least in the case of host-based IDS. The analysis of event logs for forensic purposes is addressed in Chapter 6.

The Solaris SHIELD Basic Security Module extension (BSM) to Sun MicroSystem's Solaris provides an even more powerful logging facility. It extends the Solaris security features to Orange Book C2 level auditing [70] and produces log records with information which includes the following:

> Event description, time and date;
> User's audit ID, effective userid, effective group ID, real userid, real group ID, process ID, session ID, and terminal ID;
> Return code.

Turning to Microsoft Windows systems, we find that Microsoft Windows NT/2000/XP all have powerful event logging systems of their own. With audit policy enabled, Microsoft Windows NT/2000/XP allows for the tracking of events by recording security events in the Security Event log (there are several other logs in addition to the security event log). The security event log is then available for subsequent viewing (see Figure 2.6) and processing whether this be for intrusion detection or post hoc analysis. When configuring the audit policy, one can track either successful or unsuccessful outcomes (or both) for a wide variety of operations. The security event log entries (see Figure 2.7) show which actions have been performed, by whom, and the date and time of the action. Once again, there is a wealth of information available to both an IDS or to the forensic investigator for testing hypotheses and providing evidence for and against such hypotheses.

2.4.1.3 Time-lining

Time-lining [71] is concerned with time-tagging or the association of timestamps with each event or data item of interest. This is of course crucial

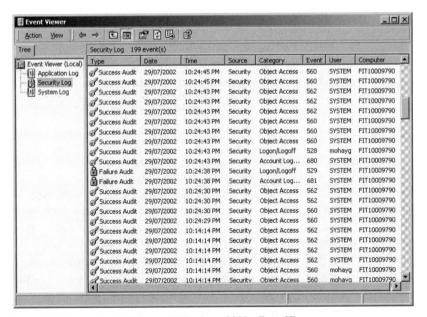

Figure 2.6 Screenshot of Microsoft Windows 2000—EventViewer.

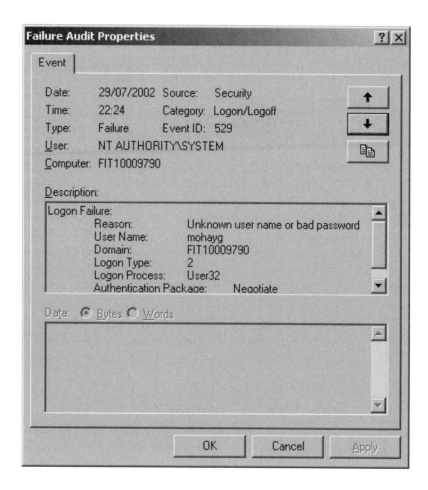

Figure 2.7 Screenshot of Microsoft Windows 2000—EventProperties.

to any investigation whether or not it must stand the test of evidentiary requirements in a court of law. The usual *MACtimes* (see 2.4.1.2) will provide time last accessed and time last written information for each file, and indeed the standard operating system commands such as "ls" in UNIX allow files to be listed in order of their time attributes so this is one of the routine things that can be done early on during a time-lining analysis. This information is then used to correlate with other sources of time-related information derived from file content and elsewhere in order to build up a time graph of activities. If imaging has occurred at the file level as opposed to at the level of disk imaging, then these time attributes must have been copied across faithfully from the original evidence. Some copy methods update that information to reflect the time of the copy operation itself which defeats the purpose of attempting to use the time information for investigative purposes. For instance, the UNIX ''cp'' (for copy) command has a ''-p'' switch to preserve the time attributes of the original source file.

It is critical that an investigation maintains a consistent view of time across the different events being analyzed during an investigation. Hence, an investigator must be able to relate the local timestamps of the captured disk image and its files to real-time (i.e., the investigator must capture the local time of the host machine at the time of disk seizure, or at the time of disk imaging if the entire computer is captured). The reason for this is simple: if file timestamps are pertinent to the case, then it must be demonstrated that the times are meaningful and consistent with the timeframes of related noncomputer evidence and with the timeframes of electronic evidence derived from other computers. In the case of investigations involving e-mail, the relative times at which different but related e-mails were posted may well be vital. In such cases, the integrity of the *time-posted* values must be demonstrable, as must evidence supporting the supposition that the possibly different local clocks used for those timestamps were in synchronization and that the relative timestamps are consistent with the case being made.

As mentioned earlier, time-lining is an important and commonly used tool in computer forensics and is discussed also in Chapters 3, 4, and 6.

2.4.2 National and international standards

It is clear and has been for some time that there is a need for standard procedures in the area of computer forensics for procedural and judicial consistency. They are also needed in order to assist LE to detect, track, and prosecute computer crime and cybercrime across jurisdictions which in turn requires standard protocols for cooperation and interworking. Both the protocols and standards require to be sanctioned and ultimately mandated across the various jurisdictions involved. A recent article whose authors include Mark Pollitt, one of the early leaders of the *International Organization for Computer Evidence* (see later) emphasizes this point [72]. The paper also makes the interesting and important observation that computer forensic science is market-driven and this has important implications for training and certification, something we return to in Chapter 3.

The best known and most influential set of guidelines relating to the seizure of electronic or digital evidence is that produced in 1994 by the Computer Crime and Intellectual Property Section (CCIPS) of the U.S. Justice Department titled ''Federal Guidelines for Searching and Seizing Computers'' [73]. Two additional supplements to these guidelines were produced in 1997 and 1999. The guidelines have been substantially revised in 2001 [74] in light of changes in the way in which computers and users make use of the Internet and in light of the growing body of law governing these activities. The guidelines emphasize the legal issues involved and a good summary of this aspect is provided by Yair Galil in *The Internet Law Journal* [75]:

> The document provides LE officials with a condensed overview of the constitutional and statutory framework applicable to the "search and seizure" of computer data and network traffic. It also provides summary descriptions of the Fourth Amendment's limitations on "search and seizure," the Electronic Communications Privacy Act, the Wiretap Act, the Cable Subscriber Privacy Act, and the Privacy Protection Act. It then explains what exceptions and loopholes exist in each protective statute, and ends by reviewing pertinent case law. In addition to providing a legal review of the statutes themselves, the document also offers practical advice on strategic planning for the "search and seizure" of computer records, touching on issues such as the optimal composition of a work team, as well as a recommended checklist of preparations.

Chapter 4 of the guidelines is particularly useful in navigating the complexities of the kinds of warrants and court orders that are required in the different circumstances that may arise:

> Two federal statutes govern real-time electronic surveillance in federal criminal investigations. The first and most important is the wiretap statute, 18 U.S.C. §§2510-22, first passed as Title III of the Omnibus Crime Control and Safe Streets Act of 1968 (and generally known as "Title III"). The second statute is the Pen Registers and Trap and Trace Devices chapter of Title 18 ("the Pen/Trap statute"), 18 U.S.C. §§3121-27, which governs pen registers and trap and trace devices. Failure to comply with these statutes may result in civil and criminal liability, and in the case of Title III, may also result in suppression of evidence.

> In general, the Pen/Trap statute regulates the collection of addressing information for wire and electronic communications. Title III regulates the collection of actual content for wire and electronic communications.

In the meantime, since original publication of the CCIPS guidelines in 1994, the United Kingdom's Association of Chief Police Officers (ACPO) established its Computer Crime Group in 1996 and published the "Good Practice Guide for Computer Based Evidence" which identifies four principles relating to the seizure and handling of computer evidence. These are [76]:

> ▶ *"Principle 1:* No action taken by the police or their agents should change data held on a computer or other media which may subsequently be relied upon in Court.

> ▶ *Principle 2:* In exceptional circumstances where a person finds it necessary to access original data held on a target computer that person must be competent to do so and to give evidence explaining the relevance and the implications of their actions.

> ▶ *Principle 3:* An audit trail or other record of all processes applied to computer-based evidence should be created and preserved. An independent third party should be able to repeat those processes and achieve the same result.

> ▶ *Principle 4:* The onus rests with the officer in charge of the case to ensure compliance with any law pertaining to the possession of, or access to information contained on a computer. The officer must be satisfied that the use of any copying device or actions of any person having access to the computer complies with these laws."

These guidelines are used widely throughout the United Kingdom.

In addition to the above two guidelines, the U.S. Secret Service and the International Association of Police have jointly produced their *Best Practices for Seizing Electronic Evidence* booklet [44] which places more emphasis on the actual seizure itself and the technology involved. It addresses seizure of not only desktop or networked computers but also other electronic storage devices, such as mobile phones, pagers, fax machines, and smart cards.

Another guide relating to the seizure of computers and computer-storage devices is the U.S. National Institute of Justice (NIJ) Guide entitled "Electronic Crime Scene Investigation: A Guide for First Responders." The decision to produce this guide goes back to May 1998, when the National Cybercrime Training Partnership (NCTP), NIST's Office of Law Enforcement Standards (OLES), and the NIJ, decided to collaborate on possible resources that could be implemented to counterelectronic crime. Crime scene investigation, nominated as the topic for the first guide was published in July 2001 and can be consulted at [77]. It is comparable to the USSS and IACP guide above in that it provides a checklist for first responders, with a greater level of technological detail but less emphasis on exposition of principles and the legal context than in the case of the ACPO and CCIPS guides, respectively. The remaining booklet topics that have been identified are as follows:

1. Managing Technology in Law Enforcement;

2. Analysis of Computer Evidence;

3. Investigative Use of Technology;

4. Investigating Technology Crimes;

5. How to Setup a Digital Evidence Laboratory;

6. Courtroom Presentations of Digital Evidence.

The National Center for Forensic Science (NCFS), with NIJ support, has taken responsibility to manage the planning panel for the booklet "How to Setup a Digital Evidence Laboratory?"

One of the earliest international developments relating directly to standards for electronic evidence and computer forensics was the establishment of the IOCE. We can trace the beginnings of the IOCE back to 1993 when the first International Conference on Computer Evidence took place in the United States. The IOCE was officially instituted in 1995 coincident with the second conference in Baltimore, Maryland. The theme of the highly successful third conference in 1996 in Melbourne, Australia, was *"The World Wide Web of Crime: Who's Controlling the Traffic?"* and represented the first attempt by the IOCE to deal with the burgeoning problem of crimes committed on the Internet. The fourth conference was held in The Hague in 1998 and heralded a renewed commitment to foster cooperation between members and to act as an international voice for the computer forensics community. Membership in the IOCE had grown considerably in the preceding years and at that stage included 45 agencies representing 25 countries. In 1997, the IOCE and the G8 group of nations determined independently to develop standards relating to digital evidence. This was followed in 1998 by the G8 requesting the IOCE to undertake this task and the U.S. Scientific Working Group on Digital Evidence (SWG-DE) was formed in the same year to be a focus for U.S. participation. See [78] for a joint publication by the IOCE and SWG-DE on computer evidence. The next, seventh conference in the series was held in Orlando, Florida, United States from May 5 to 10, 2002 [79].

Some of this history of the IOCE and of the SWG-DE is described in [78].

2.5   Computer crime legislation and computer forensics

2.5.1   Council of Europe Convention on Cybercrime and other international activities

One of the most significant happenings in recent times regarding the future of computer crime and cybercrime legislation around the world and implications for the investigation and prosecution of transnational computer crime has been the development and then acceptance on November 23, 2001 of the Council of Europe *Convention on Cybercrime* [1, 80]. The convention or treaty was on that date signed by 26 member states of the Council and by 4 nonmember states—Canada, Japan, South Africa, and the United States—and the CoE portal [81] indicates the legal obligations that devolve from such conventions on states [82]. The treaty will enter into force when it is ratified by five states, at least three of which must be members of the Council of Europe. Ratification by states will typically require legislative changes at the national level although the changes may in some cases be relatively minor as a result of antiterrorism legislation passed in the interim in reaction to the events of September 11, 2001.

Countries that ratify the convention will be obliged to adopt laws that are consistent with it in dealing with the areas with which the convention is concerned:

> ... crimes committed via the Internet and other computer networks, dealing particularly with infringements of copyright, computer-related fraud, child pornography and violations of network security. It also contains a series of powers and procedures such as the search of computer networks and interception.

> Its main objective, set out in the preamble, is to pursue a common criminal policy aimed at the protection of society against cybercrime, especially by adopting appropriate legislation and fostering international cooperation. [80]

As a result, legislative action arising out of the convention will focus on the following:

▸ Dealing with crimes relating to infringements of copyright, computer-related fraud, child pornography, and violations of network security;

▸ Development of procedures for dealing with computer crime in general and the collection of electronic evidence;

‣ Achievement of interworking and cooperation between agencies which are separated by national boundaries and which work within separate jurisdictions and operational environments.

These issues have been increasingly and repeatedly raised in international fora and in the literature. The G8 group of nations has identified such issues as amongst the most important in prosecuting computer crime and cybercrime in its oft-quoted and influential list of principles and actions arising out of the G8 Meeting of Justice and Interior Ministers in Washington, District of Columbia in 1997 [83]. These principles and actions were endorsed by the respective heads of state in the G8 Communique on Hi-Tech Crime issued at the following 1998 Birmingham summit meeting.

The reader is referred to Michael Sussmann's detailed account [84] of some of the significant challenges presented by escalating transnational computer crime and the steps being taken by the major multilateral organizations such as the CoE, the EU, and the G8 to deal with it. Since its 1998 Birmingham summit, the G8 has been active in developing principles for *Transborder Access to Stored Computer Data* (as reported in [85]) for the purposes of facilitating computer crime and cybercrime investigation and prosecution across national jurisdictions. Arising out of this has been the successful realization of a global network of 24/7 "one-stop shop" national cybercrime contacts, a particularly significant operational outcome that also lays the foundation for increased cooperation between nations and jurisdictions in the future.

Not unexpectedly, there has been opposition to the CoE treaty from civil liberties groups which will no doubt also oppose the second of the two additional protocols already planned for the treaty. Two protocols are likely to present problems for some states, which have already accepted the main body of the treaty are:

1. The *First Protocol* deals with the criminalization of acts of a racist or xenophobic nature committed through computer systems [86].

2. The *Second Protocol* deals with antiterrorism measures and particularly how to deal with the use of computers and the Internet by terrorist for their communication [87].

There have also been technological criticisms leveled at the convention. Considerable efforts have been made by those drafting the convention to attempt to resolve some of those issues (e.g., including exemptions from the cracking restrictions in the case of security testing). On the other hand, the detailed nature of the requirements relating to divulging encryption keys, concerns relating to intellectual property, and the nature of the as yet unfinalized second protocol still needs to be addressed.

In the meantime, there is still great variation across nations regarding the computer crime and cybercrime legislation they have in place and the extent to which their legislation deals with such crimes and allows them to cooperate with other nations when dealing with transnational crime issues. For instance, it was the shortcomings of Philippine law that prevented prosecution of the alleged author of the Love Bug virus in 2000. A report prepared by McConnell International *Cyber Crime . . . and Punishment? Archaic Laws Threaten Global Information* [88] analyzed the computer crime legislation of 52 countries and found that only 10 (ironically including the Philippines) had substantially revised their legislation to address some forms of cybercrime: Australia, Canada, Estonia, India, Japan, Mauritius, Peru, Philippines, Turkey, and the United States. Another nine had embarked upon some amendments. The report provides a summary of how the 19 countries address 10 broad categories of computer crime but presents no further detail. A detailed discussion of the report is provided by Illena Armstrong in "Legislators Turn up the Heat on Cybercrime" [89]. That article also discusses a more detailed report on computer crime legislation around the world by Stein Schjolberg [90].

Computer crime legislation and computer crime investigation techniques have not surprisingly moved more or less in parallel. Changes in computer usage brought about by increased Internet connectivity have resulted in the definition of new crimes or new variations of old crimes in order to assist successful prosecution of criminal behavior involving a computer, behavior which may have been difficult to prosecute under previous legislation. Some of the notable new crimes that have been defined include hacking into a computer system, and data and identity theft, and there will no doubt be other new computer crimes defined in the future. As an aside, we note that it was still true in the United Kingdom until recently that the use of a stolen credit card or credit card number to make a purchase in which the financial part of the sale is conducted entirely by machine or computer was not a crime. The reason for this is that it was at that time impossible in law to *deceive a machine,* something that came as a surprise to the lawmaker who was quizzing a police officer on the topic of computer crime [91].

The evidence for crimes involving computers—old crimes or new—consists of digital information which is often temporary and which must be exposed in a manner and form which leads to the provision of plausible and preferably admissible evidence. It is partly a need to expose such information that has prompted governments in the United States and the United Kingdom to introduce procedures and legislation (see Section 2.5.2) to enable LE and national security agencies to acquire traffic and content information from ISPs in a more systematic manner, and more easily, than previously. A more recent and pressing push has come from the terrorist attacks of September 11, 2001 which have resulted in the passage of antiterrorist legislation around the world (see Section 2.5.3). Data retention legislation in particular will ensure an even greater availability of such information than had ever been expected before September 11. Investigative techniques rely inevitably on such information in order to establish a digital trail, either to lead to a conviction or to yield intelligence information. The richer the information the more likely a successful outcome, but by the same token the more sophisticated must be the process of forensic investigation.

Other multilateral bodies with significant interests in computer crime and cybercrime investigation and prosecution which have been instrumental in significant developments in this regard are the United Nations and the European Commission. The UN in particular has had a long involvement in addressing computer-, telecommunications-, and network-related crime and in developing a harmonized approach to transnational and computer crime. In 1994, it produced a comprehensive manual on the prevention and control of computer crime and cybercrime titled *International Review of Criminal Policy—United Nations Manual on the Prevention and Control of Computer-related Crime*. More recently, it has produced, in May 2001, a timely report titled *Conclusions of the Study on Effective Measures to Prevent and Control High-Technology and Computer-Related Crime, Report of the Secretary-General* [92]. The report includes the following:

> It further recommends that the Commission at its 11th session consider a series of options for further action, including the possible drafting of an international instrument against computer-related crime and options for a short-term strategy, including the establishment of a United Nations global programme against high-technology and computer-related crime. It also provides information about the activities of other relevant international and intergovernmental organizations and seeks to respond to some of the concerns raised by individual Member States.

The report is a succinct yet comprehensive account of recent developments at the international level and of the arguments for and against greater control of the Internet. It concludes by setting the scene for greater controls on the Internet and greater international cooperation to combat transnational computer crime.

The European Commission has been active in dealing with computer crime concerns through its member organizations. It has historically focused primarily on individual rights and safeguards with regard to privacy and personal data and has been active also in the e-commerce area. Prior to September 11, it has been engaged in developing proposals for specialist computer crime police units in member countries and proposals for harmonizing penalties for serious computer hacking offences [93], thus reflecting some of the recommendations expressed by the UN (see earlier). As in the case of the United States and the United Kingdom, it has recently been preoccupied with antiterrorism initiatives and this is discussed later.

2.5.2 Carnivore and RIPA

The trend towards increased monitoring, towards further restriction of what constitutes legitimate Internet activity and towards increased controls on ISPs and communications service providers (CSPs) was already in place well before September 11. At the national level there were two particularly noteworthy developments in the English speaking world:

1. The introduction of Carnivore (or DCS1000 as it is now titled though the previous name seems hard to shake off) in the United States.

2. Enactment of the RIPA legislation in the United Kingdom.

Existence of Carnivore (initially called Omnivore, now called DCS1000 for Digital Collection System) and its use by the U.S. Federal Bureau of Investigation was revealed to the public in 2000. Its use has attracted considerable controversy and opposition in much the same way (and for similar reasons), as has enactment in the same year of the RIPA legislation in the United Kingdom. In both cases there have been strongly voiced concerns that privacy rights of the individual will be threatened and that there are insufficient safeguards to prevent that happening.

Subject to a number of caveats, there are in both cases safeguards built into the regulated procedures, which if properly observed will be effective in protecting privacy rights. The two caveats are that the systems put in place by those procedures operate correctly according to specification and that they are correctly deployed. Carnivore is a combination of hardware and software and allows for the FBI to monitor ISP and CSP communications by intercepting their traffic. In principle, such a system can monitor anything and everything including packet and e-mail headers and packet and e-mail content, that is, it is able to access both *traffic data* and *communications content*. (Somewhat misleadingly, the word *data* in this context does not mean the data or content contained in a communication, rather it means the data *about* a communication namely, the information about how the communication is directed and handled and in the case of e-mail, header information.) Some of the controversy surrounding both Carnivore and RIPA is to do with the distinction between these two types of surveillance, and the extent to which access to *traffic data*, which is more easily permitted than access to *communications content*, may provide LE and government agencies the means for intentional or unintentional *back door* access to *communications content*. RIPA uses the term *communications data* as a synonym for *traffic data*.

At the instigation of the U.S. Congress, the FBI arranged for Carnivore to be evaluated by the Illinois Institute of Technology. The draft [94] and final [95] evaluation reports by the IIT Research Institute (IITRI) gave Carnivore a clean bill of health apart from the need for better auditing of its use and improved data security, recommendations that the FBI has accepted. The draft report [94] linked to the U.S. DOJ FBI Carnivore page notes that Carnivore is used to achieve court-ordered surveillance of electronic communication when other means are unsuitable either because of the needs of the investigation or because of restrictions imposed by the court. It goes on to emphasize that Carnivore can be used to access either content or *traffic data* and that rigorous procedures are followed in seeking and obtaining proper authorization of either, and furthermore that installation is carried out by a separate team from the case agents who establish and justify the need for the surveillance in the first place. Interestingly, the report also notes that at that date all installations of Carnivore had been carried out by one small team.

This puts to rest some of the doubts that have arisen from some very confusing and conflicting reports on Carnivore and its intent. Carnivore can monitor and record both traffic and content, if configured to do so. Two key sentences in this regard appear in the draft report under "ES.5 CONCLUSIONS:"

> In response to the DOJ's four questions, IITRI concludes when Carnivore is used correctly under a Title III order, it provides investigators with no more information than is permitted by a given court order. When Carnivore is used under pen trap authorization it collects TO and FROM information, and also indicates the length of messages and the length of individual field within those messages possibly exceeding court-permitted collection.

> While the system was designed to, and can, perform fine-tuned searches, it is also capable of broad sweeps. Incorrectly configured, Carnivore can record any traffic it monitors.

This reinforces the point made earlier: if properly implemented, and properly configured and deployed on a case-by-case basis—and for many critics these are sticking points—Carnivore should be no different in effect to previous procedures.

Kevin Di Gregory, Deputy Assistant Attorney General, U.S. Department of Justice, described the Carnivore system and its intended use in a statement to a committee of U.S. Congress on July 24, 2000. His statement "Carnivore and the Fourth Amendment" appears in [2], and the FBI describes Carnivore in [96]. The Electronic Privacy Information Center (EPIC) contains information regarding Carnivore obtained through provisions of Freedom of Information Act (FOIA) in [97]. The FBI has stated that it will not release the software source code for public scrutiny.

The U.K. RIPA [3] was enacted in 2000 and repeals sections of the Interception of Communications Act 1985. RIPA is intended to update existing law covering interception of communications to take into account technological change, in particular the development and widespread use of the Internet, and to align more closely with the Human Rights Act and European Commission on Human Rights. The act comprises five parts:

1. Interception of communications and the acquisition and disclosure of communications data;

2. Surveillance and covert human intelligence sources;

3. Investigation of electronic data protected by encryption;

4. Scrutiny of investigatory powers and codes of practice;

5. Miscellaneous and supplemental.

Passage of the bill was surrounded by considerable controversy and it is still controversial with uncertainty regarding much of its operation. In late 2001, the Home Office published a draft Code of Practice on Accessing Communications Data [98], which refers to the collection of communications (i.e., traffic) data and not content, seeking public consultation with LE and industry at large. The safeguards prescribed by RIPA to ensure that the surveillance provisions of the bill are not abused in regard to *communications data* are as follows [98]:

▸ "The Act (5.1) provides two different ways of authorizing access to communications data; through an authorization under section 22(3) and by a notice under section 22(4). An authorization would allow the relevant public authority to collect or retrieve the data itself. A notice is given to a postal or telecommunications operator and requires that operator to collect or retrieve the data and provide it to the public authority which served the notice. A designated person decides whether or not an authorization should be granted or a notice given."

‣ A *relevant public authority* means the NCIS, the National Crime Squad, HM Customs and Excise, the Inland Revenue, the Security Service, the Secret Intelligence Service, and the Government Communications Headquarters.

‣ The *designated person* must be of the rank of Superintendent or equivalent, except for billing information where Inspector or equivalent is sufficient.

With regard to interception of *content*, there are strict procedures to be met in applying for an interception warrant and as previously the warrant has to be approved personally by the Secretary of State ("the Secretary of State must have given personal consideration to the application") [99]. To give some idea of the level of such applications, although one speculates that the number has risen sharply since 1998, the U.K. Home Office notes the following [100]:

‣ "In 1998, 2031 interception warrants were authorized by the Home Secretary and the Secretary of State for Scotland.

‣ On and average, one in every two interception warrants that I issue results in the arrest of a person involved in serious crime. (Jack Straw, Home Secretary)"

Concerns relating to implementation of the bill are highlighted in an independent report prepared for the British Chambers of Commerce June 12, 2000 on RIPA entitled "The Economic Impact of the Regulation of Investigatory Powers Bill" [101]. The report criticized both the potential erosion of civil liberties and the potential costs represented by the bill, and commented

> The practical implications of RIP will depend to a great extent on the provisions in secondary legislation, and the scope of the anticipated Code of Practice. The fact that the government has failed to provide details of either has placed U.K. business at a great disadvantage in assessing the legislation.

It is the third part of the act relating to encryption that has come up against most criticism. For instance, this part of the Act can require ISPs and CSPs (when served with a notice) to reveal encryption keys in order to enable decryption of information. In the bill as enacted, the onus is on LE to demonstrate that the recipient of the notice is in possession either of the key or the protected information but this was absent from an early draft. Concerns have also been expressed about the following:

▸ The threat to the privacy of individual e-mail (these concerns relate to uncertainty about the extent to which traffic and content surveillance will be kept separate in practice).

▸ The burden placed upon the recipient of a notice (e.g., an ISP) by the potential offence of *tipping-off* since complying with this provision, that is, to say divulging an encryption key to the authorities without notifying the key's owner, may expose them to legal action from non-U.K. jurisdictions [101].

▸ The unreasonable burden potentially placed upon ISPs and the like by the possibility of further regulation to do with data retention. For instance, a recent report indicated that the U.K. government was considering requiring communications traffic data to be retained for a period of 7 years [102].

Clearly developments such as RIPA and Carnivore will assist the computer forensics investigator and aid the prosecution of computer crime. In particular, developments with respect to ratification of the CoE treaty, implementation of RIPA in the U.K., and developments within the EU will be important in the near future. It is clear that ISP regulatory requirements, and corresponding liability protection measures, are needed in order to make the Internet safer. This may be achieved through consultation and voluntary codes. The latter is already happening to some extent and a combination of regulatory and self-imposed controls will likely become the norm. One problem for the future which will persist is dealing with encryption, government will continue to seek strategic ways of dealing with this such as the use of key escrow and similar solutions, while forensic scientists will continue to attempt to utilize cryptanalysis techniques where this is appropriate and economic. We return to the issue of encryption in Chapter 7.

2.5.3 Antiterrorism legislation

2.5.3.1 The Patriot Act

The U.S. Patriot Act was enacted in the aftermath to the events of September 11, 2001, and took effect as of October 26, 2001. Among other things, the Act legislates for increased electronic surveillance powers of LE and national security agencies and increased penalties for certain computer crimes. It has resulted in some significant changes to related Federal Statutes, which address the searching and seizing of computers and the seizing of electronic evidence. The Computer Crime and Intellectual Property Section (CCIPS) of the U.S. Department of Justice has subsequently issued a memorandum [103], which provides an overview of these changes, including their rationale and intended effect. That document lists 11 sections of the Act relating to such *New Authorities* of which the following five are especially noteworthy:

1. *Section 210 scope of subpoenas for electronic evidence:* Foremost amongst the provisions of this section are that the list of records that investigators can obtain with subpoena will in future be allowed to include financial records such as credit card number or the details of other forms of payment for a communication service. This will materially assist the determination of a network computer user's true identity, given that other user identification information maintained by a service provider is often not very useful as it can be deliberately falsified.

2. *Section 212 emergency disclosures by communications providers:* The memorandum notes that the law in this regard had previously been inadequate for two reasons. On the one hand, the law had not previously explicitly safeguarded ISPs, for instance, from civil action in the event that they disclosed account or communications information in good faith on suspecting intent to commit a terrorist act. On the other hand, the law had not previously explicitly allowed a provider voluntarily to provide noncontent information to LE for their own self-protection.

   The provisions of this section address the above issues.

3. *Section 216 pen register and trap and trace statute:* [3] The memorandum notes as follows: "The pen register and trap and trace statute (the pen/trap statute) governs the prospective collection of noncontent traffic information associated with communications, such as the phone numbers dialed by a particular telephone. Section 216 updates the pen/trap statute in three important ways:

---

3. These terms refer historically to the recording of telephone-related information, in particular to the recording of numbers dialed (Pen Register) and to the recording of the telephone numbers of incoming calls (Trap and Trace).

(1) the amendments clarify that LE may use pen/trap orders to trace communications on the Internet and other computer networks, (2) pen/trap orders issued by federal courts now have nationwide effect, and (3) LE authorities must file a special report with the court whenever they use a pen/trap order to install their own monitoring device (such as the FBI's DCS1000) on computers belonging to a public provider."

4. *Section 220 nationwide search warrants for e-mail:* The provisions of this section relate to the issuing of search warrants for the disclosure of unopened e-mail. Previously, jurisdictional aspects of the law have led to some courts declining to issue search warrants for the disclosure of such e-mail, if the e-mail was located outside the district of the court. This section allows courts with jurisdiction over an investigation "to compel evidence directly, without requiring the intervention of agents, prosecutors, and judges in the districts where major ISPs are located." This will reduce the administrative burden of the courts and LE when following a transjurisdictional trail and will expedite investigations where time is of the essence.

5. *Section 814 deterrence and prevention of cyberterrorism:* The memorandum notes that Section 814 makes a number of changes to improve 18 U.S.C. §1030, the Computer Fraud and Abuse Act. Amongst other things:

   ‣ It increases penalties for hacking offences as they relate to *protected computers* (a *protected computer* is defined as one which is used by financial institutions or the U.S. government or in interstate or foreign commerce or communication).

   ‣ It criminalizes hacking acts with a general intention to damage (previously there had been the requirement that there be intent to damage in a particular fashion).

   ‣ It criminalizes explicitly the damage caused to national security and criminal justice computers.

   ‣ It expands coverage of the statute to include damage to computers in countries outside the United States if United States commerce is affected.

The net effect of the above changes is to strengthen the powers of the U.S. LE and national security agencies in gathering electronic evidence and to strengthen the position of CSPs against possible civil action when divulging evidence to and soliciting assistance from authorities in the case of suspected abuse. The changes also serve to strengthen the situation of individual computer owners confronted by computer hacking by facilitating and clarifying measures that can be used by LE in dealing with such hacking. The changes will as a result assist in the prosecution of crime that involves electronic evidence in general and will in particular assist in identifying and tracking computer abuse.

Several of the above provisions have a sunset clause and are set to expire on December 31, 2005.

2.5.3.2   Data retention

The events of September 11 have not surprisingly prompted nations to press for increased powers of surveillance and monitoring. The tension between attempts to secure the public good via legislation which permits greater surveillance on the one hand, and the rights of the individual on the other, has nowhere been more acute than in the area of national initiatives intended to ensure data retention by CSPs.

Data retention is the term used to mean the wholesale and a priori storage and retention of all *traffic data* for a set period well beyond the normal (billing) requirements of the CSP. With data retention procedures in place, it is possible for investigators in the United Kingdom, for example, to invoke the provisions of RIPA to access the information they require, knowing that the data is available, that is, that it has not been deleted. In this case, data retention ensures that the data survives long enough for investigators to seek legal access to that data. The United Kingdom's RIPA authorizes provision of access while the recent U.K. data retention act, the Antiterrorism, Crime and Security Act 2001 (ATCSA), authorizes the retention of the recorded data to which access may then be allowed under RIPA. The U.S. Patriot Act—perhaps surprisingly—has no specific data retention requirements, it includes no new requirements on CSPs to configure their systems in a manner which will allow them to store and retain *traffic data* (see Section 222 of the Act). U.S. legislators have been reported as favoring *data preservation* which signifies the less onerous and perhaps less intrusive procedure of CSPs retaining (but not disclosing) individual records only, on a case-by-case basis, followed by subsequent disclosure on authorization to do so. At the same time, there have also been reports that the U.S. administration is in the process of drafting new legislation focused on data retention similar to ATCSA, in which case the administration appears to be on a collision course with the legislature.

The United Kingdom and the EU have recently passed data retention legislation. The United Kingdom's ATCSA 2001 which took force as of December 14, 2001 consists of 14 parts, part 11 relates to data retention by CSPs. The essential points in this regard are as follows:

1. The legislation provides for the voluntary retention (a period of a year has been discussed) by CSPs of communications data in accordance with a code of practice to be developed by government in consultation with the industry.

2. It allows CSPs to retain such data, exempts them from the obligation to erase such data once no longer needed for business (billing) purposes and thus safeguards them against actions brought against them under other legislation (e.g., the Data Protection Act).

3. The legislation includes provision for mandating data retention and a code of practice via statutory authority if the voluntary arrangements appear not to be working.

Access to such data by government agencies is then subject to RIPA.

In May 2002, the European Parliament which had previously been expected to vote against the kinds of data retention measures advocated by both the CoE and the G8, surprisingly passed legislation that will allow EU governments to force CSPs to implement data retention [6]. The data to be retained includes: call records and cell site data (for mobile phones), and login/logout records, Web cache information, e-mail header information and IP addresses for ISPs.

The wealth of information available in such *traffic data* is in principle relatively easily processed so as to identify a person's time-line of activity and communication so it is not surprising that civil libertarians and privacy advocates are opposed to the measures. Documents emerging from the May 2002 meeting of G8 Justice and Interior Ministers include an Annex listing "log details related to some services that may be available to an Internet Service Provider" [104], there are over 30 individual fields listed under the following services:

1. Network Authentication Systems;

2. E-mail servers;

3. FTP;

4. Web servers;

5. Usenet;

6. IRC.

It may very well be that this wealth of information is what is needed to regulate—or if not regulate then monitor—the anarchic Internet in order to attempt to limit its abuse. The question at this stage still remains: will the data retention measures be successful and useful in those cases where it is needed or will it be a case of little gain for a very significant financial cost accompanied also by a potentially huge loss of privacy and associated potential for abuse.

2.6 Networks and intrusion forensics

The investigation of networked systems, perhaps in the laboratory or more likely at the site of an investigation and possibly involving seizure of components, is not surprisingly more complicated by far than the investigation of an individual computer. In addressing this issue, Sommer points out in *Digital Footprints: Assessing Computer Evidence* [105]:

> There are two principal situations to be considered: where the offence is concentrated on an individual's use of the Internet and where a remote site holds evidence of an offence.

It is unlikely that all the networked information is necessarily available from the scene of an on-site investigation, in many situations there will be evidence not only on the host computer but also on network servers, ISPs and the like. In the case of an ISP there will be both legal and practical considerations that will require direct contact with the ISP, and possibly a warrant, and even with local network servers the required information is likely to be available only via a privileged account on the server itself and may require a separate warrant. In such a situation the investigator needs to have regard for the following considerations:

1. Possession of the appropriate authorization.

2. Investigating the network on site or seizing (parts of) the network must not compromise the rights of the organization or business running the network (see *Best Practices for Seizing Electronic Evidence* by the U.S. Secret Service and the International Association of Chief of Police [44]).

3. All relevant evidence from sites presently or previously connected to the computer in question is gathered.

4. All relevant evidence from the computer in question is gathered.

Gathering the network-related information referred to in the penultimate point will rely upon a person experienced in computer networking and in the particular platforms and software in use on that network, and possibly upon assistance from nontargeted personnel on-site, such as a reliable employee not implicated in the investigation. In the latter case, it is necessary that the employee be carefully instructed so that no compromise of the evidentiary reliability of the information may occur.

We return to these and some related topics in Chapter 6.

References

[1] Council of Europe, "Convention on Cybercrime," http://conventions.coe.int/Treaty/en/Treaties/Html/185.htm, signed Nov. 23, 2001, visited July 2002.

[2] Di Gregory, K., "Carnivore and the Fourth Amendment," delivered to a subcommittee of U.S. Congress on July 24, 2000, http://www.usdoj.gov/criminal/cybercrime/carnivore.htm, visited July 2002.

[3] The Home Office U.K., "Regulation of Investigatory Powers Act 2000," http://www.homeoffice.gov.uk/ripa/ripact.htm, 2000, visited July 2002.

[4] Plesser, R. L., J. J. Halpert, and E. W. Cividanes, "USA Patriot Act for Internet and Communications Companies," *Computer and Internet Lawyer,* March 2002, http://eon.law.harvard.edu/privacy/Presser%20article–redacted.htm, visited July 2002.

[5] Her Majesty's Stationary Office U.K., "Explanatory Notes to Anti-Terrorism, Crime and Security Act 2001," http://www.legislation.hmso.gov.uk/acts/en/2001en24.htm, visited July 2002.

[6] Grossman, W., "A New Blow to Our Privacy," http://www.guardian.co.uk/online/story/0,3605,727644,00.html, visited July 2002.

[7] Sommer, P., "Intrusion Detection Systems As Evidence." In *RAID 98,* Belgium: University of Louvain-la-Neuve, Sept. 1998; "Intrusion Detection and Legal Proceedings," http://www.raid-symposium.org/raid98/Prog_RAID98/Talks.html#Sommer_09, visited July 2002.

[8] Johnson, T., "Storage Trends Disk, Optical, Tape," http://www.dcs.napier.ac.uk/~vldb99/IndustrialSpeakerSlides/johnson.pdf, 1999, visited March 2002.

[9]   van der Knijff, R., "Embedded Systems Analysis." In *Handbook of Computer Crime Investigation*, Chapter 11, E. Casey (ed.), London: Academic Press, 2002.

[10] NIST, "Hard Disk Write Block Tool Specification," http://www.cftt.nist.gov/WB-spec-jan-07-1.pdf, visited July 2002.

[11] Digital Intelligence Inc., "Software and Hardware Solutions for the Computer Forensics Community," http://www.digitalintel.com, visited July 2002.

[12] Guidance Software, "FastBloc," http://www.encase.com/products/hardware/fastbloc.shtm, visited July 2002.

[13] Holley, J., "Computer Forensics," *SC Magazine*, Sept. 2000, http://www.scmagazine.com/scmagazine/2000_09/, visited May 2002.

[14] Farmer, D., and W. Venema, "Bring Out Your Dead," *Dr. Dobb's Journal*, Jan. 2001.

[15] ForensiX, Fred Cohen & Associates, http://www.all.net/ForensiX/, visited June 2002.

[16] White Glove, http://www.all.net/WG/PLAC/tools.html, visited July 2002.

[17] Dittrich, D., "Basic Steps in Forensic Analysis of Unix Systems," http://staff.washington.edu/dittrich/misc/forensics/, University of Washington, visited June 2002.

[18] Seglem, K., M. Luque, and S. Murphy, "Unix Systems Analysis." In *Handbook of Computer Crime Investigation*, E. Casey (ed.), London: Academic Press, 2002.

[19] Sammes, T., and B. Jenkinson, *Forensic Computing—A Practitioner's Guide*, Berlin: Springer-Verlag, 2000.

[20] Holley, J., "Meeting Computer Forensic Analysis Requirements," *SC Magazine*, March 2001, http://www.scmagazine.com/scmagazine/sc-online/2001/article/016/article.html, visited March 2002.

[21] Computer Forensics Tool Testing (CFTT) Project Web Site, "Disk Imaging Tool Specification Version 3.1.6," http://www.cftt.nist.gov/DI-spec-3-1-6.doc, Oct. 12, 2001, visited March 2002.

[22] Schneier, B., *Applied Cryptography*, New York: Wiley, 1994.

[23] NIST, "Announcing the Standard for SECURE HASH STANDARD SHA-1," *Federal Information Processing Standards Publication180-1*, http://www.itl.nist.gov/fipspubs/fip180-1.htm, 1995, April 17, visited March 2002.

[24] Tripwire Inc., http://www.tripwire.com, visited July 2002.

[25] WetStone Technologies Inc., "SMART Watch," http://www.wetstonetech.com, visited March 2002.

[26] Power Quest Corporation, "PartitionMagic 7," http://www.powerquest.com/partitionmagic/, visited March 2002.

[27] Sanderson, P., http://online.securityfocus.com/archive/104/254394, 04 Feb. 2002, visited Jan. 2003.

[28] National Committee on Information Technology Standards, "Technical Committee T13 AT Attachment," http://www.t13.org/, March 8, 2002, visited March 2002.

[29] McDonald, A., and M. Kuhn, "StegFS: A Steganographic File System for Linux", *Lecture Notes in Computer Science*, Vol 1768, 2000, pp. 463–477.

[30] Johnson, A., "Steganography for DOS Programmers," *Dr. Dobb's Journal*, Jan. 1997.

[31] DIBS USA Inc., "DIBS Mobile Forensic Workstation," http://www.dibsusa.com/, visited July 2002.

[32] Guidance Software Inc., "EnCase," http://www.guidancesoftware.com/, visited March 2002.

[33] New Technologies Inc. (NTI), "Law Enforcement Computer Evidence Suite," http://www.forensics-intl.com/, visited March 2002.

[34] AcessData Inc., "The Forensic Toolkit," http://www.accessdata.com/, visited March 2002.

[35] Ontrack Data International Inc., "DataTrail FacTracker," http://www.ontrack.com/factracker/, visited July 2002.

[36] Vogon International Ltd., "GenX," http://www.vogon-international.com/, visited March 2002.

[37] Bryson, C., and M. R. Anderson, "Shadow Data the Fifth Dimension of Data Security Risk," NTI, http://www.forensics-intl.com/art15.html, visited March 2002.

[38] Guttmann, P., "Secure Deletion of Data from Magnetic and Solid-State Memory," Proc. *6th USENIX Security Symposium*, San Jose, CA, July 22–25, 1996.

[39] Gross, A., "Analysing Computer Intrusions," Ph.D. Thesis, San Diego Supercomputer Center, University of California, San Diego, 1997.

[40] The Houston Chronicle, "Computer Sleuths Sniffing Out Deleted Enron E-mail," Jan. 16, 2002.

[41] Computerworld New Zealand, "Handhelds Give up Secrets," http://www.idg.net.nz/webhome.nsf/nl/63E5E17F5D96FBFACC256B0500727855, Monday, Nov. 19, 2001, visited March 2002.

[42] Paraben Inc., "PDA Seizure," http://www.paraben-forensics.com/, visited March 2002.

[43] Gibbs, K. E., and D. F. Clark, "Wireless Network Analysis." In *Handbook of Computer Crime Investigation*, Chapter 10, E. Casey (ed.), London: Academic Press, 2002.

[44] The U.S. Secret Service and the International Association of Chief of Police, "Best Practices for Seizing Electronic Evidence," http://www.infowar.com/law/00/e-evidence/e-evidence.shtml, visited July 2002.

[45] Jensen, D., "Prospective Assessment of AI Technologies for Fraud Detection: A Case Study." In *AI Approaches to Fraud Detection and Risk Management, Collected Papers from the 1997 Workshop*, Technical Report WS-97-07, Menlo Park, CA: AAAI Press, http://eksl-www.cs.umass.edu/~jensen/papers/aaaiws97a.html, 1995, visited March 2002.

[46] Jensen, D., "Link Analysis," http://eksl-www.cs.umass.edu/aila/link-analysis.html, visited Jan. 2002.

[47] i2 Inc., "Analyst's Notebook," http://www.i2.co.uk/home.html, visited July 2002.

[48] Operational Information Security, Information Networks Division, Defence Science and Technology Division, *CFIT® User Manual*, 2001.

[49] i2 Inc., "Case Studies, Analyst's Notebook," http://www.i2.co.uk/applications/casestudies/, visited March 2002.

[50] NetMap Analytics LLC, "NetMap," http://www.netmapsolutions.com/, visited Feb. 2002.

[51] Xanalys, "Watson," http://www.xanalys.com/watson.html, visited July 2002.

[52] SPSS Inc., "Clementine," http://www.spss.com/products/, Feb. 2002.

[53] IBM, "Intelligent Miner for Data/Text," http://www-3.ibm.com/software/data/iminer/, visited March 2002.

[54] Socho, G. J., Fios Inc., "Once You Have the Evidence—Then What?" In *IP Litigator*, Vol. 5, No. 2, http://www.fiosinc.com/wp-once.html, March/April 1999, Aspen Law Business, visited March 2002.

[55] Holley, J., "Getting the Hard Facts," *SC Magazine*, April 2001, http://www.scmagazine.com/scmagazine/2001_04/, visited May 2002.

[56] ACES, Statement of Janet Reno Attorney General of the United States Before the United States Senate Committee on Appropriations, http://www.senate.gov/~appropriations/commerce/testimony/reno229.htm, Feb. 29, 2000, visited March 2002.

[57] ILook Investigator, http://www.ilook-forensics.org/, visited July 2002.

[58] U.S. DOJ National Drug Intelligence Center, "Hashkeeper Database," http://www.hashkeeper.org/, visited July 2002.

[59] Vais, M., "Law Enforcement Tools and Technologies for Investigating Cyber Attacks: A National Needs Assessment Report," *Institute for Security Technology Studies*, Dartmouth College, NH, June 2002.

[60] Fisher, G., "Computer Forensics Guidance," *NSRL ITL Bulletin,* Nov. 2001, http://www.nsrl.nist.gov/itlbulletin.html, visited March 2002.

[61] Patzakis J., "The EnCase Process," In *Handbook of Computer Crime,* Chapter 3, E. Casey (ed.), London: Academic Press, 2002.

[62] Guidance Software, Encase3 distribution, June 2001.

[63] Bahl, L., F. Jelinek, and R. Mercer, "A Maximum Likelihood Approach to Continuous Speech Recognition," *IEEE Trans. on Pattern Analysis and Machine Intelligence,* Vol. 5, No. 2, 1983, pp. 179–190.

[64] Church K., and P. Hanks, "Word Association Norms, Mutual Information and Lexicography," *Computational Linguistics,* Vol. 16, No. 1, 1990, pp. 22–29.

[65] Huang X., et al., "The SPHINX-II Speech Recognition System: An Overview," *Computer, Speech and Language,* Vol. 2, 1993, pp. 137–148.

[66] http://rr.sans.org/incident/IRCF.php, visited July 2002.

[67] Guidance Software, *EnCase Legal Journal,* 2nd ed., 2001.

[68] Law Foundation of NSW, Database of "Federal Court of Australia Cases," http://www.austlii.edu.au/au/cases/cth/federal_ct/index.html, visited July 2002.

[69] Farmer, D., "What are MACtimes?" *Dr Dobbs Journal,* Oct. 2000.

[70] Department of Defense, U.S., "Department of Defense Trusted Computer System Evaluation Criteria," (The Orange Book), August 15, 1983.

[71] Hosmer, C., WetStone Technologies Inc., "Time-Lining Computer Evidence," www.wetstonetech.com/timpaper.htm, visited March 2002.

[72] Noblett, M. G., M. M. Pollitt, and L. A. Presley, "Recovering and Examining Computer Forensic Evidence," *Forensic Science Communications,* Vol. 2, No. 4, Oct. 2000.

[73] Computer Crime and Intellectual Property Section (CCIPS), Department of Justice, "Federal Guidelines for Searching and Seizing Computers," http://www.usdoj.gov/criminal/cybercrime/search_docs/toc.htm, 1994, visited June 2002.

[74] Computer Crime and Intellectual Property Section (CCIPS), Criminal Division, United States Department of Justice, "Searching and Seizing Computers and Obtaining Electronic Evidence in Criminal Investigations," http://www.usdoj.gov/criminal/cybercrime/searchmanual.htm, Jan. 2001, visited June 2002.

[75] Galil, Y., "New Federal Guidelines for Searching and Seizing Computers— From Servers to PDAs," *Internet Law Journal,* Feb. 5, 2001, http://www.tilj.com/content/litigationheadline02050102.htm, visited July 2002.

[76] Jones, N., "Law for Systems Administrators Conference—Working with the Police," http://www.ja.net/conferences/security/january01/N.Jones.pdf, Jan. 30, 2001, visited July 2002.

[77] NIJ (U.S. National Institute of Justice) Guide, "Electronic Crime Scene Investigation: A Guide for First Responders," http://www.ncjrs.org/txtfiles1/nij/187736.txt, visited March 2002.

[78] U.S. Department of Justice Federal Bureau of Investigation, "Digital Evidence: Standards and Principles," *Forensic Science Communications*, Vol. 2, No. 2, April 2000.

[79] International Organization on Computer Evidence, *International Conference on Digital Evidence,* http://www.ioce2002.com/index.cfm, visited March 2002.

[80] Council of Europe, "Convention on Cybercrime (ETS No. 185), Summary," http://conventions.coe.int/treaty/en/Summaries/Html/185.htm, visited July 2002.

[81] Council of Europe, "Council of Europe," http://www.coe.int/, visited July 2002.

[82] Council of Europe, "About Conventions and Agreements in the European Treaty Series (ETS)," http://conventions.coe.int/, visited March 2002.

[83] G8, "G8 Meeting of Justice and Interior Ministers in Washington, DC in 1997," http://www.g8summit.gov.uk/prebham/washington.1297.shtml, visited Match 2002.

[84] Sussmann, M., "The Critical Challenges from International High-Tech and Computer-Related Crime at the Millennium," *Duke Journal of Comparative & International Law,* Vol. 9, No. 2, Spring 1999, p. 451.

[85] Parliament of the Commonwealth of Australia, Parliamentary Joint Committee on the National Crime Authority, "The Law Enforcement Implications of New Technology," Aug. 2001.

[86] Council of Europe, "Convention on Cybercrime First Protocol," http://www.legal.coe.int/economiccrime/cybercrime/AvProjetProt2002E.pdf, visited March 2002.

[87] Council of Europe, Convention on Cybercrime "Second Protocol," http://conventions.coe.int/Treaty/en/Treaties/Html/182.htm, visited March 2002.

[88] McConnell International, "Cyber Crime ... and Punishment? Archaic Laws Threaten Global Information," Dec. 2000, http://www.mcConnellinternational.com, visited Feb. 2002.

[89] Armstrong, I., "Legislators Turn up the Heat on Cybercrime," *SC Magazine*, April 2001, http://www.scmagazine.com/scmagazine/2001_04/feature.html, visited Feb. 2002.

[90] Schjolberg, S., "The Legal Framework—Unauthorised Access to Computer Systems. Penal Legislation in 43 Countries," http://www.mossbyrett.of.no/info/legal.html, visited Feb. 2002.

[91] Select Committee on European Union, Minutes of Evidence, "Examination of Witnesses (Questions 404–419)," http://www.parliament.the-stationery-office.

co.uk/pa/ld199900/ldselect/ldeucom/95/0031503.htm, Wednesday March 15, 2000, visited July 2002.

[92]   United Nations Economic and Social Council, "Conclusions of the Study on Effective Measures To Prevent and Control High-Technology and Computer-Related Crime, Report of the Secretary-General," http://www.odccp.org/adhoc/crime/10_commission/4e.pdf, visited July 2002.

[93]   European Commission, "Creating a Safer Information Society by Improving the Security of Information Infrastructures and Combating Computer-Related Crime," Brussels, Belgium 2001.

[94]   U.S. Department of Justice, "Independent Technical Review of the Carnivore System Draft Report," http://www.usdoj.gov/jmd/publications/carnivore_draft_1.pdf, visited March 2002.

[95]   IIT Research Institute, "Independent Technical Review of the Carnivore System Final Report," http://www.epic.org/privacy/carnivore/carniv_final.pdf, visited March 2002.

[96]   Federal Bureau of Investigation, "Carnivore Diagnostic Tool," http://www.fbi.gov/hq/lab/carnivore/carnivore2.htm, visited July 2002.

[97]   The Electronic Privacy Information Center (EPIC), Information Regarding Carnivore Obtained Through FOIA, http://www.epic.org/privacy/carnivore/foia_documents.html, visited March 2002.

[98]   The Home Office U.K., "Accessing Communications Data Draft Code of Practice," http://www.homeoffice.gov.uk/ripa/pcdcpc.htm, October 25, 2001, visited March 2002.

[99]   The Home Office U.K., "Explanatory Notes to Regulation of Investigatory Powers Act 2000, chapter 23," http://www.hmso.gov.uk/acts/en/2000en23.htm, Aug. 15, 2000, visited July 2002.

[100] The Home Office U.K., "Mass Surveillance?" http://www.homeoffice.gov.uk/ripa/mass.htm, Aug. 1, 2000, visited March 2002.

[101] Brown, I., S. Davies, and G. Hosein (eds.), "The Economic Impact of the Regulation of Investigatory Powers Bill," http://www.britishchambers.org.uk/newsandpolicy/downloads/lsereport.pdf, June 12, 2000, visited July 2002.

[102] Privacy Digest, "Home Office Backs Seven-year Retention Laws," http://www.privacy.digest.com/2001/10/03, Oct. 3, 2001, visited March 2002.

[103] Computer Crime and Intellectual Property Section (CCIPS), Department of Justice U.S., "Field Guidance on New Authorities that Relate to Computer Crime and Electronic Evidence Enacted in the U.S. Patriot Act of 2001," http://www.usdoj.gov/criminal/cybercrime/PatriotAct.htm, visited July 2002.

[104] G8 Justice and Interior Ministers, "Principles on the Availability of Data Essential to Protecting Public Safety," http://www.g8j-i.ca/english/doc3. html, visited July 2002.

[105] Sommer, P., "Digital Footprints: Assessing Computer Evidence," *Criminal Law Review*, Special Edition, Dec. 1998, pp. 61–78, http://www.giustizia.it/ cassazione/convegni/dic2000/sommer_6.pdf, visited July 2002.

Contents

3.1 The origins and history of
computer forensics

3.2 The role of computer forensics
in law enforcement

3.3 Principles of evidence

3.4 Computer forensics model for
law enforcement

3.5 Forensic examination

3.6 Forensic resources and tools

3.7 Competencies and certification

3.8 Computer forensics and
national security

References

Computer Forensics in Law Enforcement and National Security

3.1 The origins and history of computer forensics

Chapter 2 provided an overview of the current state of the practice of computer forensics and described Encase and ILook Investigator, two of the best known forensic suites or toolsets. In this chapter, we consider the origins of computer forensics, provide a detailed examination of its role in Law Enforcement and National Security, and take a detailed look at principles, procedures, and tools adopted by computer forensic examiners in these communities.

As Carrie Morgan Whitcomb, director of the National Center for Forensic Science in the United States puts it:

> Computer forensic science is largely a response to a demand
> for service from the law enforcement community. [1]

The first known employment of computer forensic techniques was, however, by the U.S. military and intelligence agencies in the 1970s. Very little is known about these activities due to their occurrence in classified environments (Michael R. Anderson, private communication, March 23, 2002). However, it is logical to assume that they had a counterintelligence focus using mainframe computer systems.

Some of the first government agencies with an overt and publicly visible requirement to carry out forensics on external systems relating to criminal offences were taxation and revenue collection agencies including the U.S. Internal Revenue Service Criminal Investigations Division (IRS-CID) and Revenue Canada.

In looking at the state of computer forensics in law enforcement today, or as it should be more correctly termed now digital evidence recovery, it is useful to examine its beginning and its progression. It was not until the 1980s that the advent of the IBM PC and its many variants introduced new problems into the world of investigation: volume of data, ability to alter data without trace, and the ability to hide or delete data. Computing was made available to the masses that naturally included the criminal fraternity. It became apparent that a level of specialist knowledge was needed to investigate this new technology and thus was born the science of "Forensic Computer Examination."

As previously mentioned, in North America, organizations that were initially most active in the computer forensic field from the mid-1980s to the early 1990s were the IRS-CID and Revenue Canada. In 1984, the FBI had established the Computer Analysis and Response Team (CART), based out of FBI Headquarters in Washington, District of Columbia, to provide computer forensic support, however, it did not actually become fully operational until 1991 [2].

No specific forensic tools existed in the 1980s, so existing data protection and recovery suites of utilities, such as Peter Norton Inc. *The Norton's Utilities*, Central Point Software *PC Tools* and Paul Mace Software *Mace Utilities* were used. As of January 1990, there were 100,000 registered users of *Mace Utilities* and as most people would know, *Norton's Utilities* has become probably one of the most popular PC utility suites available. Due to the lack of specific forensic software, personnel from IRS including Michael R. Anderson, Andrew Fried and Dan Mares, and Stephen Choy from Revenue Canada, later developed their own suites of MS DOS−based (Microsoft Disk Operating System−based) forensic utilities, many of which have been refined and updated, and persist in use to this day.

Initially, the only method available to the forensic examiner to preserve evidence was to take a logical backup of files from the evidence disk to magnetic tape, hopefully preserving appropriate file attributes, restore these files to another disk and then examine them manually using command line file management software, such as *Executive Systems Inc., Xtree Gold, The Norton Commander,* and appropriate file viewing software.

Many early mainframe and minicomputer backup packages used the "sector imaging" method that was described in detail in Chapter 2.

By the mid-to-late 1980s, however, the image backup had been replaced by logical backup, which copied the file and directory structure of a disk to the backup media that allowed the user or administrator to selectively backup and restore files from the system. This was a leap forward as far as the user was concerned, but was less useful from an evidentiary perspective.

Logical backup software operates only at the operating system or file system level and consequently does not duplicate free and slack space (ambient data) making the backup copy incomplete from an evidentiary perspective. Deleted files and any other relevant information that may have temporarily been written to the disk, such as encryption passwords, were therefore unrecoverable.

The next step was therefore to examine the original media using a disk editor, such as *Norton's Disk Editor* (DE). The threat of unintentionally altering the original evidence makes this a hazardous task with the potential for disaster. This was one reason for the development of logical write blocking software, such as Revenue Canada's *Disklok* that blocked interrupt calls to write to the hard drive. Many hours have been spent with DE examining hard drives for evidence, or more recently carrying out raw examinations of image files, only to have to defend later, an allegation that the original evidence had somehow been tampered with and rendered inadmissible through incorrect acquisition process and/or unproven forensic software.

In the United States, the requirement for forensically sound bit stream image duplication of hard drives was identified by a small, informal group of like-minded U.S. federal, state and local computer forensic practitioners way back in late 1989 during the development of the first computer forensic science training courses at the Federal Law Enforcement Training Center (FLETC). The first specific forensic program created to perform this task was named *IMDUMP*, developed by Michael White, who was employed by Paul Mace Software at that time. That program proved to be useful until approximately 1991, when most of the Paul Mace utilities were sold to another software company.

Lacking continued support for *IMDUMP*, the group went to Charles P. (Chuck) Guzis, President of Sydex, Inc. in Eugene, Oregon and presented him with the dilemma posed by the loss of support for *IMDUMP*. Guzis, who had previously worked in leadership roles for Control Data Corporation, Durango Systems Inc., Stellar Software Systems, and Peritus International Inc., had been a friend of the U.S. law enforcement computer specialists for years and after some persuasion, he agreed to develop a specialized program that would meet bit stream backup needs from an evidence standpoint. Some people think of Guzis as the father of electronic crime scene preservation and the resulting program, SafeBack, which was first distributed in 1990, mentioned in Chapter 2, became the de facto worldwide standard for sector disk imaging.

Besides Safeback, Guzis and Sydex went on to develop other low-level data recovery and analysis tools including *Anadisk* (a low level diskette analysis tool), *Teledisk* (a diskette imaging tool), and *CopyQM* (a diskette analysis tool that images and duplicates FAT and non-FAT formatted floppy disks) [3].

Forensic imaging requirements in the United Kingdom developed during research work on computer viruses in the mid-to-late 1980s. Bit stream cloning of hard drives infected by viruses allowed the exact effect of the virus to be examined through the actual execution of the virus code. These requirements led to the development of the original Disk Image Backup System (DIBS™), a forensic hardware and software solution using a parallel port connected magneto-optical drive (MOD), which was first sold commercially in 1991 [4].

In another part of Europe, the Dutch National Forensic Institute was working away developing leading edge forensic technologies, with a particular lead in the area of embedded digital devices and PDAs.

In the meantime, Interpol had formed a Computer Crime Working Group in Europe, chaired by then Detective Inspector John Austen from the U.K. Metropolitan Police Computer Crime Squad, to look at developing standards and training within the European community.

Law enforcement in Australia, which had always had a close working relationship with the U.S. and Canadian law enforcement, heard of and acquired forensic tools including Safeback, Mares and Fried Utilities, and similar tools. Rod McKemmish, one of the coauthors of this book, has the distinction of being the primary developer of forensic tools in Australia and his *Fixed Disk Image* (FDI) software provided functionality almost identical to Safeback but free to Australian law enforcement.

Besides the lack of specific forensic tools, the second major deficiency was the lack of specific training for computer search, seizure, and forensic analysis. The same people in the United States and Canada who identified the deficiencies with respect to tools also began to identify the training requirements. Michael R. Anderson, then a special agent with the IRS-CID, developed the Seized Computer Evidence Recovery Specialist (SCERS) curriculum for the FLETC and was a cofounder of the International Association of Computer Investigative Specialists (IACIS®).

IACIS®, the oldest and probably best known computer forensic organization in the world, was formed in 1990 in Portland, Oregon to provide training and certification for law enforcement computer forensic examiners [5]. IACIS training was the forum in which the United States, Canadian, Australian, and many other countries computer forensic specialists, first became acquainted with the principles, techniques, and tools of computer forensics, many of which are still valid to this day.

3.2 The role of computer forensics in law enforcement

Early computer crime cases involving forensic examinations to recover evidence, related typically to fraudulent activity where the computer either facilitated the crime or stored evidence relating to the commission of the crime. Later, as computer networking became prevalent, computers became targets themselves for criminal activity in the form of computer intrusions (hacking).

These differing circumstances provide two distinct scenarios that need to be considered, each a little differently:

- Computers as the facilitators or repositories of evidence relating to a more traditional form of crime, such as fraud.

- Computers themselves as targets of a crime, such as system cracking (hacking).

This distinction will be important later when we discuss the nuts and bolts of forensic process. From a criminal law perspective, it is reasonable to state that the general objective of the physical forensic sciences is, through the application of rigorous scientific method, to be able to circumstantially reconstruct a series of events linking a suspect to a crime using the available trace evidence.

The objective of computer forensics is therefore similar by providing the means whereby a series of events surrounding a crime with manifestations in a digital environment is reconstructed. There are many techniques for carrying out these reconstructions and it should be recognized that, due to the quantity of information that may need to be duplicated, extracted, processed, and analyzed, this task could potentially be a very time consuming one.

The FBI Handbook of Forensic Services [6] has identified the following types of computer examinations and recovery processes that can be conducted:

1. *Content:* Examinations can determine what type of data files are in a computer.

2. *Comparison:* Examinations can compare data files to known documents and data files.

3. *Transaction:* Examinations can determine the time and sequence when data files were created.

4. *Extraction:* Data files can be extracted from the computer.

5. *Deleted data files:* Deleted data files can be recovered from the computer.

6. *Format conversion:* Data files can be converted from one format to another.

7. *Keyword searching:* Data files can be searched for a word or phrase and all occurrences recorded.

8. *Passwords:* Passwords can be recovered and decrypted.

9. *Limited source code:* Source code can be analyzed and compared.

10. *Storage media:* Storage media used with standalone word processors (typewriters) can be examined.

In addition to these, which appear to be somewhat dated, the following other types of examination can also be conducted:

› *Network history:* Internet browser history, e-mail, and other network related activities on a system may be reconstructed to provide a picture of the activity on a network, in the majority of cases, the Internet.

› *Graphics and multimedia file identification:* Graphics and multimedia files related to illegal activities, such as child pornography, may be identified and recovered.

Some illustrative examples of recent and well-documented cases where computer forensics played a key role include the following:

1. *Fraud:* Enron is possibly the most well-known case of alleged fraud where computer forensics will likely yield the definitive evidence in the matter due to the destruction of potentially incriminating hard copy records [7].

2. *Homicide:* In 1998, New York State Police Computer Crimes Unit personnel assisted State Police Highland with the examination of

two computers believed to contain information on a homicide in Plattekill (Ulster County). The primary suspects in the killing were the victim's wife and a man from Jacksonville, Florida. Forensic analysis of the home computer revealed the presence of Internet chatroom transcripts detailing murder threats by the Florida suspect against the victim. When confronted with this evidence at the trial in October, he pleaded guilty of murder in exchange for a sentence of 18 years [8].

3. *Narcotics:* In a U.K. case prosecuted under the Misuse of Drugs Act 1971, which involved smuggling of millions of British pounds worth "class A narcotics" (which include heroin, cocaine, ecstasy, and LSD), the origin of one particular document created on a computer was central to the case. Although the printed version of the document was signed and dated in 1997, forensic analysis of the computer established that the electronic version of the document had in fact been created in 1999, nearly 2 years after it was purported to have been signed. When presented with the facts of this discovery, the suspect claimed to have originally deleted the document and when he realized that it would be central to his defense, had recreated the document from the printed version. Comparison of the printed version and the electronic version revealed that, had this in fact been true, the suspect had somehow faithfully included in the electronic version, every single typing mistake on the document, and the use of a rather obscure font at one point. The suspect was later convicted and sentenced to 10 years imprisonment [9].

4. *Pedophilia:* In 1998, a major raid of a global child pornography ring known as the "Wonderland Club" resulted in raids of nearly 200 persons in the United States and 13 other countries including the United Kingdom and Australia. Search warrants were served on 90 addresses that had been identified in the United States and the U.S. Customs Service seized computers from 32 suspects in 22 states. British authorities conducting the investigation retrieved over 100,000 images of children as young as 18 months engaged in sexual acts. Forensic analysis of the computer seized revealed many members of the club had over 10,000 child porn images. Bill Anthony, a special agent with U.S. Customs said during an interview that some members of the club may even have used their own children in the images or accepted money for having used their children. (*USA Today*, 3/9/98). Suspects are also believed to

have sent live video feeds of child molestation across the Internet (*Washington Post*, 3/9/98) [10].

5. *Organized crime:* In January 1999, the FBI was told by confidential informants that Nicodemo Scarfo Jr., son of jailed Philadelphia Mob boss "Little Nicky" Scarfo, was running a gambling and extortion operation in New Jersey. The FBI agents subsequently obtained authority to serve a search warrant on Scarfo's office and forensically examine Scarfo's computer, which was suspected to contain incriminating records of Scarfo's operations. It was found that Scarfo was cognizant of the threat to the information on his computer and had encrypted the particular file that the FBI were interested in using *Pretty Good Privacy* (PGP), a publicly available data encryption program. The FBI was unable to crack the encryption, and agents returned with a covert search warrant and placed a keystroke-monitoring device, referred to by the FBI as a key logger system (KLS), on the computer. The KLS recorded every keystroke made by every user of the computer. After nearly two months of monitoring, the FBI covertly returned and retrieved the KLS. They found Scarfo's PGP password in the KLS logs: *nds09813-050*. This was later confirmed to be Scarfo Sr.'s federal prison identification number, which the FBI had known all the time [11].

6. *Computer hacking:* Kevin Mitnick, sometimes referred to as "America's Most Wanted Computer Outlaw," had eluded police, U.S. Marshals, and FBI for over 2 years after vanishing while on probation for previous computer crime offences. He had previously been convicted in 1989 for federal computer and access device fraud under 18 U.S.C. §§1029 and 1030. While on the run, Mitnick continued to break into numerous computers, intercepted private e-mail communications, and copied personal and confidential materials from a number of computer systems he had compromised. He stored the illegally obtained material, including personal e-mail, stolen passwords, and proprietary software, in various sites around the Internet. This fact posed significant jurisdictional problems for investigators. Amongst the stolen data was a large amount of software that contained proprietary source code for key products into which companies had invested many millions of dollars for developmental efforts in order to maintain their competitive edge. Mitnick caused a great deal of disruption on the systems he compromised, often altering information, corrupting system software,

and monitoring system users, sometimes preventing or impeding a legitimate use of the systems. To evade capture, Mitnick used cloned and stolen cellular telephones, and stolen Internet services to carry out many of his intrusions. In December 1994, Mitnick made the serious error of breaking into systems operated by Tsutomu Shimomura of the San Diego Supercomputer Center. Shimomura took the compromise personally and set out to track Mitnick down. Less than 2 months later, Shimomura had tracked him down to Raleigh, North Carolina, where on February 15, 1995, the FBI arrested him [12].

3.3 Principles of evidence

Chapter 2 mentioned some of the legal and operational issues affecting the current state of practice and it is worthwhile here to highlight the core principles of evidence. A unique feature of computer forensics that sets it apart from any other area of computer technology is the requirement that the application of the technology must be carried out with due regard to the requirements of the law. Failure to do so can result in the digital evidence being ruled inadmissible or at the very least being regarded as tainted.

Computer evidence needs to meet the same legal requirements as any form of evidence to be produced successfully in a court. It needs to be:

1. *Admissible:* It must conform to certain legal rules before it can be put before a jury.

2. *Authentic:* It must be possible to positively tie evidentiary material to the incident.

3. *Complete:* It must tell the whole story and not just a particular perspective.

4. *Reliable:* There must be nothing about how the evidence was collected and subsequently handled which causes doubt about its authenticity and veracity.

5. *Believable:* It must be readily believable and understandable to members of a jury.

However, due to the digital and transitory nature of computer evidence, it does require special consideration and the challenge for any person or organization searching, seizing, or analyzing computer evidence is to secure it and retrieve relevant information in a manner that ensures its authenticity and veracity.

There are a number of obstacles that are faced by those who seek to carry out the computer forensic process, namely:

1. Computer evidence can be readily altered or deleted.

2. Computer evidence can be invisibly and undetectably altered.

3. Computer evidence can appear to be copied while in fact it is undergoing alteration.

4. While in transit, computer evidence can share the same transport pipeline as other data.

5. Computer evidence is stored in a different format to that when it is printed or displayed.

6. Computer evidence is generally difficult for the layman to understand [13].

The problems identified above can be summarized in two terms, *mutability* and *interpretation*.

Within the first issue identified, the mutability of computer-based evidence, there are two subdivisions. Transitory, real-time system events are even more sensitive to alteration than magnetic disk media and are sometimes referred to as *volatiles*. *Volatiles* can be defined as active, transient information reflecting the system's current operational state including registers, caches, physical and virtual memory, network connections, shares, running processes, media mount points, floppy, tape, CD-ROM, and printing activity [14].

Nonvolatile refers to files and information stored in a semipermanent or permanent form on media, such as a disk or CD-ROM.

The second issue is that information stored in binary form normally needs a degree of interpretation (unless you happen to think in binary or hex) before it becomes intelligible to the human eye. During normal system operation, data interpretation is carried out automatically by the controlling programs, however, during forensic analysis this is not normally the case and the data is typically reinterpreted from file header information using multiformat viewer programs such as *Outside In* and *Quick View Plus*. This reinterpretation of the information from its binary form requires special consideration when formulating procedures by which computer-based information can be collected and presented as reliable evidence in a court of law.

3.3.1 Jurisdictional issues

In an age of global computer networking, there will be issues with respect to sovereignty and jurisdiction in dealing with criminal offences and evidence that can easily cross national boundaries. As with most other laws across international boundaries, there is very little consistency with respect to the rules governing the acquisition and admission of evidence. Unfortunately, unlike many other forensic disciplines, computer forensics currently lacks an international accepted standard against which the legal and judicial fraternity can measure the competencies and procedures of computer forensic practitioners.

Evidence obtained in one jurisdiction and which is perfectly acceptable to the legal system may be completely inadmissible in the judicial system of its immediate neighbor. This lack of an international accepted standard for obtaining computer evidence has been an issue since the late 1980s, and has been considered since then by such organizations as the Council of Europe, the Organization for Economic and Cooperative Development (OECD), the International Criminal Police Organization (Interpol), and since 1995, the IOCE which was described in Chapter 2.

The SWGDE, also mentioned in Chapter 2, developed a draft document that proposed the establishment of standards for the exchange of digital evidence between sovereign nations, and this document has subsequently been adopted as a draft standard by some U.S. law enforcement agencies [15].

3.3.2 Forensic principles and methodologies

Having a standard methodology, that is one which is thorough, logical and provides appropriate protection for the original evidence is crucial for computer forensics [16]. As mentioned in Chapter 2, the U.K. Association of Chief Police Officers (ACPO) prepared a document titled "Good Practices Guide for Computer Based Evidence" to assist in the acquisition, analysis, and presentation of computer-based evidence. It is the only accepted current standard of practice for digital evidence and has been adopted outside the United Kingdom by various private and government forensic organizations. Despite its role in providing a standard which computer forensic examiners should seek to follow, the guide states up front that "Noncompliance with this guide should not necessarily be considered as grounds to reject evidence." During the International Hi-Tech Crime and Forensics Conference (IHCFC) of October 1999, an IOCE working group reviewed the U.K. Good Practice Guide and the SWGDE Draft Standards. The working group proposed the following principles, which were voted upon by the IOCE delegates present with unanimous approval:

1. Upon seizing digital evidence, actions taken should not change that evidence.

2. When it is necessary for a person to access original digital evidence, that person must be forensically competent.

3. All activities relating to the seizure, access, storage, or transfer of digital evidence must be fully documented, preserved, and available for review.

4. An individual is responsible for all actions taken with respect to digital evidence while the digital evidence is in their possession.

5. Any agency that is responsible for seizing, accessing, storing, or transferring digital evidence is responsible for compliance with these principles.

IACIS® has stated that there are three essential requirements for the conduct of a competent forensic examination:

1. Forensically sterile examination media must be used.

2. The examination must maintain the integrity of the original media.

3. Printouts, copies of data, and exhibits resulting from the examination must be properly marked, controlled, and transmitted [17].

These principles are mainly common sense and consistent with the standard evidentiary requirements as detailed previously. These principles can be synthesized into four rules that are fundamental to ensure admissibility of the evidence presented in a court of law.

Rule 1
Minimal handling of the original—the application of computer forensic processes during the examination of original data shall be kept to an absolute minimum.
     This can be regarded as the single most important rule in computer forensics. Any examination of original evidence should be conducted in such a way so as to minimize the likelihood of alteration. Generally, this rule is adhered to by the application of various preservation techniques. Essentially the original is, where possible, duplicated and the examination takes place on the duplicate data. The Best Evidence Rule, which was established to deter any alteration of evidence, either intentionally or unintentionally, states that the court prefers the original evidence at the trial, rather than a copy; however, duplicates will be acceptable under the following conditions:

> ► Original lost or destroyed by fire, flood, or other act of God, which has included such things as careless employees or cleaning staff;

> ► Original destroyed in the normal course of business;

> ► Original in possession of a third party who is beyond the Court's subpoena power.

This rule has been relaxed in many jurisdictions to allow duplicates to be tendered in Court unless there is a question as to the original's authenticity, or admission of the duplicate would, under the circumstances, be unfair [18].

As stated in the IACIS® forensic requirements detailed later, appropriately sterile media should be utilized to ensure that the duplication processes are conducted to uncontaminated media.

The duplication of evidence has a number of advantages. Firstly, it ensures that the original is not subjected to alteration in the event of an incorrect or inappropriate process being applied. Secondly, it allows the examiner to apply various techniques in cases where the best approach is not clear. Consequently, if during such trials the data is altered or destroyed, it simply becomes a matter of working on a fresh copy. Thirdly, it permits multiple computer forensic specialists to work on the same data, or parts of the same data, at one time. This is especially important if specialist skills (e.g., cryptanalysis—password breaking) are required for various parts of the analysis process. Finally, it ensures that the original is in the best state possible for presentation in a court of law.

Unfortunately, while there are advantages to duplicate evidence, there are also a number of disadvantages. Firstly, the duplication of evidence must be performed in such a manner, and with such tools, so as to ensure that the duplicate is a perfect reproduction of the original. Failure to properly authenticate the duplicate will result in questions being raised over its integrity. This in turn can lead to questions being raised over the accuracy and reliability of both the examination process and the results achieved. Secondly, by duplicating the original, we are adding an additional step into the forensic process. This in turn has resourcing and procedural implications. Additional resources are required to accommodate the duplicated data, and extra time is required to facilitate the duplication process. Furthermore, the methodology being employed must be expanded to include the duplication process. Finally, the restoration of duplicated data in an effort to recreate the original environment can, by its very nature, prove to be difficult. In some instances, in order to recreate the original environment, specific items of hardware may be required. This again adds further complexity and time to the forensic process.

Best practice within the computer forensic community also indicates that two copies of the duplicated evidence should be maintained, one of which is never touched and the other that is used as the working copy on which further analysis is performed.

Rule 2
Account for any change—where changes occur during a forensic examination, the nature, extent, and reason for such change should be properly accounted for.

During an examination, despite all intentions to the contrary, it may be inevitable and/or necessary for either the original or duplicate to be subjected to alteration. This applies both at a physical and logical level. In such cases, it is essential that the examiner both fully understands the nature of the change, and is the initiator of the change. Additionally, the examiner must be able to correctly explain the extent of any change and give a detailed explanation as to why such change was necessary. As stated earlier, this includes any examination whether it is conducted on the original or on a duplicate. Essentially, this applies to any evidentiary material that is derived from a forensic process in which the change has occurred.

This is not to say that change shall not occur, but rather in situations where it is inevitable, the examiner has a responsibility to correctly identify and document change.

The ability of the examiner to correctly describe the change is directly attributable to his/her skills and knowledge. While during the forensic examination this point may seem insignificant, it becomes a critical issue when the examiner is presenting his/her findings during any legal proceedings. While the evidence may be sound, questions regarding the examiner's skills and knowledge can affect both his/her credibility as well as the reliability of the process used. Hence, given sufficient doubt, the results of the forensic process can in the worst case be ruled inadmissible and therefore be disregarded.

Many computer forensic practitioners believe that any change in the original evidence will rule the entirety of evidence on the system inadmissible. However, it should also be noted that in the other, more mature forensic sciences, such as fingerprint identification, in many cases, the examination to recover the evidence is in fact destructive to the original evidence. For example, in the case of a burglary, one does not normally produce the pane of glass from which the suspect's fingerprint was recovered.

Rule 3
Comply with the rules of evidence—the application or development of forensic tools and techniques should be undertaken with regard to the relevant rules of evidence.

One of the fundamental concepts of computer forensics is the necessity to ensure that the application of tools and techniques is carried out in such a manner, so as not to lessen the admissibility of the final product. It therefore follows that the type of tools and techniques used, as well as the way they are applied, is important in ensuring compliance with the relevant rules of evidence.

Another critical factor in complying with the rules of evidence is the manner in which the evidence is presented. While this is very much dependent upon the existing legislation, it is nevertheless necessary to ensure that the method of presentation does not alter the meaning of the evidence. Essentially, the information should be presented in a manner that is as indicative of the original as possible [19].

Rule 4

Do not exceed your knowledge—the forensic computer specialist should not undertake an examination that is beyond their current level of knowledge and skill.

It is essential that the computer forensic examiner is aware of his own limitations with regard to his current level of skills and knowledge. In effect, the examiner must be able to recognize at what point the examination requires knowledge and skill beyond their own capabilities. On reaching this point, the examiner has a number of options. The first is to cease any further examination and to seek the involvement of more experienced and skilled personnel. The second is to conduct the necessary research to improve his knowledge to a point that permits a continuation of the examination. The third is to continue with the examination in the hope that all goes well.

The final option without doubt, is the most dangerous. It is imperative that the forensic examiner be able to describe the processes employed during an examination correctly. Additionally, the examiner should be able to explain the underlying methodologies for such processes. Failure to competently and accurately explain the application of a process or processes can result in the expertise and credibility of the examiner being called into question in any subsequent judicial proceedings.

Another danger with continuing an examination beyond one's skills is the increased likelihood for damage. All too often in these situations, changes that the examiner is not aware of or does not understand take place. Consequently, such changes are usually ignored. This in turn becomes a ticking time bomb, waiting to explode back on the examiner's face. When it does, it usually occurs when the examiner is giving his/her evidence.

Essentially, properly skilled and qualified staff should undertake complex computer forensic examinations. The actual level of skill and knowledge will determine the complexity of the examination. To ensure that these conditions are met, it is imperative that the examiner has undergone the appropriate level of training. Additionally, given that technology is continually advancing it is important for the examiner to partake in ongoing training [20].

3.4   Computer forensics model for law enforcement

It is useful to draw the evidentiary requirements, legal considerations, and principles together into a framework or model that provides coherency and consistency for all aspects of conducting computer forensics.

In developing such a framework, it is important to focus on the challenges that may be presented to the examiner in applying the model to carry out examinations and in the presentation of the resulting evidence in such a way that it is subsequently accepted in court.

1. *Expertise test:* Obviously a key test will be to challenge the expertise and credibility of the computer forensic examiner who conducts the forensic analysis and presents the resulting evidence. This test essentially seeks to establish the strength and reliability of the expert's knowledge as applied to the IT environment in which the electronic evidence is extracted.

2. *Methodology test:* The methodology test probes the processes and procedures adopted by the computer forensic examiner during the computer forensic examination. The adoption of poorly constructed methodologies can lead to erroneous analysis results, and may even lead to the destruction of, or alterations to, potential evidentiary data.

3. *Technology test:* The technology test examines the technology used during the forensic examination process, and aims to test the accuracy, reliability, and relevance of the technology as applied in the computer forensic analysis.

3.4.1   Computer forensic—secure, analyze, present (CFSAP) model

The CFSAP (computer forensic—secure, analyze, present) model essentially combines the four key elements of computer forensics (identification, preservation, analysis and presentation) into three distinct steps. Each step combines a number of processes to achieve three key objectives:

1. The securing of potential evidence;

2. The analysis of secured data;

3. The presentation of the analysis results.

The CFSAP model (Figure 3.1) provides a framework within which detailed individual forensic processes and procedures may be developed. It is of a sufficiently high level that it can be used to develop procedures for any of the different types of computer forensics as detailed throughout this book [19].

3.4.1.1   Secure—securing potential evidence

The securing of evidence encompasses both the identification of potential sources of evidence as well as the preservation of data residing within each source. The development of a suitable methodology to secure electronic evidence will be dependent upon the rules of evidence and the technology available at the time. The primary focus of this stage is to ensure that all available evidence is identified and captured in such a way that its integrity and value is not diminished.

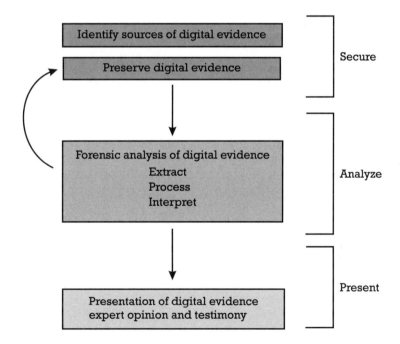

Figure 3.1   The CFSAP model.

Identification   The identification of data requires a comprehensive understanding of both the nature of the IT environment as well as the underlying technology. Failure to understand both of these can result in the key evidence being missed. Once potential evidence is located, and before it is preserved, the forensic examiner must ensure that it is relevant to the facts under investigation. Depending on the circumstances and grounds on which the evidence is being acquired, failure to determine relevance could see it ruled inadmissible in any future legal examination.

Preservation   Once potential evidence has been identified it will be necessary to either preserve the original data in the state in which it is found or to make an exact duplicate of the data. Essentially computer forensic rule 1 (minimal handling of the original) and rule 3 (comply with the rules of evidence) are critical in the securing stage. The preservation of data under these circumstances involves two distinct steps:

1. Duplication;

2. Authentication.

While it is preferred that the original source of evidence be preserved, in reality this may not be possible. Electronic evidence may reside on a computer system that is critical to the ongoing operations of a business, or alternatively it may reside on a computer geographically removed, yet remotely accessible. In either case, securing of the original is not realistic. In such instances, it is desirable to duplicate the data by making an exact copy through the use of forensically sound duplication techniques. Similarly, where data of evidentiary value is being collected in real time, as in the case of live monitoring of system logs during unauthorized network activity, it would be unrealistic to take the receiving system off-line for the purposes of preserving data captured. Interestingly in some instances, such as some criminal investigations, retention of the original data by the systems owner may constitute a continuation of an offence, thereby necessitating the seizure of the original computer system(s).

After duplicating the data it is necessary to authenticate the copy by applying some means of comparison with the original. This is particularly a problem if the original data is resident on a live system that is constantly subjected to change. This raises the question, why would you need to authenticate a copy sometime after the duplication process has occurred? The simple answer is that in some instances, it could be alleged that the copy has been altered, either deliberately or inadvertently, and as such is not reliable. The best way to authenticate data is to fingerprint the files by generating a OWHF—discussed in Chapter 2, of both the original and copy data at the time of duplication. If the duplication process is accurate, the fingerprints should match up. Additionally, if it is alleged that the data has been tampered with, the retaking of a mathematical fingerprint from the copy data should yield the same result as that derived at the time of duplication.

3.4.1.2   Analyzing data

The analysis of potential digital evidence essentially encompasses three steps:

1. The preparation of data;

2. The processing of extracted data;

3. The interpretation of data.

Preparation   This is the preparatory process in which captured data is made ready for processing. Whether the original data is seized, or an authenticated copy of the original is obtained, it is essential that the forensic examiner possess a master copy of the data to be examined. The master copy is simply an authenticated copy of the original that is preserved for future reference. To alleviate possible changes, it is not uncommon for the master copy to be stored on some form of permanent storage media (e.g., CD-ROM or DVD). The master copy forms the benchmark upon which the forensic process may proceed. To this end, it is regarded as standard practice to work from a secondary copy of the master copy. If during the examination process changes to the data occur, or some form of research and development on the data is required to overcome a problem, the computer forensic examiner still has, by way of the master copy, an authenticated duplicate to recommence the examination.

Processing   The processing of data essentially encapsulates the application of computer technology, in the form of data recovery and analysis tools, to the retrieval of relevant electronic evidence. Simply put, it is the finding of the proverbial needle in the haystack. The processing of data entails two key steps:

1. The search for relevant data;

2. The extraction of relevant data.

The search for relevant data involves scanning through all preserved data, searching for information that matches a predetermined criterion. The predetermined criteria can encompass things such as key words, recorded events or activities, system changes or anomalies, or disguised or encrypted data. In searching for relevant data, the forensic examiner will not only examine current files, but also consider searching for deleted material or residual data. Additionally, the computer forensic examiner may apply various pattern matching or data analysis techniques in an effort to identify relationships between data that may afford valuable evidence of an event or course of conduct.

The extraction of data can only take place when relevant data has been located. The extraction process simply involves the isolation and duplication of the relevant items of data from the copy undergoing examination. These extracted copies form the basis of the electronic evidence for the particular matter under investigation.

Interpretation   The interpretation stage relies heavily on the knowledge and skill of the computer forensic examiner, rather than the capabilities of the forensic technology as relied upon in the processing step. Once the computer forensic examiner has isolated electronic evidence, he/she must be able to interpret it to establish its meaning and, therefore, its bearing within any investigation or inquiry. The interpretation of data is undertaken to establish key issues, such as relevance, context, ownership, and identity (these are discussed in more detail later on in the book). It is in the interpretation stage that the computer forensic examiner may express an opinion or belief regarding things such as the following:

• How the data came to be on the computer system?

• The accuracy and reliability of the data.

• The possible identity of the owner.

• The purpose of the data.

In expressing an opinion that may be used in subsequent legal proceedings, the computer forensic examiner must possess sufficient knowledge regarding the IT environment from which the data is derived to satisfy the expertise test.

3.4.1.3 Presentation of results

The presentation of the results of a computer forensic examination is the final step in the computer forensic process. It is at this point that all relevant data should have been identified, preserved, and extracted. In presenting the results of an examination, it is critical for the computer forensic examiner to be able to clearly and concisely convey both the results obtained and the meaning of those results. To this end, it is essential that the computer forensic examiner be able to explain complex technological concepts and techniques in easy-to-understand terms.

This is important given that in some instances the results of the computer forensic examination may end up being tendered in evidence before a court of law. Consequently, the computer forensic examiner must be able to convey the significance of any results to persons who may have little or no understanding of the technology employed.

To assist the computer forensic examiner in the presentation stage, it may be necessary to employ various visualization tools, such as flow charts and link analysis charts, in an effort to explain underlying concepts and relationships. While such visualization techniques may assist, it should be remembered that they are merely an aid to, and not a substitute for, the actual evidence.

In presenting the results, the computer forensic examiner is faced with the possibility of being challenged on his/her findings based on the following:

▸ The tools used;

▸ The methodology employed;

▸ The examiner's expertise.

A failure to satisfy any challenge can result in the electronic evidence being regarded with suspicion, and may ultimately result in the computer forensic examiner's credibility being challenged.

3.5 Forensic examination

While all computer forensic cases are unique, examiners should develop their own documented standard operating procedures (SOPs) and follow them consistently. These SOPs will help protect the integrity and authenticity of evidence by ensuring that all data is acquired, analyzed, and preserved in a systematic and consistent manner.

The *prime directive* in carrying out any forensic examination on a *live* or a *dead* system is

> Strive to capture as accurate a representation of the system(s), as free from distortion and bias as possible [20].

3.5.1 Procedures

Procedures will vary to some extent based on the IT environment, the type of case, the status of the system at the time the examiner confronts it (is it live and functioning, on a network, or is it shutdown) and what resources they have to safely acquire and analyze the evidence. Procedures will also be dictated to some extent by what evidence is actually being sought and the context surrounding it. Due to the amount of information that is contained on a modern disk drive, the more information the computer forensic examiner has about the context of the case, the easier it is to provide the analysis in a timely manner.

A computer forensic examiner should seek to obtain the following information prior to conducting the analysis:

1. What is suspected or needs to be proven?;

2. Any specific information about times and dates to support time-line analysis of activities;

3. Any specific keywords and text strings;

4. Access to any other supporting computer evidence already in possession of the investigator to support evidence correlation, such as proxy logs (logs of Internet browser activity from firewalls and proxy servers), and pen register logs (records of incoming and outgoing phone calls from a suspect's phone line);

5. A description of the computer skill level of the suspect;

6. If the system is used for business rather than a personal computer, as detailed a description as is available about the network environment in which the system was located and what the system's primary function was.

Searching without this information can be like looking for a needle in a haystack.

Live System Processing

Many computer forensic examiners are used to dealing only with static magnetic media (hard drives and floppy disks) from a computer system that has been seized in a shut down state. Intrusion forensics (a focus of Chapter 6) conducted on a live, compromised system in a computer security incident response scenario is much more demanding of the forensic examiner. Many transitory system events occurring on the system, such as which network connections are open and what processes are running, may constitute critical evidence about the compromise that needs to be appropriately preserved and acquired. Every forensic examiner should therefore have an understanding of the protocols for safely acquiring volatile data from live systems, not just analyzing static file system structures from magnetic media.

Any system being examined live should be considered to be *hostile* until proven otherwise. There are circumstances that need to be considered in these cases such as the "continued presence of [the] intruder on the system, possible 'booby traps', impact of system compromise on continued operations [and] involvement of law enforcement" [14].

Data acquisition in these cases requires that processes followed need to address the order of volatility of information resident on a system. The order of volatility for system events, and therefore the order in which they need to be acquired during forensic processing, is as follows:

1. Registers, peripheral memory, caches;

2. Memory (virtual, physical);

3. Network state;

4. Running processes, open files, media mount points;

5. Logical file system;

6. Physical hard drive, floppies and backup tapes;

7. CD-ROMs and printouts [14].

It should be noted in all cases that protection of the system, and therefore evidence, is paramount so the computer forensic examiner should always err on the side of caution if anything appears to be amiss. That is, always have the power plug nearby just in case you need to pull it because the disk drive light starts whirring unexpectedly.

Prior to carrying out forensic examinations, the following should be considered:

▸ *On live systems avoid tools that use a graphical user interface.* Command-line utilities, and in particular, statically linked binary files, are best utilized as they are more likely to leave little or no footprint on the evidence system if they are properly utilized. Command line tools are also much easier to use if you have run your own known, trusted command shell.

▸ *Validate your tools.* Only utilize tools from trusted sources and personally verify their actions and that they work as advertised. This supports not only the evidence acquired during the examination of the system but also supports your credibility, should you be called upon in court to validate the processes followed and tools used. Generate checksums for all of the tools and store the list of checksums with the toolkit.

▸ *Keep copies of the tools on removable media.* Create a CD-ROM or a number of floppy disks that contain the trusted operating system kernel(s) and the data acquisition tools. Write-protect the media as necessary.

▸ *Document, document, document.* Documentation of exactly what is done and when it is done during every facet of an investigation cannot be overemphasized. Testimony may take place as much as a year or more later and the more comprehensive your notes, the easier it is to provide accurate and less refutable testimony.

The following evidence processing guidelines should not be considered to be definitive, proscriptive process but as the best practice. The process has been derived from the U.S. Secret Service *Best Practices for Seizing Electronic Evidence* [21], New Technologies Inc. *Computer Evidence Processing Steps* [22], IACIS® *Forensic Examination Procedures* [17], RFC 3227 *Guidelines for Evidence Collecting and Archiving* [23], and Foundstone's forensically sound initial response processes [24]. The guidelines implement the three phases of the CFSAP model (secure, analyze, present) with each phase comprising several steps.

3.5.1.1 Securing evidence

1. *Establishing forensically sterile conditions:* All media utilized during the data acquisition and analysis process are to be freshly prepared, completely wiped of nonessential data using an appropriate sanitization program, scanned for viruses and verified before use.

2. *Following a complete, documented, logical process in acquiring evidence from the system:* While each forensic case is unique, all examiners should develop documented SOPs and follow them. SOPs help ensure consistency in the manner that all data collected is preserved, acquired, analyzed, and presented. Documentation should describe exactly what was observed on the system when it was examined, what actions were carried out and at what time. Litigation may be a lengthy process and accurately describing the situation will be difficult if appropriate notes are not kept. Accurate

documentation differentiates a professional, scientific approach to an examination from an ad hoc, unprofessional one. Such documentation should include photographs and video where possible.

3. *Using a known trusted command shell and tools for acquiring data from a system:* The operating system and applications on a machine potentially containing evidence should never be trusted. Information gathered in the examination cannot be trusted if the tools cannot be trusted. The computer forensic examiner, should, if possible, carry out all actions using a known, trusted kernel and applications that he/she can be sure has not been compromised or modified. Except in dire circumstances where no other option is available, no information from the examination should be written to the system being examined. Where this does occur, accurate notes of the actions required and the reasons for them should be kept to justify any alterations made to the system. The objective is to preserve the state of the system as far as possible at the time the acquisition and examination take place.

4. *Data acquisition—volatiles:* Where appropriate, acquire evidence using the order of volatility listed earlier to ensure that resulting output is written to the sanitized media previously prepared. Prosise and Mandia [24], and Romig [14] describe in-depth acquisition processes for volatile data in Microsoft Windows–based and UNIX environments.

5. *Copying system files for analysis:* In system compromise cases, it may be necessary to copy system log files or binaries for further analysis. Where this is necessary, the copy process should be authenticated. All efforts must be made to ensure that these processes are as noninvasive as possible.

6. *Logical volume imaging on live systems:* In some cases, it may not be possible to shutdown a system in order to image it. This may particularly be the case with business systems where the system is critical to the functioning of the company, and/or the system is a server running a redundant array of inexpensive disks (RAID) array where multiple physical disks appear as one logical volume. In these cases, the most reasonable option is probably to carry out a logical image of the volume either to plug-and-play removable storage media, such as a tape drive or over a network connection to another system. More and more systems like these will be encountered, so

the computer forensic examiner must have procedures developed, should this scenario be encountered. Another circumstance where logical volume imaging may be employed is where a cryptographically secured filesystem is encountered on a live system. Unless keys and passwords are provided then the encrypted data may be irretrievable once the system is shutdown. The alternative here is to logically image the encrypted filesystem while it is open to ensure that no potential evidence is lost or later unavailable.

7. *Shutting down the computer:* There is much discussion about whether or not to carry out a normal system shutdown due to flushing of caches, overwriting of swap files and such. The determination of whether a system should be normally shutdown or a hard shutdown with power removed will be a case dependent judgment call for the examiner at the time. If it appears all appropriate volatile information has been acquired, and if there are no booby traps or malicious programs apparent, the system should be shut down to allow imaging of the hard disk drives and seizure of the system. The system should be observed closely during this process with the ability to "pull the plug" on system power in case of emergency.

8. *Documenting the hardware configuration of the system:* This step should take place on site if the system is not being seized, so that an accurate record of the system hardware configuration, BIOS settings including boot order, and drive translation settings are maintained. Serial numbers for all components should be noted before any removal of the hard drive or the system itself. Where possible, photographs or video should be taken of the systems surrounding environment, cabling, and configuration. Network environmental information, if appropriate, should describe network type, topology, and relevant physical or media access control (MAC) addresses. Reviewing the BIOS configuration of a hard drive and comparing it to the manufacturer's physical parameter details listed on the label is important to ensure that no concealed data areas are being maintained on the disk through BIOS manipulation.

9. *Documenting the system date and time:* Time is one of the single most important attributes used in computer forensics. Establishing the dates and times associated with the modification, access, and creation (commonly referred to as MACtimes) of computer files is extremely important in reconstructing the sequence of events on a system. The accuracy of the dates and times may be critically

important and the system BIOS date and time settings and its variation from the true date and time need to be accurately recorded at the time the computer is seized or examined.

10. *Continuity of evidence (chain of custody):* If the computer is being seized, it should not be left unattended unless it can be secured from unauthorized access. For transport, it should also be appropriately protected with bubble-wrap or other protective packaging. In vehicles, computer evidence should be kept as far away as possible from magnetic fields caused by stereo speakers and two-way radio equipment.

It is a common practice in some circumstances to seize only the hard drive and if this is the case it should be protected using an antistatic bag and appropriate padded packaging for transport.

In all cases documentation should be prepared which has an adequate description of the item including any serial numbers, the sequence of handling and control of the hardware until any potential legal proceedings are finalized. All potential and actual evidence should be secured in a limited access filing cabinet or locker with one person identified as the custodian controlling access to it. Any time access is made to the evidence, it should be appropriately noted with time, date, person, reason, and a signature.

11. *Data acquisition—magnetic media:* Noninvasive sector image backups should be made of all hard disk drives and floppy disks. Where possible, a hardware write blocking system, such as Intelligent Computer Solutions Inc.'s *Drive Lock* [25] or Digital Intelligence Inc's *FireChief* [26] should be used to ensure no inadvertent writes are made to the evidence drive. Drive Lock, shown in Figure 3.2, is commercially available and employs an IDE-to-IDE interface unlike other tools which employ IDE-to-SCSI or IDE-to-IEEE 1394

Figure 3.2 Drive Lock. (Reprinted with permission, 2002.)

controller translation. Where this type of hardware is not available, a controlled boot disk specially modified to ensure the operating system on the disk makes no probes to drives attached to the system, should be used [24].

12. *Authentication of copied and imaged media:* If the imaging utilities used to make the bit stream backup do not do so automatically, manual authentication of the imaging should be carried out to validate the duplication process. Tools used should employ an accepted one-way hash function—see Chapter 2 (OWHF) algorithm over 128 bits, such as MD5 or SHA-1.

13. *Malicious code protection:* All reasonable precautions must be taken during any copying or imaging process to ensure that there is adequate protection from malicious code including viruses, destructive programs or other programs that could potentially corrupt or compromise the original evidence media.

14. *Archiving media images:* Once imaging has been completed, two archive copies, a master copy and a working copy should be made to preserve the image files before analysis. As with all other copying and imaging processes, the archive copies are compared to the original image files to ensure the integrity of the archive process. All analysis should then be conducted on the working copy of the image rather than on the original, seized media or the master copy.

3.5.1.2 Analyzing secured data

1. *Logical analysis of the media structure:* An examination of the volumes, partitions, and file systems located on the image is conducted to identify the data structure of the original media. The characteristics and configuration should be noted in detail.

2. *Operating system configuration information:* Details from the boot record and information on the operating system configuration, user-defined system configuration such as *CONFIG.SYS, AUTOEXEC.BAT, WIN.INI, SYSTEM.INI,* and registry are examined and findings are noted.

3. *Document file names, dates, and times:* From an evidence standpoint, file names, ownership, and MACtimes can be extremely important [27]. Therefore, it is important to fully catalog all active and deleted files. Some forensic analysis programs, such as Access Data's *Forensic*

Toolkit (FTK), will do this automatically and allow export of the results into a spreadsheet [28].

4. *File signature recognition:* Many systems these days have upwards of 20,000 active files located on them so any technique that reduces the amount of data needed to be examined is very useful. Comparative analysis of existing operating system and application files using signatures generated by OWHFs can significantly reduce the number of active files that need further examination. Some forensic analysis tools such as Guidance Software's EnCase (see Chapter 2) and FTK, can do this automatically as file signatures are prepared as part of the evidence processing.

   Tools such as the *Hashkeeper* database [29] discussed later and the *National Software Reference Library* [30] discussed later and in Chapter 2 can also assist the computer forensic examiner by providing specialized signature databases to identify child pornography images and hacking related software.

5. *Identifying file content and type anomalies:* Some file types including encrypted, compressed, and graphic files may be stored in a binary format that a standard string/text search program will not search. Suspects may also purposely alter file extensions or employ steganographic techniques to conceal incriminating information. Evaluation of the file headers, the leading bytes of a file, may also identify files with inappropriate file extensions.

6. *Evaluating program functionality:* Depending on the application software involved, running captured programs to learn their purpose may be necessary. Appropriate backup and isolation measures should be employed to ensure disconnection of the system from a network to prevent potential dissemination of malicious software. Use of an operating system emulator such as *VMWare* from VMWare Inc. can allow safe analysis of the suspect software. For programs of uncertain function, employment of debugging programs that intercept system calls, and decompilation programs, which extract the original high level programming code from executable binary code, may be required.

7. *Text string and key word searching:* Normally an investigation will require some searching for the presence of information in a textual form and there are many tools available to assist in the search for text-based evidence. Prior to conducting this search, the information described previously should be collected to assist the analysis

process. FTK, a forensic analysis suite, incorporates a complete text indexing capability that allows almost instantaneous searches for most text strings [28].

8. *Evaluating virtual memory:* Virtual memory, in the case of Windows, the swap or page file, and in the case of UNIX, the swap partition, is a potential source of evidence and investigative information. Some forensic analysis tools, such as FTK, automatically extract swap file information and process it for textual information in the same manner as any other file.

9. *Evaluating ambient data:* Ambient data (file slack and unallocated space) is described in Chapter 2. Forensic acquisition and analysis tools can extract deleted files and relevant fragments from these areas on disk to support the investigation process. Deleted files may be particularly important and MACtime attributes should be obtained for recovered files to support the time-line analysis (Section 3.5.2.3).

3.5.1.3   Presenting the results of analysis

1. *Document, document, document:* The importance of comprehensive and accurate documentation cannot be stressed enough. As indicated previously, documentation should be contemporaneous, that is, notes should be taken at the time, not prepared from memory, hours or days later. Documentation should include chain of custodial information, dates, times, forensic software details including version numbers and details of evidence located. Some forensic analysis tools also maintain detailed electronic reports in HTML or document format and these should be treated as the same as other forms of digital evidence and secured appropriately. Documentation should be printed and physically signed where possible or a jurisdictionally acceptable digital signature method is employed where this is not possible.

2. *Retaining copies of software used:* As part of the documentation process, copies of the software used to carry out the imaging and analysis should be retained with the output of the forensic tool involved. With software being updated so regularly, duplication of results may prove difficult or impossible if the original version used has not been retained.

3.5.1.4 Limited examinations

In some circumstances it may be legally or operationally impractical to carry out a complete forensic acquisition and examination. This may be due to

1. Physical equipment limitations requiring examination of the original evidence on the premises, with appropriate precautions.

2. Sheer quantity of data to search due to size of the media (RAID arrays terabytes in size are now prevalent in the corporate environment).

3. Considerations with respect to business impact of shutting down the system as are present in some jurisdictions such as under Federal Law in Australia.

4. Legal constraints that may be applied to the "search and seizure" process, particularly in the case of civil rather than criminal matters.

5. So much corroborative evidence has already been identified that makes further evidence redundant and further search unnecessary.

6. Circumstances beyond the examiner's control prevent the examination from being conducted fully.

The IACIS® procedures [17] identify these potential limitations and as with all other aspects of the computer forensics process, the reasons for these limitations should be fully documented should the issue arise in future.

3.5.2 Analysis

We will now briefly examine some of the analytical techniques that can be carried out in support of the forensic process, some of which are mentioned in the guidelines earlier. Some of these techniques have been mentioned previously and some will be mentioned in the later chapters, however, they should be identified as part of the process as the majority were developed to support criminal and national security investigations. The techniques include but are not limited to

1. System usage analysis;

2. Internet usage analysis;

3. Time-line, or temporal, analysis;

4. Link analysis;

5. Password recovery and cryptanalysis.

3.5.2.1 General system analysis

This seeks to identify user(s) of the system, the system name, its apparent primary function, and any other characteristics that can be determined from the operating system and application configurations. Usernames and passwords will be important particularly if encrypted files are identified during the analysis.

In the case of Microsoft Windows 9x systems, the *system.dat* and *user.dat* files should be examined in detail as information on the hardware and software configured and their use by a user are stored there. Likewise on Microsoft Windows NT and 2000 systems the hidden *ntuser.dat* file(s) should likewise be examined for similar information. *Dr. Watson* error log files can also reveal a great deal of information, particularly in intrusion cases, where in many cases an intruder will crash a program and a *drwatson.log* file will be created. Computer forensic examiners should become very familiar with the format and functions of the Microsoft Windows registry file in all its variations, as it is a repository of a large amount of information about the systems, its use and its external connectivity.

Application software may also maintain logs and where available these should also be examined.

3.5.2.2 Internet usage analysis

It is relatively obvious what this type of analysis sets out to achieve; however, it is useful to list the types of information sought during this type of analysis. The following types of information that should be sought:

1. *Network host and connectivity information:* This should include the last IP address allocated to the system, NetBIOS name, network card MAC address, dial-up applications used, and any relevant log or registry entries showing addresses, times, and dates of connection. The Microsoft Windows registry, Microsoft Windows NT, and Microsoft Windows 2000 event logs should be examined if applicable.

2. *Internet browser history:* This includes recovery and analysis of Internet browser history files, such as *index.dat* for Internet Explorer and *netscape.hst* for Netscape Navigator as well as

recovery, identification, and examination of browser cache files
and downloaded files, such as images, movies, and sounds.

3. *E-mail use:* This includes identification of e-mail applications
   utilized, application configuration including e-mail addresses and
   names, recovery of the e-mail data file(s) with content and analysis
   of any relevant e-mail messages based on the information contained
   in the e-mail headers.

4. *Other network applications:* This includes other applications that may
   be used for communication, file sharing, or transfer such as ICQ,
   Internet relay chat (IRC) clients and similar software should be
   identified and cataloged.

3.5.2.3 Time-line, or temporal, analysis

Law enforcement personnel often construct time-lines of criminal activity as
an aid in other types of investigation and computer forensic examinations
are no different. Time-line analysis comes up also in Chapters 2, 4, and 6, but
it is useful here to highlight its importance in a law enforcement examination
because dates and times are so critical in criminal investigations.

Time lines of computer usage can provide valuable information about
the computer user and the sequence of events affecting the computer.

Analysis tools employed for criminal intelligence can be utilized in
conjunction with the MACtimes and times obtained from system logs to
construct event time lines which may be able to be corroborated with other
external sources of evidence, such as telephone call records and physical
surveillance logs.

3.5.2.4 Link analysis

Link analysis, another criminal intelligence analysis tool, is discussed in
detail elsewhere in the book (see, for example, Sections 4.5 and 7.6); but its
utility in visualizing information obtained during analysis needs to be
recognized and considered.

3.5.2.5 Password recovery and cryptanalysis

Password protection and encryption pose unique problems for law
enforcement. Password recovery using brute force password attack using
specialized programs, such as AccessData's Password Recovery Toolkit
(PRTK) and New Technology Inc.'s Advanced Password Recovery Software
Toolkit may yield results depending on the strength of algorithm employed

to encrypt the data. In the case of high strength algorithms, cryptanalysis using supercomputing systems, such as those employed by National Security agencies or advanced distributed processing techniques, such as that employed by AccessData's Distributed Network Attack (DNA) offer the only solutions.

Chapter 7 addresses this topic in more detail in the context of recent and expected future developments in the area of encryption technologies.

3.5.3 Presentation

Presentation, the final phase of the forensic process, ties all the previous activities together to provide the overall picture of what has occurred. Information presented may include what is termed the *real evidence*, such as the following:

1. The output from the forensic tools utilized;

2. Printed command line histories and monitor snapshots;

3. Handwritten notes and checklists;

4. Audio and video recordings;

5. Diagrams and manuals.

Real evidence is a physical thing, the existence or characteristics of which are relevant and materialistic. As with the original evidence, these derived forms of evidence need to be appropriately handled and protected using continuity of evidence protocols.

Direct evidence, which can incorporate real evidence, is the evidence that stands on its own to prove an alleged fact, and includes oral testimony from the computer forensic examiner about what they personally saw, heard, or did.

3.5.3.1 Supporting props

Due to the complexities of a computer forensic examination, it may be necessary to attempt to simplify the information, and present it in a format more readily understandable to a Court in the form of demonstrative evidence. The adage ''a picture is worth a thousand words'' could not be more correct in the complex world of information technology. Demonstrative evidence including graphs, charts, diagrams, and more recently computerized simulations, can be effectively used to illustrate or explain a witness's testimony to assist the jury to better grasp a particular aspect of the case. When such an exhibit is computer-generated, it is founded on a complex set of data that either underpins or undermines its reliability. To adequately present and defend a digital exhibit, as with other aspects of the forensic process, you need to understand the underlying science, technology, and engineering of the computational process.

3.6 Forensic resources and tools

We will now briefly examine some, but not all, of the forensic resources and tools that are employed in the law enforcement community. The forensic aspects of the major operating systems will also be discussed. The tools will be examined on the basis of their major functional category including duplication, authentication, search, forensic analysis, and file viewing tools. The lists are indicative, not all inclusive and the examiner should never rely on just one tool, but have a toolbox of options.

3.6.1 Operating systems

All operating systems are not created equally from a forensic perspective and it is useful to discuss issues surrounding the employment of various operating systems during various stages of the forensic process.

3.6.1.1 MS-DOS

The MS-DOS, in its various guises, is one of the most widely known operating systems in existence. Depending on the version, its strength is its relative simplicity and the fact is that only three files are really required to have a functional operating system—*COMMAND.COM, MSDOS.SYS,* and *IO.SYS.* It is very easy to modify older versions of MS-DOS to be forensically noninvasive and for this reason it is used for controlled boot disks for the major imaging utilities, such as Safeback and the EnCase (see Chapter 2) imaging client.

3.6.1.2 Microsoft Windows

Microsoft Windows, in all its versions since Microsoft Windows 3.1x, is probably the most forensically invasive operating system. The latest versions of Windows are very complex and automatically probe, and where possible mount, every local Windows compatible file system that is detected during the system boot process. Windows versions have only recently been utilized for forensic duplication due to the development of hardware write blocking devices that prevent the operating system from altering the evidentiary magnetic media.

Microsoft Windows, strength lies in its market pervasiveness and the fact that comprehensive forensic analysis tools like FTK and EnCase have been developed to run on it. Most people, and particularly law enforcement officers, feel comfortable using the graphical user interface and the operating system in general. From a forensic duplication perspective, however, due to its complexity and invasiveness, it is less than ideal.

3.6.1.3   BeOS

BeOS [31] is a relatively little known operating system, developed by Be Inc., designed for digital media applications and Internet appliances. It is a high performance operating system similar in some ways to Linux that provides professional users and enthusiasts with a high performance environment to quickly and easily develop applications and content and is designed to facilitate the integration of new technologies.

BeOS does not make invasive probes or automatically attempt to mount magnetic media that is connected to it which means it can be used for media acquisition. BeOS also includes many text and file utilities from the UNIX world such as *dd* and *grep*, which can be employed for forensic purposes.

3.6.1.4   Linux

Linux, developed by Linus Torvalds, is a very popular, modular freeware operating system that is highly compatible with the UNIX operating system that was developed in the 1960s at Bell Laboratories.

From an acquisition perspective, like BeOS, Linux does not make forensically invasive probes or automatically attempt to mount magnetic media that are connected to it. This means that Linux can be utilized for media acquisition. Linux also includes many powerful low level and file utilities that can be employed for forensic purposes. Linux modularity also means that distributions that exist are able to provide a controlled boot environment on a single floppy disk and CD-ROM with various forensic utilities for forensic acquisition and analysis. Work is being conducted to develop and distribute forensic specific versions of Linux, and Forensic and Incident Response Environment (FIRE) and the Portable Linux Auditing CD (PLAC), freely available on the Internet, are two examples of this.

From a forensic analysis perspective, Linux is a very powerful operating system. It natively incorporates support to be able to mount and analyze many different types of file systems both attached locally and over a network using a capability known as *network block device*. Linux is able to be employed on large clustered multiprocessor computing systems, known as Beowulf clusters, supporting high-speed text searching and where necessary cryptanalysis. NASA's Computer Crime Division and the Defense Computer Forensic Laboratory (DCFL, part of the Defense Cyber Crime Center, DCCC) in the United States are two organizations that procured high end Linux Beowulf clusters (see Figure 3.3) for forensic processing [32].

A good discussion on the basic employment of Linux as a forensic platform is the paper "The Law Enforcement Introduction to Linux: A Beginner's Guide" by Grundy [33].

3.6.2 Duplication

As stated in Chapter 2, unlike normal backup programs, sector imaging preserves all data contained on the hard disk and unlike in the late 1980s and early 1990s, there are now many sector-imaging and duplication tools available. The safe acquisition of potential evidentiary data from magnetic media is one of the most critical aspects of computer forensics and so examiners should use their own judgment and independently evaluate and

Figure 3.3 A forensic Linux Beowulf cluster.

verify the duplication tools and processes and not rely on marketing rhetoric about a tool's capability.

3.6.2.1 Safeback

Safeback, as previously mentioned, was designed from the ground up as an evidence-processing tool with error-checking built into every phase of the evidence backup and restoration process [34]. A command-line-based utility executed from a controlled boot disk has not changed all that significantly over the past 12 years and continues to be in use with many law enforcement and government agencies worldwide.

3.6.2.2 Snapback DatArrest

This is another commandline-based imaging utility that is used by some law enforcement agencies primarily because of its ease of use [35]. Snapback, due to the background of its developers, has particular strength in imaging SCSI disk drives as described in *SC Magazine*'s September 2000 market survey of forensic utilities [36].

3.6.2.3 EnCase and FastBloc

To provide an imaging capability with EnCase (see Chapter 2), Guidance Software makes a restorable disk image of a Microsoft Windows 98 SE boot disk with a MS-DOS imaging client that supports both software-based write blocking and seamless authentication [37]. FastBloc is a hardware write-blocking device that allows forensic acquisition of an IDE hard drive using EnCase in the Microsoft Windows environment which provides greatly increased acquisition speed. It operates through conversion of IDE-to-SCSI bus signals and supports hot swapping of IDE drives during the acquisition process [38].

3.6.2.4 ByteBack

ByteBack is another command line forensic duplication utility [39]. On examination it appears to be derived from or related to the data recovery program Media Tools by ACR Data Recovery that is marketed through a number of data recovery companies [40]. ByteBack's data recovery heritage is apparent in the number of data recovery features including the ability to rebuild lost data structures including partition and FATs.

3.6.2.5 Disk Image Backup System (DIBS™)

DIBS, originally developed in the United Kingdom by Computer Forensics Ltd., is an integrated hardware and software imaging and analysis system, first marketed in 1991. Unlike other forensic systems, it employs a SCSI-MOD system to store evidentiary images [41]. DIBS is used predominantly in the United Kingdom and Europe.

3.6.2.6 VOGON evidential hardware

Vogon, another U.K. company, markets another integrated hardware and software imaging and analysis solution. The Vogon hardware adopts a different approach to other imaging systems in that it utilizes high capacity, 200 GB Hewlett Packard LTO Ultrium SCSI tape drives as the imaging media [42].

3.6.2.7 Norton Ghost

Norton Ghost, an acronym for General Hardware Oriented System Transfer, is a widely utilized commercial system backup and recovery program from Symantec [43]. In standard use Ghost does not meet forensic requirements due to the fact that it does not produce a true image but instead interprets information from the master boot record and partition tables. With the employment of certain command line switches, particularly the image raw (IR) switch, however, Ghost can be utilized to create forensically sound clones and images [44].

3.6.2.8 *dd*

dd is a low-level file utility and potentially the lowest-cost forensic imaging utility that is included with most distributions of UNIX and Linux. The U.S. DCFL has enhanced *dd* by integrating md5 authentication [45]. It has also been ported across the Win32 command line environment [46]. *dd* has since been utilized in a forensic capacity due to its noninvasiveness and flexibility, and achieved highest marks as a forensic imaging utility in *SC Magazine*'s September 2000 Market Survey [36].

3.6.2.9 ICS Image MASSter Solo 2 forensic systems

The suite of *MASSter Solo 2* forensic systems from Intelligent Computer Solutions (see Figure 3.4) is an integrated hand-held duplication system that is in use with the U.S. Secret Service and other law enforcement

Figure 3.4 ICS Image MASSter Solo 2 hand-held forensic unit (Reprinted with permission, 2002.)

agencies around the world. It is capable of imaging and cloning multiple IDE and SCSI drives and maintains an audit trail of all device activities [47].

3.6.3 Authentication

Authentication is a critically important element of the forensic process and should take place at many stages. There are many authentication tools available, we list here some of the most important ones.

3.6.3.1 Hash

Hash is a command line program developed by Dan Mares, mentioned previously, that calculates a 32-bit cyclic redundancy check (CRC), 128-bit md5 or 160-bit SHA-1 hash of a file supporting file signature analysis [48].

3.6.3.2 md5sum

md5sum is a GNU implementation of the md5 algorithm for the UNIX and Linux operating system. Like *dd*, it has also been ported across to the Win32 environment [46].

3.6.3.3 Hashkeeper

Hashkeeper is a Microsoft Access database developed by the National Drug Intelligence Center (NDIC) to maintain a record of md5 hash sets for forensic use. Besides normal operating system and application software hash sets, NDIC also maintains specialized hash sets related to child pornography and narcotics. Unfortunately, the database is available only to law enforcement authorities [29].

3.6.3.4 National Software Reference Library

National Software Reference Library (NSRL) from the U.S. National Institute of Standards and Technology (NIST) is similar to Hashkeeper in that it provides a set of OWHF reference data derived from md5 that can be used to reduce the number of files that have to be reviewed or examined during an investigation. NSRL is available on a set of CD-ROM for purchase [30]. NSRL is discussed in Chapter 2.

3.6.4 Search

In looking at search tools, we will encompass not only text searching utilities but also those tools that carry out advanced file recovery through raw, binary search of a hard disk or image file(s) for hexadecimal file headers.

3.6.4.1 dtSearch

dtSearch by dtSearch Corporation is a full text search and retrieval engine for this Windows environment that, due to its use of an index, allows extremely fast search over gigabytes of data [49].

3.6.4.2 DiskSearch Pro

DiskSearch Pro is a command line text search engine from New Technologies Inc. that is able to search through both active files, and free and unallocated space employing fuzzy logic technology [50]. It is able to deal with embedded and encoded text formats and is able to search on up to 250 keywords simultaneously.

3.6.4.3 Net Threat Analyzer

Net Threat Analyzer, previously called IPFilter, is also from New Technologies Inc. It is a command-line search tool designed to detect text strings specifically related to Internet usage including e-mail, Web browsing and file downloads [51]. It is free to qualifying law enforcement agencies.

3.6.4.4 String Search

SS is a command-line text search engine from Dan Mares that is specifically designed to search media at the logical file system level for keywords [52].

3.6.4.5 *grep*

grep is a UNIX/Linux low-level, regular expression text string search utility that is extremely powerful. It is able to search through active files, unallocated space or a hard drives at the raw device level [53]. In conjunction with *dd*, *grep* obtained the highest rating for identifying each and every occurrence of concealed test data during the *SC Magazine*'s September 2000 Forensic Tool Market Survey.

3.6.4.6 File Extractor

File Extractor is a Microsoft Windows–based advanced file recovery utility from the Datalifter suite of forensic support tools by Stepanet Communications. It is specifically designed to search through unallocated space on hard drives or contained in forensic image files at the binary level for hexadecimal values that represent specific file headers of interest to the computer forensic examiner [54]. File Extractor is then able to sequentially extract an arbitrarily specified amount of data past the file header and write it to a file of the same type as the detected header. It is very useful for recovering deleted, partially overwritten files where the header is still intact, particularly graphics files.

3.6.4.7 Foremost

Foremost is a new advanced file recovery utility developed by the U.S. Air Force Office of Special Investigations that provides a similar type of functionality as File Extractor, but for Linux. It is available as a separate package or as part of the FIRE forensic Linux distribution [55].

3.6.5 Analysis

There are many individual tools available to assist in the forensic analysis process but in this section we shall only discuss integrated software suites.

3.6.5.1 Expert Witness™

Expert Witness, by ASR Data Acquisition and Analysis LLC, was the first fully integrated forensic data acquisition and analysis program designed based on the specifications and requirements of the law enforcement community. It was initially developed for the Macintosh platform but was then ported over to the Microsoft Windows environment in 1997/1998. Expert Witness was an ancestor of EnCase.

3.6.5.2 Forensic Toolkit™

FTK (see Figure 3.5) developed by AccessData as mentioned previously is a relatively new and fully integrated forensic data acquisition and analysis program that integrates a number of extremely powerful features not found in other forensic analysis suites including integrated dtSearch® technology, supporting full text indexing of image files without having to restore them to a hard drive, and Stellent's Outside In® file viewer technology. FTK's interoperability with AccessData's PRTK is also very capable when

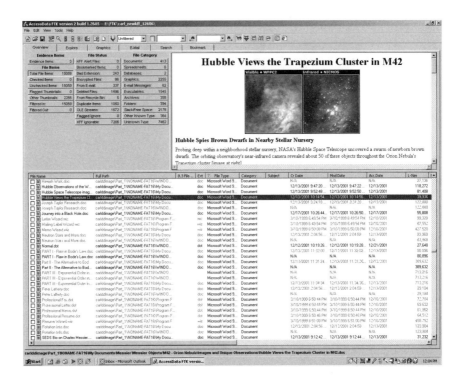

Figure 3.5 AccessData's FTK. (Reprinted with permission, 2002.)

encrypted and password protected files are detected during the analysis process [28].

3.6.5.3 EnCase

EnCase, from Guidance Software, described in detail in Chapter 2, is a fully integrated forensic data acquisition and analysis program widely used in commercial forensics.

3.6.5.4 ILook Investigator©

As described in detail in Chapter 2, *ILook Investigator*, developed by Elliot Spencer from the United Kingdom and supported by the U.S. Department of the Treasury, is a freeware, law enforcement integrated forensic data analysis program [56]. ILook is designed to examine image files of seized computer systems that have been made with *Safeback*, *dd*, *EnCase* or any other utility that makes a straight sector dump image. ILook is widely used in the U.S. law enforcement community and training on its use is carried out at FLETC. As a free tool, it offers a very viable alternative to those smaller law enforcement agencies unable to afford the expense of a tool such as EnCase, although it does require the user to be an experienced and knowledgeable computer forensic examiner in order to be able to utilize it fully.

3.6.5.5 WinHex

No forensics toolkit is complete without a powerful hex editor program for low-level file analysis and WinHex, by Stefan Fleischmann from X-Ways AG, fills this role admirably [57]. WinHex (see Figure 3.6) is extremely flexible and a white paper on its employment as a forensic utility is available.

3.6.5.6 [N] Curses Hexedit

A powerful hex editor program for the UNIX/Linux environment is [N] Curses Hexedit, by Rogoyski [58]. It has a very similar interface to DE.

3.6.5.7 Automated Computer Examination System

Automated computer examination system (ACES), a multimillion dollar FBI funded project for CART, is a law enforcement only integrated forensic data

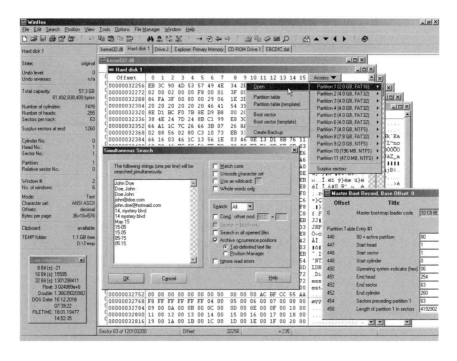

Figure 3.6   WinHex. (Reprinted with permission, 2002.)

analysis program designed for the Microsoft Windows NT4 platform [59]. ACES was to be made freely available to state and local law enforcement in the United States, however, its deployment at these levels was cancelled in favor of supporting ILook.

3.6.5.8   ForensiX

ForensiX, developed by the well-known U.S. forensics practitioner Fred Cohen, is a law enforcement only integrated forensic data acquisition and analysis program, designed for the Linux operating system [60].

3.6.5.9   Storage Media Archival and Recovery Toolkit

Storage Media Archival and Recovery Toolkit (SMART) from the developer of Expert Witness Andy Rosen, is a very powerful integrated forensic data acquisition and analysis program designed for the Linux and BeOS operating systems [61]. SMART (see Figure 3.7) combines sanitization, acquisition, authentication, and analysis capabilities into an intuitive X Window System

Figure 3.7   SMART. (Reprinted with permission, 2002.)

(X11) graphical user interface on powerful forensically sound operating system platforms. SMART is commercially available.

3.6.5.10   Datalifter v2.0 forensic support tools

Datalifter, previously known as Datasniffer, from StepaNet Communications Inc., is a suite of 10 tools supporting recovery and analysis of data from both cloned drives and sector image files. Tools include disk content cataloging, Internet cache and history viewer, file signature comparison, e-mail recovery for the major e-mail programs, such as Outlook Express, Outlook, recycle bin recovery, Microsoft Windows network ping and trace route, and a screen-capture utility [62].

3.6.5.11 NetAnalysis

NetAnalysis, by Craig Wilson from digital-detective.co.uk, is a forensic Internet history analysis tool currently in BETA testing. It supports analysis of browser use, file downloads, and any function that is carried out from browser interface through the examination of the Internet Explorer *index.dat* and Netscape *netscape.hst* files [63].

3.6.6 File viewers

A number of file viewing programs have already been mentioned; however, it is useful to list here some of the more useful utilities.

3.6.6.1 Quick View Plus

Quick View Plus for Microsoft Windows, from Jasc Software, incorporating Stellent's Outside In technology, is probably the best known general file viewing utility available [64]. It has support for almost all documents, presentations, and graphic formats making it an invaluable tool for the computer forensic examiner.

3.6.6.2 IRFanView 32

IRFanView 32 by Irfan Skiljan is a very fast 32-bit graphics viewer that supports almost all image formats that are in use on the Internet and plugins available that support many movie formats [65].

3.6.6.3 Resplendent Registrar

Resplendent Registrar, from Resplendence Software, allows detailed examination of Microsoft Windows registry files with more advanced features than those offered by the native Microsoft Windows registry editors, *regedit.exe* and *regedt32.exe*. It supports searching, bookmarking, and printing details of relevant keys [66].

3.6.6.4 GUIDClean

GUIDClean is a freeware program that allows detection and display of the Global Unique Identifiers (GUID) that Microsoft Windows 98 and some versions of Microsoft Word and Excel, prior to MS Office 2000, placed in documents [67]. The GUID is based on the MAC address of the systems network card, if one is present, allowing tracking of documents to the system on which they were authored [68].

3.6.6.5 Unmozify

Unmozify, from Info Evolution, is an Internet browser offline viewer program that can be used to examine and reconstruct Web pages from browser history files and the cache directories of Internet Explorer and Netscape Navigator [69].

3.7 Competencies and certification

Determining what is an acceptable level of competency in order to present computer forensic evidence in a court has always been a complex question that has many jurisdictional implications. Acceptance of evidence from expert witnesses varies widely across jurisdictions and the complexity of computer forensic examination is exacerbated by the lack of international standards as described previously. Fortunately, the IOCE and SWGDE have risen to the challenge and are attempting to address this inadequacy. The issue of acceptable training and more to the point, certification, is much more complex than just the lack of an international standard of practice. Key Computer Service in Florida, a computer forensic training and examination company has come up with a series of questions to ask to determine the level of competency of a computer forensic examiner to undertake a commercial litigation examination [70]. The questions that they deem relevant are as follows (permission by John Mellon, Key Computer Service Inc. to reproduce this material is gratefully acknowledged):

What are the examiner's qualifications?

1. Can the examiner testify in court as an expert if necessary?

2. Has the examiner testified in court previously?

3. How many forensic examinations have they conducted in the past?

4. Does the examiner hold any certifications in computer forensic examination?

5. Where did the examiner receive his training?

6. How long has the individual been conducting examinations?

Does the examiner understand all of the techniques and issues described below to conduct an examination or is he/she relying on one software suite to conduct the examination?

1.   It is the examiner who must qualify as an expert witness, not the software.

Is the examiner familiar with the particular operating system that you wish to be examined?

1.   What type of operating system are you dealing with?

2.   Is it a standalone computer?

3.   Is it MS-DOS, Microsoft Windows or UNIX?

4.   Is it a network?

5.   If so, what kind of network?

Is the examiner knowledgeable about acquiring magnetic data and can the individual advise about processes to be followed during the original acquisition of the media?

1.   Is this a voluntary or an involuntary collection of data?

2.   What procedures does the examiner recommend to preserve the original data during acquisition?

3.   Will the recommended procedures reduce the potential of someone trying to destroy evidence while it is being collected?

What does the examiner do to preserve the original media from accidental writes, viruses, and booby traps?

1.   Will these procedures prevent the introduction of viruses and prevent the accidental destruction of data?

2.   Does the examiner work from a forensic duplicate or bit stream copy?

3.   If so, what software is used?

4.   If not, completely avoid them!!!

Does the examiner have the knowledge, skill, and software to recover deleted files?

1.   Has the individual simply explained how files are stored, deleted, and recovered?

2. Has the examiner explained how Microsoft Windows long file names are stored and recovered. Ask them if they must be recovered?

Does the examiner have the knowledge, skill, and software to recover a formatted drive or diskette?

1. Has the individual simply explained what happens when a drive or diskette is formatted and how this data is recovered?

Does the examiner have the knowledge, skill, and software to find and recover hidden files?

1. Has the individual explained some common methods used to hide files?

Does the examiner have the knowledge, skill, and software to recover password-protected files?

1. Has the individual explained the two basic methods used to password protect files or data?

2. Does the individual use software solutions?

3. If so, what software?

4. What approach is adopted for RSA, PGP, or other difficult to break password protection schemes?

Does the examiner have the knowledge, skill, and software to find, access, and translate the Microsoft Windows swap, temporary, cache, and similar files?

1. What is the exact file name of the Microsoft Windows swap file?

2. Where is it normally stored? (two places)

3. Is it dynamic and how big can it become?

4. Has the examiner explained what general types of applications keep temporary files?

5. Has the individual discussed Internet cache files?

6. Has the individual explained cookies?

Does the examiner have the knowledge to provide sound opinions on file creation, access, deletion dates, and similar topics?

1. What dates and times are stored in all Microsoft Windows file entries?

2. Were all of these entries stored in the MS-DOS 6.22 (or below) file entries?

Does the examiner have the knowledge, skill, and software to recover data in unallocated space that cannot be linked to a directory entry?

1. How does the examiner do this?

2. What software is employed?

3. How thorough is this search and recovery of data from unallocated space?

How will the data be presented?

1. Printouts?

2. CD-ROM?

3. Can the examiner convert the format of the data to a format that will be useful in legal proceedings (i.e., convert proprietary database or spreadsheet data into something like Excel)?

What controls will be in place to ensure the proper *chain of custody* of any potential evidence recovered?

1. The examiner should fully understand the *Rules of Evidence* as they relate to storage of evidence and *chain of custody*. The case could be lost here, if the *Rules of Evidence* are not followed.

3.7.1 Training courses

There are many computer forensic courses offered these days and unfortunately it is very much a case of caveat emptor (buyer beware), particularly with respect to those courses that propose a complete forensic methodology around only one tool. A computer forensic examiners training course should be broad and encompass core technology and forensic methodology training which is not specific to any one tool, particularly those with just a point and click interface. The questions above can be equally applied to a computer forensic training course curriculum to determine whether it adequately covers the necessary material.

To address the training issues in the context of increasing law enforcement and national security requirements, the U.S. government set up the National Cybercrime Training Partnership (NCTP) to provide guidance and assistance to local, state, and federal law enforcement agencies in an effort to ensure that the law enforcement community is properly trained to address electronic and high-technology crime. NCTP sponsors free computer forensic training for U.S. law enforcement through the National White Collar Crime Center (NWC3). Other U.S. organizations involved in training for law enforcement include FLETC, SEARCH (The National Consortium for Justice Information and Statistics), and the High-Tech Crime Investigation Association (HTCIA).

In Europe, NATO's Lathe Gambit Information Security program has a computer forensic program that is open to NATO member countries, military, national security, and law enforcement personnel. Interpol has similarly conducted training programs under the auspices of its regional working parties.

In the Asia–Pacific region, the Australasian Center for Policing Research (ACPR, previously the National Police Research Unit, NPRU) has conducted a number of seminars and training courses for state and federal law enforcement from Australia and New Zealand. A number of academic institutions are also looking at collaborative, tertiary recognized forensic training programs.

3.7.2 Certification

With respect to certification, other than tool specific certifications offered by software vendors, there are currently only two independent computer forensic certifications, the IACIS Certified Forensic Computer Examiner (CFCE) and the High-Tech Crime Network (HTCN) Certified Computer Forensic Technician. The IACIS certification is the oldest but is unfortunately restricted to law enforcement only.

3.8 Computer forensics and national security

In the national security arena, computer forensics has likewise been applied to the investigative subjects of counterintelligence and counterterrorism. National Security matters may encompass threats from internal or external sources relating to hostile intelligence collection, sabotage, and terrorism. Like everyone else, spies and terrorists have been caught in the information revolution and therefore, intelligence and military agencies have been forced to develop technological countermeasures to safeguard national security.

Computer Security is an integral part of a modern military capability and in particular, Computer Network Operations (CNO), which is an element of a relatively new warfighting doctrine, termed Information Operations (IO). According to draft British military IO doctrine, CNO comprises the following:

- Computer network exploitation (CNE), that is, "the ability to gain access to information hosted on information systems and the ability to make use of the system itself";

- Computer network attack (CNA), that is, the "use of novel approaches to enter computer networks and attack the data, the processes or the hardware";

- Computer network defense (CND), that is "protection against the enemy's CNA and CNE and incorporates hardware and software approaches alongside people-based approaches."

Computer forensics in this environment is obviously part of the CND mission, recovering evidence to attribute attacks to a particular party for the purposes of retaliatory action be it criminal, military, economic, or diplomatic. None of the techniques differ between the law enforcement and the national security arenas and in many cases identical tools may be employed. The only significant difference between the two relates to the quantity of resources that may be brought to bear the investigation as the implications of a national security investigation may be more serious (from some perspectives) than those of a criminal matter.

3.8.1 National security

As in the case of law enforcement, the employment of computer forensics in national security matters is on the increase. Also as with law enforcement, the first national security related case, at least publicly acknowledged, dates back to the mid-1980s.

3.8.1.1 The Cuckoos Egg

The investigation during 1986 and 1987 which became known as *The Cuckoo's Egg* related to a computer intrusion investigation conducted by an astrophysicist-turned-hacker-tracker, Clifford Stoll, who from a computing utilization discrepancy found the first acknowledged use of the Internet to carry out intelligence collection by agents operating in Germany on behalf of the KGB. Along the way, the U.S. National Security Agency and the FBI become heavily involved [71].

3.8.1.2 Solar Sunrise

In February and March 1998, U.S. military, government, and research and development systems on the Internet experienced a large number of systematic network intrusions that were subsequently determined to be related. Code-named *Solar Sunrise*, the timing of these activities was very suspicious since it coincided with another buildup of U.S. military personnel in the Middle East in response to tensions with Iraq over United Nations weapons inspections. The unidentified intruders penetrated many unclassified U.S. military computer systems, including those on Air Force bases and Navy installations, DOE National Laboratories, NASA installations, and university sites. The timing of the intrusions, and the apparent origin of some activities from the Middle East, led many government officials to suspect that this could be an instance of Iraqi CNA timed to cause disruption to U.S. military build up in the region.

Subsequent investigation by NASA Computer Crime Division, U.S. Naval Criminal Investigation Service Computer Crime Division, the Air Force Office of Special Investigations (AF-OSI) and the FBI's Computer Investigations and Infrastructure Threat Assessment Center (CITAC), predecessor to the National Infrastructure Protection Center (NIPC), working closely with Israeli law enforcement, determined that after several days that two teenagers in Cloverdale, California, and an Israeli individual with several accomplices were the perpetrators [72].

3.8.1.3 Moonlight Maze

If Solar Sunrise set the scene, the next series of incidents code-named Moonlight Maze delivered the wake-up call to the U.S. government. Dr John Hamre, the Deputy Secretary of Defense at that time, described the Moonlight Maze events to a congressional committee in 1999 stating bluntly, "We are in the middle of a cyber war." The Moonlight Maze operation was enormous and officials have publicly stated that the intruders systematically accessed and exploited hundreds of unclassified but sensitive computer networks used by the DOD, DOE, NASA, various defense contractors, and several universities. A large amount of technical data related to defense research was copied and transferred to Russia. One defense technician trying to track the computer intruder is said to have watched in amazement, as a document from a naval facility was hijacked from a print queue to a location in Moscow right in front of him. The first Moonlight Maze attack was detected in March 1998. Three months later, U.S. agencies were able to monitor a series of intrusions as they occurred and traced them back to seven dial-up Internet connections located near Moscow.

The FBI has yet to publicly make the determination that the United States was subjected to CNE over the Internet conducted by Moscow's prestigious Russian Academy of Sciences. Moonlight Maze has so far been the most insidious and focused assault yet on sensitive U.S. computer networks [72].

3.8.1.4   September 11, 2001, and the war on terrorism

The tragic events of September 11, 2001, weigh heavily on the minds of many people the world over and computer forensics has had a critical role to play in the investigations of that event and the subsequent war on terrorism.

On August 15, 2001, Zacarias Moussaoui was apprehended by the FBI in Minnesota while attempting to obtain flight training on a Boeing 747 simulator under suspicious circumstances. Moussaoui was in possession of a laptop computer, and FBI agents in Minneapolis sought to obtain a Foreign Intelligence Surveillance Act (FISA) search warrant authorizing a search of it. Unfortunately, attorneys at FBI Headquarters Counterterrorism Division in Washington believed there was insufficient probable cause, and the case was being further investigated when the incidents of September 11 occurred. No information has been released postmortem as to whether Moussaoui's laptop in fact contained information that could have assisted in the prevention of the attack other than the fact that it contained large amounts of information about crop dusting aircraft, which prompted an alert about the potential for spreading chemical or biological agents using such aircraft [73]. There is also significant evidence that Al Qaeda members were using the Internet to communicate and carry out their plans. At the time of writing, there was a Congressional Hearing underway to discuss the issue and intelligence failures prior to September 11.

As far back as 2000, convicted Al Qaeda operative Khalil Deek, a naturalized U.S. citizen born in the Middle East, was arrested in Pakistan, extradited to Jordan and charged as a suspected ringleader in Al Qaeda. FBI officials stated a laptop computer taken from Deek had been exploited in the hunt for evidence of an alleged conspiracy to facilitate foiled attacks against tourist sites in Jordan.

3.8.2   Critical infrastructure protection

In recognition of the growing threat from hostile CNA and CNE against not only the U.S. military and government but also against the social and economic fabric of the country in 1998, the then President Bill Clinton issued U.S. Presidential Decision Directive 63—the Clinton Administration's Policy on Critical Infrastructure Protection (CIP). The policy stated that critical infrastructures are those physical and cyber-based systems essential to the minimum operations of the economy and government. They include, but are not limited to, telecommunications, energy, banking and finance, transportation, water systems and emergency services, both governmental and private. This expanded the perspective of national security to encompass those private industry services that were nationally critical. In the United Kingdom, Canada, and Australia similar studies were being conducted and findings identified.

The role of computer forensics in CIP is identical to that with respect to its employment in other aspects of national security, system recovery, and attack attribution. The expansion of scope of national security interest in these areas has meant a much greater level of information sharing that has previously been possible and has included the opening up of training at organizations like the NW3C to private individuals involved in CIP activities.

3.8.3   National security computer forensic organizations

Many national security agencies have been forced to develop computer forensic capabilities to support their operational mission due to the move from a paper-based environment to a digital one. It is left up to the imagination of the reader to determine which agencies have undisclosed computer forensic programs and mention those organizations that do have a declared computer forensic capability and a national security mission.

3.8.3.1   FBI CART

In the United States, the FBI fulfils both law enforcement and national security roles and consequently the CART as mentioned previously has to deal with examinations in support of both roles.

3.8.3.2   DCFL

DCFL as an agency of the U.S. Department of Defense quite obviously has a very visible role in national security matters supporting DOD investigative and counterintelligence agencies with their missions.

3.8.3.3   Air Force Research Laboratory

The U.S. Air Force Research Laboratory (AFRL), like DCFL, an agency of the U.S. Department of Defense has a computer forensic research and development oriented mission in support of Air Force and DOD activities.

3.8.3.4 U.K. Defense Evaluation and Research Agency

The U.K. Ministry of Defense Research and Development Agency has both an operational computer forensic team, and a research and development team. The teams are able to provide support to both law enforcement and national security missions as required.

3.8.3.5 NATO's Lathe Gambit

Previously mentioned, the Allied Command Europe—Counter Intelligence (ACE-CI) organization runs the Lathe Gambit program for member countries. The program provides both operational and support activities in support of the ACE counterintelligence mission.

References

[1] Morgan-Whitcombe, C., "A Historical Perspective of Digital Evidence: A Forensic Scientist's View," http://www.ijde.org/carrie_article.html, visited April 2002.

[2] Federal Bureau of Investigation, "FBI in Brief—History of the FBI—Rise of International Crime (1980s)," http://www.fbi.gov/libref/historic/history/rise.htm, visited April 2002.

[3] New Technologies Inc., "Charles P. Guzis, About Us," http://www.forensics-intl.com/guzis.html, visited April 2002.

[4] Rendall, J., "Computer Forensics—Electronic Evidence Techniques for Now, Problems for the Future," http://www.bcs.org.uk/branches/cov/prog0001/report/oct.htm, visited April 2002.

[5] International Association of Computer Investigative Specialists (IACIS), "About Us," http://www.cops.org/about_us.htm, visited April 2002.

[6] Federal Bureau of Investigation, *Handbook of Forensic Services*, http://www.fbi.gov/hq/lab/handbook/intro.htm, visited April 2002.

[7] Godoy, M., "Cybersleuths Seek Truth About Enron," Tech Live, http://www.infowar.com/law/02/law_021202a_j.shtml, visited April 2002.

[8] New York State Police, "NYSP Annual Report 1998—Forensic Investigation Support Services," http://www.troopers.state.ny.us/Intro/Annual/98Ann/98FICServ.pdf, visited April 2002.

[9] Sanderson Forensics, "Case Studies," http://www.sandersonforensics.co.uk/html/case_studies.html, visited April 2002.

[10] Concerned Women for America, "Cracking Down on Child Pornography—Internet Child Pornography Ring Is Raided Globally," http://www.cwfa.org/library/pornography/1998-09-19_porn-child.shtml, visited April 2002.

[11] Black, J., "Needed: Wiretap Laws for a Wired World," http://www.businessweek.com/bwdaily/dnflash/aug2001/nf20010823_686.htm, visited April 2002.

[12] Shimomura, T., "Takedown," http://www.takedown.com, visited April 2002.

[13] Collie, B., "Intrusion Investigation and Post-intrusion Computer Forensic Analysis," http://www.mirrors.wiretapped.net/security/info/papers/law-enforcement/intrusion-investigation-and-post-intrusion-forensic-analysis.pdf, visited April 2002.

[14] Romig, S., "Forensic Computer Investigations," http://www.net.ohio-state.edu/security/talks/2001-10_forensic-computer-investigations/notes-pdf/04-evidence1.pdf, visited April 2002.

[15] Federal Bureau of Investigation, "Digital Evidence: Standards and Principles," http://www.fbi.gov/hq/lab/fsc/backissu/april2000/swgde.htm # Proposed % 20 Standards%20for%20the%20Exchange%20of%20Digital%20Evidence, visited April 2002.

[16] McMillan, J., "Importance of a Standard Methodology in Computer Forensics," http://rr.sans.org/incident/methodology.php, visited April 2002.

[17] International Association of Computer Investigative Specialists (IACIS), "Forensic Examination Procedures," http://www.cops.org/forensic_examination_procedures.htm#Forensic Examination Procedures, visited April 2002.

[18] Information Systems Security, "Computer Crime Investigation Computer Forensics," http://telecom.canisius.edu/cf/computer_crime_investigation.htm, visited April 2002.

[19] McKemmish, R., "Computer Forensics: Building a Computer Forensic Model and Confronting Key Issues," Sept. 2001.

[20] Stanczewski, Y., "ComputerForensics101," http://www.share.org/proceedings/sh96/data/S1753.PDF, visited April 2002.

[21] United States Secret Service, "Best Practices for Seizing Electronic Evidence," http://www.secretservice.gov/electronic_evidence.shtml, visited April 2002.

[22] New Technologies Inc., "Computer Evidence Processing Steps," http://www.forensics-intl.com/evidguid.html, visited April 2002.

[23] Brezinski, D., and T. Killalea, "RFC3227: Guidelines for Evidence Collection and Archiving," http://www.faqs.org/rfcs/rfc3227.html, visited April 2002.

[24] Mandia, K., and C. Prosise, *Incident Response: Investigating Computer Crime*, CA: Osborne, 2001.

[25] Intelligent Computer Solutions Inc., http://www.ics-iq.com/show_item_188. cfm, visited April 2002.

[26] Digital Intelligence Inc., http://www.digitalintel.com/firechief.htm, visited April 2002.

[27] Farmer, D., "What Are MACtimes?" http://www.ddj.com/documents/s=880/ ddj0010f/0010f.htm, visited April 2002.

[28] AccessData Corporation, "Forensic Toolkit," http://www.accessdata.com/ Product04_Overview.htm?ProductNum=04, visited April 2002.

[29] National Drug Intelligence Center, "Hashkeeper," http://www.hashkeeper. org, visited April 2002.

[30] National Institute of Standards and Technology, "National Software Reference Library," http://www.nsrl.nist:gov/, visited April 2002.

[31] Bebits, "What is Beos?" http://www.bebits.com/whatisbeos, April 2002.

[32] Stutz, M., "NASA Greets Beowulf," http://www.wired.com/news/print/ 0,1294,14450,00.html, visited April 2002.

[33] Grundy, B., "The Law Enforcement Introduction to Linux: A Beginner's Guide," http://home.columbus.rr.com/bgrundy/linlaw/Linuxintro-1.8.1.pdf, visited April 2002.

[34] New Technologies Inc., "Safeback," http://www.forensics-intl.com/safe back.html, visited April 2002.

[35] Snapback, "Snapback DatArrest," http://www.snapback.com/snapback_ datarrest__main_feat.html, visited April 2002.

[36] Holley, J., "September 2000 Market Survey, Computer Forensics," http:// www.scmagazine.com/scmagazine/2000_09/survey/products_02.html, visited April 2002.

[37] Guidance Software, "Encase Technical Support—Version 3 Acquisition Support 2002," http://www.encase.com/support/v3_acquisition.shtm, visited April 2002.

[38] Guidance Software, "FastBloc," http://www.encase.com/products/hardware/ fastbloc_features.shtm, visited April 2002.

[39] Tech Assist Inc., "ByteBack," http://www.toolsthatwork.com/byte.shtml, visited April 2002.

[40] ACR Data Recovery, "Media Tools," http://www.data-recovery-software.com/ mtl.htm, visited April 2002.

[41] Computer Forensics Ltd., "Disk Image Backup System (DIBS)," http:// www.computer-forensics.com/articles/imag.html, visited April 2002.

[42] Vogon, "VOGON Evidential Hardware," http://www.vogon-computer-evidence.com/evidential_systems-02.htm, visited April 2002.

[43] Symantec, "Norton Ghost," http://www.symantec.com/sabu/ghost/, April 2002.

[44] Symantec, "Switches: Sector Copy," http://service4.symantec.com/SUPPORT/ ghost.nsf/docid/2001111413481325, visited April 2002.

[45] Biatchux, dcfldd, http://biatchux.dmzs.com/?section=tools&subsection=F, visited April 2002.

[46] Garner, G., "Forensic Acquisition Utilities," http://users.erols.com/gmgarner/ forensics, visited Nov. 2002.

[47] Intelligent Computer Solutions Inc., "Image MASSter Solo 2 Forensic Kit," http://www.ics-iq.com/show_item_187.cfm, visited April 2002.

[48] Mares and Company, LLC, "Hash," http://www.maresware.com/maresware/ gk.htm#HASH, visited April 2002.

[49] dtSearch Corporation, "dtSearch," http://www.dtsearch.com/desktop.html, visited April 2002.

[50] New Technologies Inc., "Disk Search Pro," http://www.forensics-intl.com/ dspro.html, visited April 2002.

[51] New Technologies Inc., "Net Threat Analyzer," http://www.forensics-intl. com/nta.html, visited April 2002.

[52] Mares and Company, LLC, "String Search," http://www.maresware.com/ maresware/ps.htm#SS, visited April 2002.

[53] UNIXhelp for Users, "grep," http://unixhelp.ed.ac.uk/CGI/man-cgi?grep, visited April 2002.

[54] Stepanet Communications, "File Extractor," http://www.datalifter.com/ tutorials/fe/file_extractor.htm, visited April 2002.

[55] Air Force Office of Special Investigations, "Foremost," http://sourceforge.net/ project/showfiles.php?group_id=46038, visited April 2002.

[56] Spencer, E., "iLook," http://www.ilook.fsnet.co.uk/ilook/about.htm, visited April 2002.

[57] X-Ways AG, "WinHex," http://www.winhex.com/winhex/index-m.html, visited April 2002.

[58] Rogoyski, A., [N] Curses HexEdit, http://ccwf.cc.utexas.edu/~apoc/programs/ c/hexedit/, visited April 2002.

[59] Federal Bureau of Investigation, "Automated Computer Examination System (ACES)," http://www.ncfs.org/aces.html, visited April 2002.

[60] Fred Cohen and Associates, "ForensiX," http://www.all.net/ForensiX/ index.html, visited April 2002.

[61] ASR Data Acquisition & Analysis, LLC, "Storage Media Archival and Recovery Toolkit (SMART)," http://www.asrdata.com/smart, visited April 2002.

[62] Stepanet Communications, "Datalifter," http://www.datalifter.com/, visited April 2002.

[63] Digital Detective, "NetAnalysis," http://www.digital-detective.co.uk/netan alysis.asp, visited April 2002.

[64] Jasc Software, "Quick View Plus for Windows," http://www.jasc.com/ products/qvp/qvpmore.asp, visited April 2002.

[65] Skiljan, I., "IRFanView," http://www.irfanview.com/english.htm, visited April 2002.

[66] Resplendence Software, "Resplendent Registrar," http://www.resplendence. com/registrar, visited April 2002.

[67] Jedlinski, M., "GUIDClean," http://www.lodz.pdi.net/~eristic/free/guidclean. html, visited April 2002.

[68] Murphy, D., "Microsoft Documents Secretly Track Author," http://dgl.com/ itinfo/1999/it990307.html, visited April 2002.

[69] Info Evolution, "Unmozify," http://www.evolve.co.uk/unmozify/, visited April 2002.

[70] Key Computer Service Inc., http://www.keycomputer.net/equest.htm, visited April 2002.

[71] Stoll, C., "The Cuckoo's Egg," http://www.nytimes.com/books/99/01/03/ specials/stoll-egg.html, visited April 2002.

[72] National Defense University, "Information Operations: The Hard Reality of Soft Power," *Joint Services Staff College Handbook*, Norfolk, VA, 2001.

[73] Gordon, G., "FAA Security Took No Action Against Moussaoui," http:// www.startribune.com/stories/1576/1028069.html, visited April 2002.

CHAPTER

4

Contents

4.1 Auditing and fraud detection

4.2 Defining fraudulent activity

4.3 Technology and fraud detection

4.4 Fraud detection techniques

4.5 Visual analysis techniques

4.6 Building a fraud analysis model

References

Appendix 4A

Computer Forensics in Forensic Accounting

4.1 Auditing and fraud detection

In recent years, the role of the auditor has come under increasing scrutiny. In particular, the auditor's ability to both detect and understand accounting anomalies, and then to correctly report them has gained much attention. Large corporate collapses, accounting irregularities and independence issues have all combined to make the role of the auditor more accountable particularly with regard to the identification and management of risk. Indeed, auditors "play a vital role in ensuring that an organization is efficiently run, morally sound, technologically advanced, cognizant of the environment and other areas of concern, and safe from unnecessary risk" [1]. In essence, today auditors play a significant role in ensuring adherence to good corporate governance.

As greater emphasis is placed on the role of the auditor in detecting anomalous accounting activity, so there is greater reliance on computerized fraud detection tools and techniques. In addition to the auditing function, there has emerged an accounting function designed not only to detect fraudulent behavior, but also to investigate such behavior. This function is known as forensic accounting. Forensic accounting seeks to bring together investigative, accounting, and technology skills with a view to "getting at the truth behind the numbers'' [2], and as a result plays an increasingly large part in fraud investigations and prosecutions. Combining traditional auditing functions with those of forensic accounting provides organizations with a powerful tool in both the detection and prevention of fraudulent business activity.

4.1.1   Detecting fraud—the auditor and technology

It has long been recognized that auditors, and in particular internal auditors, are well placed to identify fraud-related risks and possible fraud within organizations. Indeed, the auditor, it can be said, plays a proactive role in fraud prevention and detection by ensuring that appropriate business controls are implemented and adhered to, thereby reducing and managing fraud related risk. This function is encapsulated in the following definition of internal auditing by the Institute of Internal Auditors:

> Internal auditing is an independent, objective assurance, and consulting activity designed to add value and improve an organization's operations. It helps an organization accomplish its objectives by bringing a systematic, disciplined approach to evaluate and improve the effectiveness of risk management, control and governance processes [3].

To assist in this role, auditors utilize computer-assisted audit tools and techniques (CATTs). CATTs provide the auditor, and forensic accountant, with a series of software tools that when applied across an organization's information infrastructure facilitates the detection and investigation of anomalous activity. Essentially, CATTs provide auditors and forensic accountants with a range of capabilities that includes, among other things:

- Information retrieval and analysis;
- Fraud detection;
- Audit reporting;
- Continuous monitoring [3].

The application of computer technology, combined with statistical analysis techniques and number theory, allows the auditor to identify anomalous behaviors and trends that may provide an indication of fraudulent or unethical activity. Despite their sophistication and complexity, ultimately the success of these tools is very much dependent upon three key factors:

1. The application of the correct analysis technique(s);

2. The use of the most appropriate data;

3. An ability to correctly interpret and understand the results.

4.2 Defining fraudulent activity

With the move to electronic commerce and the increasing popularity of the Internet as a medium to effect business related transactions, it is not surprising to find a shift in the types of fraudulent behavior away from traditional business related environments to the new electronic world. Indeed, the adoption of electronic commerce brings with it an attractiveness that many fraudsters find appealing. Perhaps the greatest attraction of the on-line world is that of anonymity. No longer does a fraudster have to present himself or herself physically to perpetrate a fraud, but rather the fraud can be committed from home or from the confines of the office. In addition to the anonymity, the electronic world provides fraudsters with a global reach that they have never had before.

Given the speed of transactions, the electronic world allows a greater number of fraudulent transactions to take place in a far shorter time than was previously possible. Indeed, these advantages and disadvantages are recognized by law enforcement agencies, which in addition to the above see the volatile or transient nature of electronic evidence and the "potential for deliberate exploitation of sovereignty issues and cross jurisdictional differences by criminals and organized crime" as significant issues [4].

Not only has the electronic world provided a medium through which a higher volume of fraudulent activity can occur, but it has also facilitated a far greater variety of fraudulent activities. These activities include everything from credit card fraud to fraudulent on-line purchases and even fraudulent investment schemes. Grabosky, Smith, and Dempsey highlight the impact of an electronic world on fraud by stating "the fundamental principle of criminology is that crime follows opportunity, and opportunities for theft abound in the Digital Age" [5]. Given the broad nature of on-line fraud, it comes as no surprise to find that the very definition of fraud is constantly being challenged with more and ever-inventive techniques focused on deceptive behavior.

4.2.1 What is fraud?

The *Concise Oxford Dictionary* defines fraud as

> Criminal deception, use of false representations to gain unjust advantage; dishonest artifice or trick; person or thing not fulfilling expectation or description; deceitfulness [6].

Traditionally fraud has been viewed as one person deceiving another person for the purposes of deriving a benefit or gain. In the electronic world, the deception may not involve two persons but rather one person and a computer, or two computers one of which is controlled by a person. Regardless of the medium through which the fraudulent activity is perpetrated, for a fraud to occur there must be some form of deceptive or false representation, which results in an unjust gain for either the person perpetrating the fraudulent activity or for a third party. Such gain may not necessarily be a gain in monetary terms, but can include a benefit (e.g., delivery of a service) or the attainment of intellectual property (e.g., ownership of copyright).

Unlike traditional fraud where the victim and witness are human beings, in the electronic world it is the computer systems and on-line transaction systems that become the witness. Rather than examining paper-based evidence, in the electronic world the evidence is less tangible and more transient in nature. Consequently the need to identify, capture, and preserve potential electronic evidence through the use of computer forensics is of paramount importance.

So what drives a person to commit fraud? Essentially fraudulent behavior is driven by three key factors: *motive, opportunity,* and *benefit.* These factors combine to create what is known as the fraud triangle shown in Figure 4.1. When all the three factors are present the probability of, or opportunity for, fraudulent activity to occur is significantly increased, thereby heightening the fraud risk profile for the individual organization.

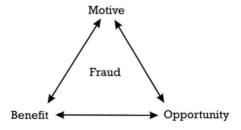

Figure 4.1 The fraud triangle.

Motive is the primary factor in fraudulent behavior. Motives for fraud includes greed, personal gain, payment of debt, and retribution. With the advent of electronic commerce, another more insidious motive has also emerged—ego. To show that electronic fraud can be committed just for the sake of establishing one's expertise is becoming more common. When a perpetrator is asked why he or she committed the fraud, sometimes the response is "to prove that I could do it."

Opportunity facilitates the means by which the fraud is perpetrated. Essentially, opportunity arises when weaknesses in controls and security are identified and exploited. Weaknesses such as poor accounting controls, lack of verification of a person's identity, lack of access controls, and poor reconciliation processes all provide opportunity.

Benefit is the perceived outcome of the fraudulent activity. In essence, frauds are committed to derive a benefit. The benefit may be gained either directly by the fraudster or by another, and can be tangible or intangible in nature. Financial advantage, acquisition of tangible property, disposal of defective or valueless property, and revenge are all examples of benefit.

While the concept of motive, opportunity, and benefit can be applied in all instances of fraud, Loebbecke, Eining, and Willingham developed a functional causal model definition to describe the likelihood of management fraud [7]. The Loebbecke, Eining, and Willingham model states

$$P(\text{MI}) = f(C, M, A)$$

where

$P(\text{MI}) =$ the probability of management fraud (MI stands for "material irregularities");

$\quad C =$ the degree to which there exist conditions favorable to management fraud;

$\quad M =$ the degree to which persons of authority have the motive to commit fraud;

$\quad A =$ the degree to which persons in authority have the attitude or ethical values that facilitate fraudulent activity.

The Loebbecke, Eining, and Willingham model classifies the warning signs (also known as *red flags*) used in the assessment of risk pertaining to management fraud by categorizing each under one of the above three elements. In addition, the model differentiates the warning signs or *red flags* by classifying them as either primary indicators or secondary indicators. The net result of this approach is to develop a causal model that can be used to develop risk management profiles useful in the detection of fraudulent activity within an organization.

Red flags are either activities occurring within an organization's business processes or behaviors exhibited by employees, each of which provide indicators of heightened levels of exposure to fraud related risk. Essentially, *red flags* are danger signs that indicate a greater likelihood of fraudulent activity taking place. In terms of business processes, the sort of *red flags* one might look for includes missing source documents, duplicate payments, missing inventory, poor segregation of duties, weak internal controls, conflicts of interest, and unexplained growth. With respect to employee-related *red flags*, the danger signs include gambling addiction, heavy personal debt, reluctance to delegate tasks, and an obsessive approach to work which seems to preclude a normal life.

4.2.2 Internal fraud versus external fraud

From a computer forensic perspective the nature of the fraud committed has a distinct impact on the identification, preservation, and analysis of pertinent electronic evidence. Fraud, whose source is external to an organization will require significantly more resources and time to investigate, principally due to the fact that little is known about the identity of the perpetrator or the potential sources of evidence. This is particularly so for on-line fraud that is not limited by geographic boundaries. Fraud that is committed within an organization, while still requiring time and effort to investigate, is, from a computer forensic perspective, far simpler to investigate. This is principally due to ease of access both physically and legally to the related electronic evidence, which resides within an organization's IT infrastructure. In addition, the jurisdictional issues that may impact on investigating external fraud are very much reduced. Before moving onto an examination of the technology underlying modern fraud detection and analysis tools and techniques, let us briefly examine the key differences between internal and external fraud from a computer forensic perspective.

4.2.2.1 Internal fraud

Internal fraud is simply fraud that originates from within an organization, and is committed or initiated by an employee or officer of that organization. Internal fraud principally arises from deficient or poorly implemented internal controls. As organizations have individuals who are in decision-making positions (e.g., chief financial officer), the need for strong and independent internal controls is essential. This is particularly the case when the individual has power to authorize expenditure. In such instances, the internal controls need to be able to discern between legitimate business activity and potentially fraudulent activity. Traditionally, this has been achieved through the use of mechanisms such as reconciliation, segregation of duties, and internal audits. Given the time and expense of implementing these processes manually, it is not surprising to find the role of automated fraud detection tools becoming more central in the fraud detection process. The application of fraud detection tools provides a level of independence and consistency that moves the fraud identification process away from interpretation errors and inconsistencies normally associated with differing human perspectives. In addition, given the time benefits and the reduced need for human interaction in the fraud detection process, fraud detection tools provide a significant cost benefit advantage to an organization.

Internal fraud can take on many guises, and as such is very much dependent on the nature and size of the business. Despite this, however, it is possible to list some of the more common types of internal fraud experienced in recent times [8], several examples of which appear in the list of 2001 U.S. computer crime convictions presented in Chapter 1:

1. *Fictitious suppliers:* Involves payment to nonexistent suppliers on receipt of a fictitious invoice by accounts receivable, for fictitious goods and services alleged to have been provided. Alternately, a fraudulent supplier can be created within the accounting system and regular payment made for nonexistent services or goods.

2. *Duplicate payments:* Overpayment is made on two separate occasions for the same instance of goods or services received. While this may simply be an accounting error, from a fraud perspective the overpayment may result from a supplier double charging or as a consequence of an employee acting in collusion with the supplier.

3. *Theft of inventory:* Inventory is stolen by employees who cover their actions by writing off the stolen inventory in the organizations stock management system as being damaged.

4. *Unauthorized manipulation of system data:* Critical accounting and inventory management data is altered or deleted with a view to covering up instances of fraudulent or unethical behavior.

5. *Tender manipulation:* Critical tender information, such as competitors' pricing and tender submissions, is leaked to a competing organization thereby resulting in an unfair advantage.

6. *False financial reporting:* The senior executive or departmental head, may misreport the financial performance of an organization or

department with a view to covering up significant shortfalls resulting from fraudulent or unethical activity. Such misreporting includes the incorrect treatment of debts and liabilities with a view to minimizing their impact on the relevant financial reports.

7. *Secret commissions:* Undeclared benefits, such as cash or gifts, are received or paid in return for maintaining or establishing the sale or purchase of goods or services.

8. *Expense account misuse:* Allocated expense account funds are used on personal, nonwork related activities, with the true purpose of the expenditure being misreported so as to minimize any suspicions.

9. *Insider trading:* Information pertaining to the future operations of an organization, are sold or leaked to third parties, or are misused by an employee with a view to gain some form of advantage. Such information can include planned infrastructure purchases, strategic takeover plans, or the pending release of profit forecasts and performance results.

4.2.2.2 External fraud

External fraud is fraud that is committed against an organization by a person or persons not employed by the victim organization. External fraud is generally perpetrated against a victim organization for some form of material gain, whether it be financial gain or unauthorized or unpaid use of a service or product. The underlying motive(s) for external fraud can be varied and include anticompetitive behavior, criminal intent, or payback by a disgruntled employee, customer, or supplier. Just as in internal fraud, the success of external fraud is very much dependent upon the strength of the organization's controls. Organizations that have poorly implemented or deficient controls run a greater risk of falling victim to fraud than those organizations that are more diligent in their fraud detection strategies and processes. As history has shown, once a fraudster has identified a control weakness, the opportunity to gain benefit provides a strong incentive to pursue fraudulent activity.

Given the growth of electronic commerce, the risks of on-line fraud being perpetrated have increased significantly. In particular, the globalization of electronic commerce provides the fraudster with a significantly larger pool of potential victims. To this end, it is not surprising to find organizations turning to automated fraud detection tools in an effort to minimize the risk of external fraud.

Whether it is traditional fraud or on-line fraud, external fraud can take on any number of forms some of which were referred to earlier and which include, but are not limited to:

1. *Credit card fraud:* Goods or services are purchased with stolen credit card numbers.

2. *Duplicate invoices:* Goods or services are double billed, on two separate occasions by a supplier with the hope that the invoices will be processed and the double payment overlooked.

3. *Forged checks:* Goods or services are purchased with forged or stolen checks.

4. *False identity:* Goods or services are received and billed to a false identity provided to the organization by the perpetrator.

5. *Bogus products or services:* An organization pays for particular products or services that are either not delivered or if delivered, are incorrect or defective.

6. *False investment schemes:* Monies are invested in bogus investment schemes that provide no return on investment and usually result in the loss of the initial funds invested.

4.2.3 Understanding fraudulent behavior

Before examining the relationship between fraud and technology, it is perhaps worthwhile spending a little time examining what underlies fraudulent behavior. While much has been written over the past few years regarding the drivers underlying fraudulent activity, from a computer forensic perspective it is essential that an understanding of such drivers be reached as that will greatly assist in the search for electronic evidence. This is particularly so in a business sense, for, depending on the size of the organization and the type of fraud, the sources and volume of evidence may be extremely large and complex.

So what drives a person to commit fraud? As has been previously stated, motive, opportunity, and gain are the underlying factors, and the first of these (i.e., motive) is fundamental. The most common motives are

1. Greed;

2. Gambling;

3. Drugs;

4. Financial pressures;

5. Challenge;

6. Retribution;

7. Psychological disorder;

8. Environmental factors.

There is one further distinction that is useful to make when considering fraud, and that is the difference between opportunist fraud and premeditated fraud. Opportunistic fraud is simply fraudulent behavior arising from a previously unforeseen and immediate opportunity to exploit a weakness in system controls. Simply put, a weakness is identified and the fraudster makes the most of the opportunity. On the other hand, premeditated fraud involves some degree of planning and forethought before the fraudulent behavior is undertaken. Just as opportunistic fraud exploits weaknesses, premeditated fraud probes system controls for those weaknesses that will facilitate, with minimal risk of being detected, fraudulent activity.

4.3 Technology and fraud detection

Applied in a fraud detection role, computer forensics utilizes a wide range of techniques to analyze system data with a view to identifying anomalous and potentially fraudulent activity. These techniques include the use of data mining techniques, visualization, and related techniques (such as link analysis, time-line analysis, and clustering), pattern matching, fuzzy logic, and statistical analysis in an effort to profile and detect fraudulent behavior. Some of these topics are sufficiently important and pervasive that they occur across the board in computer forensics. They are as a result referred to throughout the book, for instance, data mining is referred to in Chapter 2, and then again in Chapter 7 where it is discussed in the context of future developments in computer forensics, link analysis is discussed in Chapters 3 and 7, while time-lining which is a recurring theme in computer forensics is addressed in Chapters 2, 3, and 6.

The techniques can be applied either in real time or through postevent analysis. In any case, as highlighted by Richard Kusnierz and Alan Livesey: "the objective of any detection system is not only to provide indicators that a problem exists, but hopefully to also provide proof that will be admissible in court" [9]. Clearly, admissibility is a key consideration if the output of the fraud detection system is to be used in evidence. Earlier chapters have discussed in detail a number of issues surrounding the nature and admissibility of digital evidence in the forensic process.

Given the complex nature of business processes, it is very difficult, if not impossible, to develop an all-encompassing fraud detection system that can be applied across a wide range of business environments. Therefore, it is essential for any fraud detection system to be adapted or designed to meet the specific needs of the relevant business environment. Consequently, in developing such a system it is essential that an understanding of the business processes applicable to the organization be clearly defined and understood. These business processes in particular should include an understanding of where fraud risks lie, the relevant *red flags* that indicate fraudulent activity, and the sources of data that provide the basis for an accurate analysis.

By way of example, consider a retail organization that sells goods to the public. The processes surrounding the receipt of inventory may go something like this:

▸ Goods received;

▸ Shipping document and received goods notice forwarded to accounts payable;

▸ Invoice received and matched against received goods notice;

▸ Payment authorized and issued.

In this example, a number of *red flags* may exist:

▸ Poor segregation of duties;

▸ Little or no reconciliation of goods received against shipping documents and invoices;

▸ High levels of damaged inventory.

The particular mix of *red flags* should be indicative of the business process aligned to the receipt of inventory, and as such may not be ideally suited to other business processes, such as goods sold, occurring within the organization.

The process of identifying fraud risks will result in the development of fraud risk profiles that can be used for each business process to design and implement fraud rules within a fraud detection system. These rules allow the creation of automated routines designed to differentiate potentially fraudulent activity from acceptable activity [10]. Drawing on the above example, the *red flags* identified in the "receipt of inventory" process can be used as the basis for fraud rules. Essentially, a series of rules are developed that measure the level of each red flag. For example, in the case of segregation of duties a rule may be developed to measure the number of instances in which an employee performs an activity within the goods received process. If an employee performs the majority of activities then the resulting score will be high and may be indicative of poor segregation of duties. The resulting measure of what is acceptable and what is not acceptable is used to build a profile or picture of potentially fraudulent activity within the goods-received system.

Because business processes are dynamic, and given that fraudulent behavior is also dynamic, the development of a fraud profile should not be seen as a one-off activity. Rather it is an ongoing process, whereby existing rules are constantly reviewed and modified, and new rules developed and implemented. For, as fraud patterns change, so too do the fraud rules upon which the detection algorithms are based [11]. Given this dynamic nature of fraud profiling, it becomes increasingly complex to design and implement fraud detection systems on a real time basis. To this end such real time systems need to have the capacity to learn, and must therefore utilize technology that is best suited to this task (e.g., neural networks), for failure to do so would see such systems quickly become outdated [12]. However, most fraud detection systems are post-event by nature and as such are significantly less expensive to implement than real time systems.

Post-event analysis essentially involves the application of data analysis and data mining techniques to existing business information with a view to identify irregularities. Because it is post-event, the success of the analysis, from a forensic and business perspective, not only resides in the strength of the analysis techniques utilized, but also the timing. The detection of fraudulent activity from datasets derived from transactions occurring some 2 to 3 months previously, may not result in an optimal fraud detection system. Depending on the nature of the fraudulent activity, significant leakage of funds or assets may have taken place in the space of 2 or 3 months. In addition, critical primary and secondary evidence (e.g., computer logs, transaction records, and witness recollections) may have been destroyed or altered. Consequently, it is imperative that for post-event analysis to be truly effective, it must not only involve the application of the most appropriate fraud rules and analysis techniques, but must also be undertaken in a timely fashion.

This section examines the application of data mining techniques to computer forensics, and provides an overview of the use of fraud detection tools and digit analysis for fraud detection. Sections 4.4 and 4.5 address the application of statistical analysis, including regression analysis and digit frequency analysis based upon Benford's Law (Section 4.4), and visualization techniques (Section 4.5) such as link analysis, to fraud detection.

4.3.1 Data mining and fraud detection

Data mining technology has evolved over the years to provide business with a powerful tool to sort and analyze seemingly unrelated entities within datasets. In simple terms, data mining seeks to process large volumes of data by attempting to identify or extract meaningful relationships that can be used to establish and predict trends and patterns both past and future. In achieving this goal data mining involves the bringing together of data analysis techniques and "high-end technology" in the pursuit of developing useable knowledge [13]. Just as data mining tools and techniques are used to solve real world problems in business, engineering, and science, they are also being applied to the world of fraud detection [14].

Today there are numerous data mining products on the market, each using various data analysis techniques on a wide range of platforms and technologies. In fact, so diverse are the range of products that it would fill a book just to discuss their capabilities, technologies and approach to knowledge discovery and management. Therefore, rather than focusing on individual products, the later part of this chapter will instead focus on the techniques used in data mining products when applied to the detection of potentially fraudulent activity.

Given the relative maturity of data mining technology, a number of algorithms currently being implemented within data mining software have a direct application in the field of fraud detection. In particular, clustering, association rules, decision trees, and neural networks all have a role in the detection of potentially fraudulent behavior (some of these techniques are discussed also in Chapter 7).

It is worthwhile highlighting here two key issues relating to the application of data mining technology to the detection of potentially fraudulent behavior. The first relates to the accuracy of the data. As with any analysis, the accuracy of the results is very much dependent upon the reliability of the original data. Unfortunately, when it comes to fraudulent activity, particularly internal to an organization, the data itself may be corrupted or destroyed as a means of covering up the fraud. To this end, a good fraud detection system will ensure that source data is protected and preserved and not capable of being manipulated. The second issue arises from the fraud profile process. The decision as to whether behavior is fraudulent is very much dependent upon the interpretation of what is acceptable and what is not acceptable. The success of this interpretation is in turn very much dependent upon the comprehensive nature of the fraud rules that have been developed to describe unacceptable behavior. Simply put, poor rules will result in poor detection rates.

4.3.2 Digit analysis and fraud detection

Digit analysis is an auditing technique developed around the concept of analyzing the actual digits making up the numbers contained within a dataset. Essentially, it seeks to identify abnormalities within individual numbers by examining [15]:

- ▸ Digit and number patterns;
- ▸ Round number occurrences;
- ▸ Duplicate numbers;
- ▸ The relative size of numbers.

Digit analysis is based principally on the mathematical theory known as Benford's Law (see Section 4.4.1.4 for a detailed explanation of Benford's Law). Benford's Law utilizes expected frequencies of digits within numbers to identify anomalies by comparing expected frequencies with actual frequencies. In situations, where actual frequencies differ from expected frequencies, closer scrutiny is required to establish if there is fraudulent activity through number invention, manipulation, or undue biases [16]. Digit and number pattern analysis principally involves an analysis of the frequency of the first two digits occurring within a number, whereby the actual frequencies of the first two digits are compared with the expected frequencies as described by Benford's Law.

Round number occurrences seek to measure the extent and frequency of rounding thereby alerting the examiner to potential manipulation or invention. Round numbers, for the purpose of digital analysis, are defined as "numbers that are divisible by 100 or 1,000 without leaving a remainder" [17].

Duplicate numbers, as the name suggests, is a search for multiple occurrences of the same number within a dataset. The frequency and existence of duplicate numbers (particularly if the numbers are intended to be sequential) may indicate double counting or direct manipulation.

The relative size test simply examines each number to determine if it is outside the normal or expected range of like numbers [17].

4.3.3 Fraud detection tools

The discussion to date in this chapter has focused principally on the underlying principles and issues surrounding fraud and the use of technology to detect and prevent fraudulent activity. The software tools currently used to detect fraudulent behavior fall into one of two categories. There are dedicated data mining tools that have fraud detection capabilities, and there are data extraction and analysis tools the core function of which is the detection of anomalous activity.

In its Seventh Annual Software Survey, the Internal Auditor Magazine surveyed members of the Institute of Internal Auditors as to the types of software used to extract and analyze data, and detect and prevent fraud. From the survey, the following was found for data extraction and analysis [18]:

1. 40% of respondents use audit control language (ACL) by ACL Services Ltd;

2. 21% of respondents use spreadsheet software;

3. 14% of respondents use standard database software;

4. 9% of respondents use other software (e.g., Peoplesoft, SAP, and Crystal Reports);

5. 4% of respondents use internally developed software;

6. 3% of respondents use interactive data extraction and analysis (IDEA);

7. 3% of respondents use SAS software;

8. With the remaining using a mixture of Easytrieve Plus, Report Writer, Focus, and Monarch.

In terms of fraud detection and prevention, the following was found [18]:

1. 70% of respondents use ACL by ACL Services Ltd;

2. 14% of respondents use internally developed software;

3. 7% of respondents use other software (e.g., SQL, Monarch, Easytrieve, and Business Objects);

4. 5% of respondents use IDEA;

5. 2% of respondents use DATAS;

6. With the remaining using a mixture of Office-suite software and SAS.

It is interesting to note from these results that many of the mainstream data mining products did not rate in the survey results. We conclude this section by highlighting some of the high-end data mining tools currently available that incorporate fraud detection capabilities within their feature set. The top five data mining products with fraud detection capabilities, as rated by Abbott, Matkovsky, and Elder [14] include the following:

1. Clementine by Integral Solutions Ltd;

2. Darwin by Thinking Machines;

3. Enterprise Miner by SAS Institute;

4. Intelligent Miner for Data by IBM;

5. Pattern Recognition Workbench by Unica Technologies Inc.

4.4   Fraud detection techniques

As electronic commerce and transaction processing systems become faster and more capable of processing an ever-growing number of transactions, the volume of data undergoing real time processing continues to increase. This growth in turn places even greater demands on fraud detection systems, particularly with regard to the timely detection of potentially fraudulent behavior. In order to meet these increasing demands, the underlying fraud detection techniques are becoming ever more complex. No longer do organizations rely solely on one or two indicators to detect fraudulent activity, but rather it is a combination of indicators and techniques that are being applied in the fight against fraud. This in turn places even greater pressure on the capabilities of fraud detection systems to keep up with the flow of data. This is particularly a problem for fraud detection systems that operate in real time, for not only must they process existing data, but also they must be able to learn so that their fraud profiles remain current. In terms of post-event analysis, time is also a critical factor. Failure to deal with data on an ongoing basis can result in excessive historical data that may overwhelm a fraud detection system, resulting in poor detection rates.

Despite these problems, today's fraud detection techniques have been sufficiently refined to allow for timely and detailed analysis to be undertaken. What follows is a discussion of some of the key detection techniques currently being utilized in fraud detection systems.

4.4.1 Fraud detection through statistical analysis

Statistical analysis in fraud detection is the application of statistical techniques and related number theory to the analysis of information with a view to identify irregular or statistically inconsistent data. The more common statistical analysis techniques currently being applied in fraud detection systems consist of the following:

> Regression analysis;

> Correlation analysis;

> Dispersion analysis;

> Benford's Law.

4.4.1.1 Regression analysis

Statistical analysis techniques permit the forecasting of future or anticipated trends by exploring the relationships between given variables within a particular dataset. The analysis helps to determine the relationship between two or more variables and predict, with varying degrees of certainty, the future value of one variable (the dependent variable) as determined by its relationship with one or more other variables (the independent variable or variables). In simple regression analysis (also referred to as univariate regression analysis), it is assumed that the value of the dependent variable is determined by the value of just the one independent variable, while with multiple regression analysis the value of the dependent variable is determined by several independent variables acting in concert. The ability to forecast the value of a dependent variable provides the computer forensic professional with a powerful tool that can be applied to the detection of fraudulent activity by comparing forecast values with actual outcomes.

Whereas a manager or director of an organization may base future decisions on predicted outcomes, the computer forensic professional can use the same forecast techniques to compare actual outcomes, such as revenue, against the forecasted outcomes for the same period. Using historical data to build a trend line that best reflects the overall behavior of the particular variables, such as revenue per week, an expected outcome for a particular week can be computed and compared against the actual outcome for the same week. Where a significant difference emerges between expected and actual outcomes, particularly in cases where expected revenue is significantly lesser than the actual revenue, then a closer examination of revenue activity for the week is warranted. One possible cause for such a discrepancy is a leakage of income resulting from fraudulent activity. Consequently, the application of forecasting techniques based on an examination of past trends provides the computer forensic professional with a powerful tool for identifying potential instances of fraudulent activity.

The most common trend analysis technique used in fraud detection is that of regression analysis. Regression analysis allows the computer forensic professional to examine the relationship between one or more known variables (the independent variable) and an unknown variable (or dependent variable). However, given that fraud detection from a computer forensic perspective is centred on an examination of past events, the dependent variable should be seen more as an expected outcome and not as an unknown or forecast value.

Linear trend regression analysis   Linear trend regression analysis applies to relationships that can be represented by a straight line. For instance, in the case of Figure 4.2, the independent variable is a period of time (e.g., a day, week, or month) and the dependent variable is monthly expenditure. Figure 4.2 demonstrates what is known as a scatter diagram, which in this case illustrates a linear relationship between the common variables, time and revenue. Using the plot points contained within the scatter diagram, it is possible to represent the overall relationship between time and revenue by fitting a straight line between all plot points. This straight line, which indicates a direct linear relationship between the variables was derived using

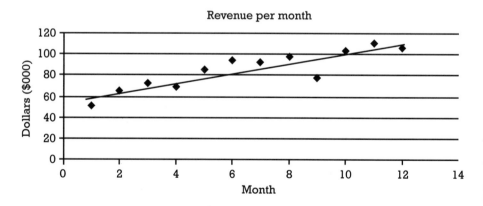

Figure 4.2 Linear scatter diagram.

the linear regression analysis equation shown in the Appendix. Use of this equation is governed by the following key factors [19]:

> There is a linear relationship between variables.

> The larger the source dataset, the greater the reliability of the forecast values.

Given that linear trend regression analysis represents relationships by mathematically examining historical trends, it is critical that the historical data used best reflects the operations of an organization over a reasonable period of time. Additionally, it is important that as much historical detail as possible is included in the source data (e.g., utilize daily or weekly figures as opposed to monthly figures) to allow *drill-down* for more detailed analysis.

> The values of the dependent variable should be derived from accurate historical data preceding the particular period for which comparisons of forecasts and actual values are to be made.

Given that linear trend regression analysis is a predictive tool that is being applied for historical analysis, it is important that the values of data upon which it is based are known to be accurate, and not influenced by fraudulent activity. Using a dataset that includes data affected by fraudulent activity can skew the regression line, thereby masking the true impact of the fraudulent activity that one is trying to detect.

To detect suspicious activity, it simply becomes a matter of calculating the forecast value of the dependent variable and comparing it with the actual value. A significant difference between forecast and actual values therefore becomes the trigger for a closer examination of the respective data with a view to identifying the reason(s) for the discrepancy. In Figure 4.2, period 9 displays a significant difference between the actual revenue earned and the expected or forecast revenue (as represented by the linear trend line).

In one particular forensic analysis, an organization's revenue stream for a given year was subjected to a linear regression analysis. Allowing for growth and seasonal variations, it was soon identified that for the second week in June, revenue for one retail outlet was expected to exceed $35,000. However, when the actual figures were examined, revenue had only reached $26,000. This in turn triggered an investigation that identified that revenue for previous weeks was above $30,000. A closer examination and subsequent analysis revealed that while the inventory movement was consistent with previous weeks, revenue earned was not. After even closer scrutiny it was discovered that an employee was stealing monies.

While regression analysis has the ability to identify instances of unexpected or irregular behavior, it cannot determine the underlying causes for such behavior. Consequently, when used in computer forensics, regression analysis should only be regarded as an alert mechanism.

Nonlinear trend regression analysis   While linear trend regression analysis can be used to detect potentially fraudulent behavior by exploiting the predictable nature of linear relationships, in reality relationships between variables may not be linear. Indeed, given the cyclical nature of business processes over any given year, the relationship between variables will often not be easily described by a linear relationship, and nonlinear regression analysis must be used.

Utilizing the same principles as in linear trend regression analysis, it is possible to predict the value of the dependent variable relative to an independent variable, by applying a suitable mathematical equation that best describes the nonlinear relationship. Figure 4.3 shows a scatter diagram that demonstrates the nature of a nonlinear relationship between two variables. This relationship, while being direct, is best described as curvilinear.

An examination of Figure 4.3 shows that units of widgets increase, over time, in a nonlinear manner. As such, if a linear trend regression analysis was applied to the underlying dataset in order to forecast the expected number of widgets for a given period, a significant disparity would emerge between forecast and actual values. This disparity could either result in a misleading indicator of the actual number of widgets, or inadvertently mask fraudulent activity. To best describe the nonlinear or curvilinear relationship shown in Figure 4.3, we apply nonlinear trend regression analysis to derive a regression

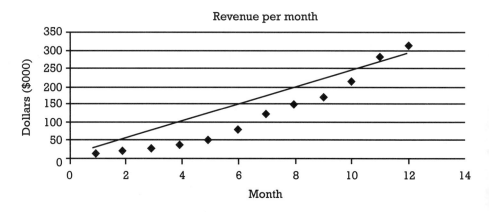

Figure 4.3 Nonlinear scatter diagram.

function that expresses the nonlinear nature of the relationship between the dependent and independent variables. Once again application of the analysis in computer forensics is governed by the following key factors [19]:

> There is a nonlinear relationship between variables.

> The larger and more detailed the source dataset, the greater the reliability of the forecast values.

> The value of a dependent variable should be derived from accurate historical data preceding the particular forecast period.

As is the case with linear trend regression analysis, the detection of suspicious activity, is simply a matter of calculating the forecast value of the dependent variable and comparing it with the actual value. A significant difference between forecast and actual values is the red flag for a closer examination of the respective data with a view to identify the causal effect of the actual results obtained.

Multiple regression analysis   The discussion on regression analysis to date has focused on the relationship between the dependent variable and one independent variable. Given that business processes comprise many interconnecting relationships, each acting on the other to influence outcomes, how then can regression analysis be employed to capture these relationships? The answer is to modify the regression analysis equations to account for the use of population samples and to include each of the independent variables (see the Appendix for the form of the regression equation derived by the use of multiple regression analysis).

The inclusion of multiple independent variables in the forecasting model greatly improves the overall accuracy of the forecasted dependent variable by reducing the size of forecasting errors. As a consequence, the use of multiple independent variables greatly enhances the computer forensic examiner's ability to both detect and define fraudulent behavior.

4.4.1.2   Correlation analysis

While regression analysis examines the relationship between two or more variables, correlation analysis is used to describe the strength of the relationship. Essentially correlation analysis, from a computer forensic perspective, is an adjunct to regression analysis. In detecting fraudulent behavior, the computer forensic professional uses regression analysis to detect potentially fraudulent activity, and correlation analysis to determine how strong the relationship is between the actual values and the regression line. Essentially if there exists a strong correlation with the regression line, then the more reliable are the forecast values derived from the regression analysis. Consequently, the more reliable the forecast value, the more reliable is its use in detecting fraudulent or anomalous behavior [20].

There are two primary measures for determining the strength of the relationship between two variables:

1. The coefficient of determination.

2. The coefficient of correlation.

It is the former that is most commonly used for correlation analysis, it is—counterintuitively—more useful for this purpose than the coefficient of correlation and we shall not consider the latter further for that reason.

The coefficient of determination   The coefficient of determination (referred to typically as r^2—see the Appendix) reflects the strength of the relationship between the dependent variable and the independent variable(s).

A coefficient of determination of 1 signifies that the regression line is a perfect fit (i.e. 100%), while a value of 0 signifies that there is no correlation between the regression line and the values of the dependent variable. The closer the coefficient of determination is to 1 the stronger the correlation. For example, if the coefficient of determination equates to 0.736, then it can be said that the independent variable accounts for 73.6% of the variation of the dependent variable.

4.4.1.3 Dispersion analysis

The earlier sections are concerned with the expected values of a variable predicted on the basis of one or more other factors (regression analysis) and the confidence to be placed in such predicted values (correlation analysis). In this section, we focus on something similar to correlation analysis, this time we focus not on *prediction* based upon known values of one or more independent variables but on *dispersion analysis*. Dispersion analysis refers to known frequency distributions, such as that presented by the 100 annual rainfall figures of the previous century, and allows us to make judgements such as "how exceptional was this year's rainfall in the context of the previous century?" Whatever the data collected, whether it is rainfall data or whether it is financial data such as sales figures or costs incurred, or simply data about inventory, it is possible in many such cases to compare individual values against other like values, with an expectation of similarity. The extent of the expected similarity depends upon the inherent dispersion of the values around some characteristic values such as the mean, mode or median, with the mean being used most commonly. The intention of making such comparisons may be to identify an emerging trend away from previous mean or simply to identify the exceptional nature of an observed value. If the value is exceptional, then it may be that new or unexpected influences are at work and the situation bears further investigation in order to identify whether something untoward has occurred.

From a computer forensic perspective, such a comparison facilitates the identification of potentially fraudulent or anomalous behavior by measuring the deviation of a given value from the expected mean of like data from previous times. This is particularly useful in cases where loss, through fraudulent behavior, results in a decreased variable value for a given period. Simply comparing a suspect variable value against all other like values is impracticable and not terribly informative in terms of providing any clear indication of exceptional values and thus associated fraudulent or suspect activity.

It is the concept of *dispersion*, that is the extent to which previously recorded values of this sort vary from each other, for example, are they closely clustered or are they widely different—that provides statisticians a means by which to provide context to a specific observed value. In particular, how does the value compare with other supposedly similar values—whether it be rainfall or dollars of revenue for a specific outlet for last week—and does it vary widely from other recorded values or not too much? There are many measures of dispersion; some common ones that can be used to provide this contextualization are

- *Range:* The highest and lowest previously recorded values—if the newly observed value is outside the min/max range then this flags that the value is highly unusual.

- *Interquartile range:* The mid 50% of previously recorded values (i.e., discard the top and bottom quartiles)—if the newly observed value is outside this range, it may be dubious.

- *Standard deviation.*

The standard deviation of a sample of observed values may be calculated using the formula appearing in Appendix 4A and reflects the extent to which the observed values vary from their mean. Chebyshev's inequality states that the proportion of values deviating more than n standard deviations from the mean will be at most $1/n^2$, irrespective of the nature of the frequency distribution of the values and no matter what the nature of the pattern of variation. For example, at least 75% of all values for a given variable within a dataset will reside within two standard deviations from the mean, while at least 89% of values will reside within three standard deviations.

From a computer forensic perspective, this relationship provides a benchmark upon which anomalous behavior may be identified. By comparing the actual figure of widgets sold on a particular day against the relevant distribution of widgets sold per day observed over many days, it is possible to establish threshold values for acceptable and unacceptable outcomes. In the case of widgets, historical data may indicate that the mean shipment of widgets per day is 3,000, with a standard deviation of 150. On one particular weekday, we note that the shipment of widgets only reached 2,450. Comparing this to the mean does not provide a meaningful picture but by taking into account the standard deviation, we see that 2,450 equates to more than three standard deviations from the mean. This in turn implies that the value is relatively low and therefore worthy of further investigation.

4.4.1.4 Benford's Law

Deriving from the work of Frank Benford in the late 1930s, Benford's Law expresses the notion that, with certain assumptions, the smaller digits (e.g., 1 and 2) occur rather more frequently as the leading digit in numbers—be they numbers from the physical universe such as the physical constants or statistics gathered by the government and social scientists—than do the larger digits (e.g., 8 and 9). The assumptions are essentially that

 ▸ The sample of numbers are reasonably large.

 ▸ There are no artifical restrictions on the values assumed by the numbers (for example, in the case of a random number generator which by definition will yield values that do not conform to Benford's Law).

In these circumstances, Benford's Law states that the probability of occurrence of a digit as the leading digit of a number is related to the log of its inverse. This behavior, it turns out, carries through to digit-combinations, so that 10 is more likely to occur than 19 as the leading two digits of a number, and also, correspondingly, to the frequency of strings of digits within numbers, so that a string of digits such as 813 is more likely to occur than 824. The following equations describe the observed outcomes [21]:

$$P(d_1) = \log\left(1 + \tfrac{1}{d_1}\right) \qquad \text{leading single digit}$$
$$P(d_1 d_2) = \log\left(1 + \tfrac{1}{d_1 d_2}\right) \qquad \text{leading two digits}$$
$$P(d_1 d_2 d_3) = \log\left(1 + \tfrac{1}{d_1 d_2 d_3}\right) \quad \text{leading three digits}$$

where

P = the probability;
d_1 = the first digit;
d_2 = the second digit;
d_3 = the third digit.

The frequencies predicted by Benford's Law are summarized in Table 4.1 [22, 23].

Benford's Law and fraud detection   It has only been in recent times that the true significance of Benford's Law has been realized with regard to its application in the detection of fraudulent activity. All too often the fraudster may select an even spread of high and low numbers in an effort to blend their fraudulent or false values into a given dataset (e.g., general ledger). Such activity has the net effect of making the distribution of numbers within the dataset more uniform by increasing the probability of higher digits and reducing the probability of lower digits. The resulting more uniform or more even distribution is at odds with the observed behaviors highlighted by Benford's Law, and as such would be a trigger for further investigation [22].

In addition to the above changes in distribution, it is not uncommon for duplicated numbers or combinations of numbers to be used by a fraudster. In such instances it is possible, with the aid of Benford's Law to calculate the probability of various number combinations, thereby allowing a comparison

Table 4.1 Benford's Law—Predicted Frequencies [23]

Digit	First Frequency (%)	Second Digit Frequency (%)	Third Digit Frequency (%)
0	–	11.97	10.18
1	30.1	11.39	10.14
2	17.61	10.88	10.1
3	12.49	10.45	10.06
4	9.69	10.03	10.02
5	7.92	9.67	9.98
6	6.7	9.34	9.94
7	5.8	9.04	9.9
8	5.12	8.76	9.86
9	4.58	8.5	9.83

to be made between the expected probability and that actual frequency. Again, a significant variation would be a trigger for further investigation.

There have been many instances where Benford's Law has provided some insight into potentially fraudulent behavior. In one such case, an organization's accounts payable system was analyzed, over a two-year period, with a view to identify possible fraudulent payments. Utilizing ACL software and Benford's Law it was soon observed that the frequency of occurrence of "1" as the first digit was 18%. A further analysis was conducted on a monthly basis revealing that the first digit frequency for "1" one was in the order of 27% for the first 8 months, dropping significantly after this period. An audit of the eighth and ninth month's payments was undertaken revealing a significant shift in the payment amounts. Further investigations identified regular payments, not matching Benford's predicted frequencies, had commenced in the ninth month to a new supplier who turned out to be fictitious.

As a word of warning, it should be noted that the application of Benford's Law might not always result in an accurate indication of fraudulent activity. This is in part due to the nature of some number distributions in some given circumstances. One particular instance is that of product pricing where the frequency of low digit numbers may be significantly less than that predicted by Benford's Law (e.g., $19.99). In such instances, the frequency of high order numbers such as 9 may be the norm and the frequency of low order numbers may be the exception. As a consequence, it is important that consideration be given to the applicability of Benford's Law in each given circumstance. Reference [24] provides some rules of thumb to be applied when determining what types of numbers would adhere to Benford's Law.

4.4.2 Fraud detection through pattern and relationship analysis

Pattern and relationship analysis provide basic fraud detection tools for the computer forensic examiner. By examining simple patterns and relationships between data items, missing or anomalous data can be readily identified. Unlike statistical analysis techniques, pattern and relationship analysis rely solely on examining past events and developing a profile of common behaviors.

4.4.2.1 Sequencing

Sequencing involves an examination of a dataset with a view to identify either missing or abnormal data elements. In some instances of fraud, it is not uncommon for the perpetrator to attempt to hide or disguise the nature and extent of their activity by manipulating the raw data so as to either remove any evidence of their activity or to disguise it. Sequencing attempts to identify such anomalies by examining the sequence of known items recorded in a dataset and identifying instances where such items are either missing or, when compared to previous trends, recorded at an abnormally high or low level.

By way of example, consider a case of inventory fraud, where daily stock levels are calculated by recording inventory received, inventory sold, and inventory damaged. To disguise theft of inventory, either the inventory-received value would need to be understated, or the inventory sold or inventory damaged levels overstated. In the case of inventory received and sold, other measurements, such as income and expenditure can be used to reconcile the actual levels of goods received and sold. However, without appropriate controls in place, theft of inventory can be disguised by increasing the quantity of damaged goods. Using sequencing techniques, it would be possible to detect the increased levels of inventory being written off by comparing values with past periods. An increase in damaged stock, without an appropriate increase in sales should be the cause for concern.

While relatively simple, sequencing provides the computer forensic specialist with a quick and easy tool for identifying anomalies without the need for complex mathematical concepts.

4.4.2.2   Duplicate investigation

Duplicate investigation is a relatively simple concept that can be used to readily identify abnormal entries and essentially involves a search for duplicate values within a given dataset. In particular, it focuses on like data contained within the same field (e.g., invoice number) and is best applied across a range of values that are intended to be unique, such as transaction numbers. In instances of fraud, it is possible for the fraudulent activity to be disguised by the replication of a previous record within a dataset. The replication may include the duplicating of unique identifiers, such as invoice or receipt numbers. In such instances, a scan of the dataset should readily identify the fact that two (or more) entries contain the same invoice or receipt number and should therefore warrant further investigation.

4.4.2.3   Historical trend analysis

Historical trend analysis, as the name implies, utilizes past trends to determine if current values are worth further investigation. However, unlike the statistical forecasting techniques described earlier, trend analysis is reliant on like periods and does not use predictive techniques (such as regression analysis) to compare expected outcomes with actual outcomes. Instead, it provides a means of detecting changes by comparing current values with past activity. To undertake historical trend analysis, it is important that a suitable quantity of historical data be available. Additionally, it is important that such historical data is obtained in a uniform manner and in a consistent and regular time frame. By way of explanation, consider the case of revenue earned. Comparing quarterly revenue against previous revenue would only be effective if the comparison was made against the same quarter for past years. Comparing current third quarter revenue against past first and second quarter revenue figures would not yield a true result, given that seasonal factors may influence a particular quarter.

Historical trend analysis simply involves comparing current figures with past trends for the same business or operational unit. Essentially, it is a way of comparing past performance with current performance with a view to detecting potential loss. In comparing past trends with current results, it is important to factor in those variables that have a direct impact on variations. Such variables include budgeted growth, special projects, and abnormal costs. Consequently, it is important not to assume straight away that a deviation from past trends is an automatic sign of fraud, but rather it is only a fraud warning when all variables have been taken into account. In doing so, it is necessary to rule out legitimate explanations prior to assuming fraudulent activity.

4.4.2.4 Ratio analysis

Just as investors use key financial ratios (e.g., profit to earnings) to determine the viability of an investment, computer forensics can use ratios to identify potentially fraudulent transactions occurring within an organization. Ratio analysis essentially targets the variance between transactions by calculating key ratios and using these to identify abnormal or suspicious numerical values. Typically, the key ratios used in determining variance consist of highest value/lowest value, highest value/next highest value, and current value/previous value [25].

Highest value and lowest value   The high low variance compares the highest value for a particular data type (e.g., price of a widget) with the lowest value for the same data type. The resulting variation gives a ratio that can then be used to compare against the same ratio for a similar product or item. The calculation of a high low ratio permits differing entities (e.g., widgets and bricks), whose overall values may vary, to be compared against one another.

A high ratio indicates significant variation between the highest value and the lowest value. For organizations that permit discretionary pricing in order to stay competitive, or in industries where price elasticity is strong, a high ratio may be acceptable. However, for those organizations that have rigid pricing structures, it would be considered abnormal to see a high ratio given that the level of elasticity in pricing policy would be low. Consequently high low variances, subject to the nature of the business environment, can identify instances where potentially fraudulent or unethical behavior (e.g., over-pricing) has occurred [25].

High value and next highest value The high next highest variance compares the highest value for a particular data type (e.g., payment) with the next highest value for the same data type. The resulting ratio provides an insight into the existence of what are referred to statistically as *outliers* or abnormal amounts and thus possibly of fraudulent activity. However, if the fraudulent activity has been undertaken for a period of time, then it is likely that the second highest value may also be fraudulent. In such instances, further differentiation between data may be required. For example, consider a situation where an accounts payable employee has, on a regular basis, been deliberately overpaying a supplier. In such instances, an alternative ratio analysis (current value and previous value) would need to be undertaken on a supplier-by-supplier basis [25].

Current value and previous value The current previous variance is similar in concept to historical trend analysis, however, rather than examining trends, it is simply a case of comparing the current value of a particular data field, with previous values for the same data field. In essence, it is a one for one comparison. Although simple in its implementation, it is prone to error under certain circumstances. Using the example of the overpaid supplier from the previous paragraph, consider the following situation. A comparison of current values for goods or services rendered is made against past values. From this analysis, it is evident that the supplier today is being paid more than previously. This of course does not necessarily indicate fraud. Before fraud can be suspected a number of other factors must be eliminated, for example, whether the increased payment is due to inflationary pressures or changes in the supplier's contract, or whether it is simply a miscalculation, an innocent mistake.

Despite the power and effectiveness of ratio analysis to detect possible instances of fraudulent behavior, it should be remembered that this technique, like all other techniques, is merely an indicator of abnormal behavior, requiring further and detailed examination.

4.4.3 Dealing with vagueness in fraud detection

Given the complex nature of both fraud and fraud profiling, it is sometimes difficult to define fraudulent activity in terms of true and false activities. For example, consider the red flag "Payment made to X was large": Is this statement true, given that the payment made to X was $10,000? Most people would hesitate to answer yes (true) or no (false), depending on the situation or context, but rather say "sort of." This does not imply uncertainty about the "world" or domain of payments (as we are sure that the payment made to X was exactly $10,000), it is more a case of vagueness or uncertainty about the meaning of the word "large." As fraud can be easily disguised within legitimate events, it is often difficult to differentiate legitimate activity from those driven by fraudulent intent. In such circumstances, the red flags making up the fraud profile are vague and not always clear-cut. Dealing with *red flags* that are not clear-cut cannot be handled solely by the techniques described above. To this end a degree of fuzzy evaluation is needed.

Fuzzy set theory treats a "large payment" as a fuzzy statement and stipulates that the truth value associated with this statement is a number between 0 and 1, rather than being simply true (1) or false (0). Fuzzy logic takes a set of compound fuzzy statements, such as "large payment quickly transferred" (involving the two component fuzzy statements "large payment" and "quickly transferred") and attempts to determine its truth value as a function of the truth value of its components.

Fuzzy logic allows the fraud detection system to deal with partial or inconclusive scenarios without having the need to reach a definite result of either yes or no. In essence, the use of fuzzy logic allows fraud detection to deal with the gray areas of fraud whereby neither a yes or no is the best answer for a given fraud rule in a particular scenario. Fuzzy logic has been used very successfully in certain commercial applications such as household appliances and video cameras. This success is thought to be due to fuzzy logic's ability to deal with a small number of fuzzy statements and yet not be involved in chains of inference (i.e., reasoning with sets of fuzzy statements).

So how does fuzzy logic help in fraud detection? If you were to ask an experienced auditor or fraud investigator what combination of *red flags* would clearly constitute fraudulent behavior, it is likely that the results would be varied. Given differing levels of experience, different business environments and variations in corporate culture, some *red flags* (e.g., revenue and cost projections) may not be viewed as significant as other red flags [26]. To this end, it is difficult to get a clear and consistent yes or no response. However, fuzzy logic allows various *red flags* to be rated not simply as strong or weak but rather in terms of the level of their effectiveness (e.g., 75% effective) for differing scenarios. For each red flag, such as "large payment," a fuzzy number or truth value ranging between 0 and 1 is used to describe its applicability. For each fuzzy number value, a level of confidence is then assigned. This level of confidence reflects the belief in the respective *red flags* ability to detect fraud in a given situation [26].

Given the dynamic nature of fraud, and the business environment in which most fraud originates, the application of rule based detection to fraud control can result in fraud profiles being out dated quickly. The ability of fuzzy logic to learn and grow therefore makes it an ideal candidate in fraud detection systems. When combined with the use of neural networks the underlying fuzzy rules can be readily adapted to changing circumstances through various inherent learning techniques. To this end, fuzzy logic provides an adaptive and intuitive approach to fraud detection that standard rules-based decision systems cannot.

4.4.4 Signatures in fraud detection

Real time detection of fraudulent behavior facilitates the timely identification of fraudulent behaviors and the timely development of suitable preventive strategies. Unfortunately, given the large volume of information being processed through on-line systems, it is often difficult to undertake a comprehensive and timely—let alone real time—fraud analysis of the huge datasets that result from these every day transactions [27].

To address this, some researchers have turned to the use of signatures that can be used to describe varying patterns of behavior. Such signatures are essentially a mathematical representation of predefined fraudulent activity or other behavior. These signatures can then be used to detect fraudulent activity through one of the following methods:

- Profile-based detection [27];

- Anomaly-based detection [27].

Profile-based detection involves the comparison of current transaction behavior against known fraudulent behavior. On the other hand, anomaly-based detection compares current transaction behavior against legitimate known past behavior for the same user or entity. In both instances, the fraudulent behavior or past behavior is represented by signatures.

To measure the degree of fraudulent activity associated with a transaction two measures of probability are first calculated. The first is the probability that the transaction is fraudulent. This is calculated by taking the signature for known frauds and mathematically comparing it against the transaction being measured. The second measure is the probability that the transaction is legitimate. This is calculated by mathematically comparing the user or entity profile, derived from previous legitimate transactions, against the particular transaction being measured. Once both probabilities have been determined an overall fraud score can then be calculated [12].

The use of signatures in real-time fraud detection holds a number of advantages over rules-based systems applied in a real time environment. Firstly, signatures require significantly less storage resources (e.g., memory) due to the fact that they are a statistical representation of the source data as opposed to a complete duplicate of the raw dataset. Secondly, signatures have the ability to evolve as more and more transactions are processed. Thirdly, significantly fewer resources are needed in the learning process, for as each new transaction is processed the current signatures are updated, resulting in little or no need to recalculate the signature from the source data. Fourthly, signatures can be transferred to new accounts without the need for large amounts of historical data. Finally, signatures are significantly quicker to process thereby reducing the time impact on the processing of real-time transactions.

4.5 Visual analysis techniques

In fraud detection through statistical analysis techniques, number theory or fuzzy logic provides a means of detecting instances of potentially fraudulent behavior, however what it does not provide is a means of identifying and demonstrating interconnected relationships that may exist between separate occurrences of anomalous behavior. Essentially, the fraud detection techniques described earlier provide limited ability to explore the relationship(s) that may exist between occurrences of fraudulent behavior. In large complex frauds, the relationships between various entities, transactions, and financial instruments can provide a clear picture as to the scope and size of the fraudulent activity. Additionally the ability to not only identify but also explore the relationships may provide additional avenues of inquiry and open up new sources of evidence not previously considered. Visual analysis techniques allow the computer forensic specialist to understand and explore the relationships between suspicious transactions.

Essentially, visual analysis involves the depiction of entities, events, and items as graphical representations graphically linked to each other according to clearly identified relationships. By visually representing each event or entity it is possible to establish, at a glance, an overall picture of the fraudulent activity or behavior. In addition, with the aid of powerful visual analysis software (such as i2 and Netmap), one is able to explore the extent and nature of the individual relationships that exist between entities. Visualization analysis is best achieved through the use of specialized software that is able to take a large dataset and analyze it for patterns and relationships, and then graphically represent those relationships.

Just as there are different analysis techniques that can be applied in the detection of fraudulent activity, so too there are a number of visual analysis techniques that can be applied in the analysis of a dataset. Principally these visual analysis techniques, as applied to fraud detection and analysis, consist of

- Link or relationship analysis;
- Time-line analysis;
- Clustering.

4.5.1 Link or relationship analysis

Link or relationship analysis explores the inherent relationships existing within a set of data by identifying links or relationships between each entity resident within the dataset (see also Chapter 7). Link or relationship analysis depicts the relationships by graphically visualizing each entity and its interconnecting relationship(s) with other entities. As a consequence of this process, a diagram is developed showing each entity and its associated events. Entities can be depicted either as a picture or icon, or as a label or tag. The associated relationships are usually depicted as lines that connect between the various entities. Each interconnecting line or relationship may be identified either by a specific color, shape or size that represents a particular type of relationship. In addition, the interconnecting lines may have a label or tag attached that describes the specific nature of the relationship. Figure 4.4 demonstrates a simple link analysis diagram depicting the relationship and frequency with which a computer of interest (center of diagram) has accessed other computers (in this case three computers). The links provide a visual representation of the relationship while the dates and times for each link depict the frequency of the connection.

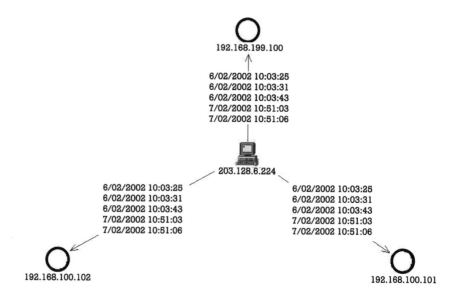

Figure 4.4 Link analysis.

Merely depicting links or relationships between entities is not sufficient. For visual analysis to work effectively, the computer forensic specialist must have the capability to explore and test the various relationships. To this end, the visual analysis software must be capable of handling customized queries and be able to depict the results according to user specifications. The ability to examine relationships with a view to identify other possible links, no matter how unexpected, is what makes visual analysis a powerful tool. The ability to depict data and the inherent relationships in a variety of ways makes data visualization a fast and powerful tool in fraud detection. In situations where thousands or even millions of transactions have taken place, the application of the right visual representation is critical. In such cases, an alternate approach to that taken in Figure 4.4 may be required. One such method is to represent the relationships in what is referred to as a *ball of string*. With this technique all entities are represented around the edge of the ball of string with the interconnecting links forming the inner core of the ball. Figure 4.5 depicts a ball of string diagram generated by the analysis software Netmap, and is representative of an analysis of insurance claims payments. In this case, the claims payments data form the outer band of the circle and the lines, within the circle, connect claims that have a common element such as claimant name or claimant address. By visually representing these relationships, it is possible to quickly identify, by means of density, instances of high activity.

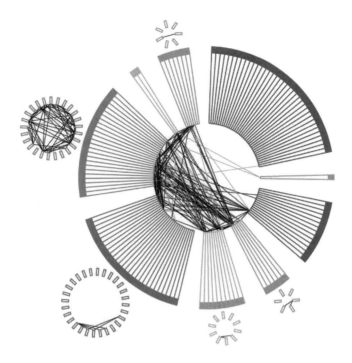

Figure 4.5 Ball of string. (Source: http://www.netmap.com/. Reprinted with permission, NetMap Analytics Pty Limited, 2002.)

4.5.2 Time-line analysis

Given that fraudulent activity, or for that matter most illegal activity, occurs over a period of time, the ability to examine and reconstruct the precise sequence of events is critical in understanding what has taken place. Not only is it needed in order to reconstruct the elapsed sequence of events, but it may also be critical as a means of identifying specific instances of vulnerability or high risk both in the past and the future. The principle technique used in this process is commonly referred to as time-lining or time-line analysis. Time-lining simply matches individual events against a time-line thereby recording the exact sequence of events, as well as depicting the relative frequency of repeated behaviors (see also Chapters 2, 3, and 6).

Time-lining essentially involves the identification of particular events, noting when and for how long they occurred, and recording them against a time-line or time scale, divided into discrete units of time. Once placed against the time-line, each event can be viewed against events from other activities, and any similarities or overlap with respect to the frequency of

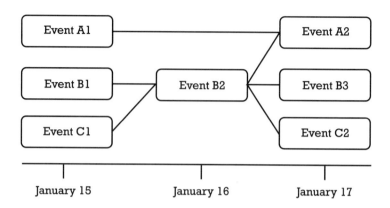

Figure 4.6 Time-line analysis.

similar events, the duration of each event, or existence of repeats can be identified and explored in more detail. In addition, time-lining provides easy identification of instances when key events may merge. In such a case, the links between merging events combine into one identifiable event on the time-line chart. Time-lining is a simple technique that facilitates a visual inspection of the sequence of events within a given period of time. Figure 4.6 depicts a simplified time-line analysis in which some events (B1 and C1) merge on a given day (January 16) and impact on a seemingly unrelated event (A2) on the following day (January 17).

4.5.3 Clustering

Clustering brings together like or similar data contained within a dataset into clearly defined groups. The similarities from which the groupings are derived is very much dependent upon the objectives or focus of the analysis. Clustering essentially provides a visual pattern detection technique that draws out similarities and irregularities in large datasets.

When data is subjected to cluster analysis techniques, a number of clearly defined rules are applied to the dataset, with the resulting matches grouped according to a common pattern. The size and density of the grouped records or clusters, provides a visual representation of the frequency and strength of each particular grouping. A small cluster may be an indication, subject to the rules applied in the clustering process, of abnormal or irregular behavior and may therefore require further analysis. Members of such small outlying clusters are sometimes referred to as *outliers*. Once clusters have been identified, they should be subjected to further testing to ensure that

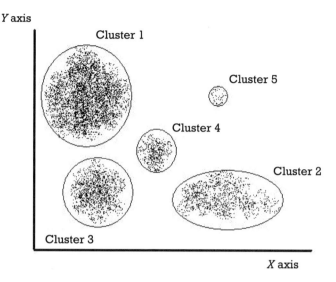

Figure 4.7   Cluster diagram.

the initial rules used in the clustering process are not flawed or were not applied incorrectly.

The resulting cluster diagram may be either two- or three-dimensional, with axes that represent the relevant measures for the underlying rules, and groups of data that represent the individual clusters. Figure 4.7 depicts a simplified cluster diagram in which naturally occurring groupings of data can be identified. It illustrates the concept with an example of orders for goods placed over time. The size of orders appears on the y-axis while the x-axis represents time. Such a cluster diagram can relate to one customer or all—in the former case, the clustering is dependent upon the nature and extent of business activity undertaken by the particular customer, in the latter case the diagram represents overall movement of stock from the point of view of the distributor.

4.6   Building a fraud analysis model

The application of CATTs in the detection of potentially fraudulent behavior is not a guarantee that such behavior will be detected. Indeed, the application of even the smartest software will not guarantee the identification of fraudulent or anomalous behavior. Auditors and audit entities the world over recognize this fact and have subsequently developed methodologies to complement the technology. It has been long recognized by the audit community that the reliability of any results is in principle derived from the accuracy and relevance of the source data.

In recognition of these issues, a number of forensic data analysis and data mining methodologies have evolved. These methodologies separate the data identification, collection and analysis processes into clearly defined and distinct phases by means of a structured data analysis model. While each phase has the ability to stand out on its own, in reality their individual inputs, processes and outputs are all very much dependent upon the interrelationships existing between all phases within the resulting model. In addition to the clearly defined stages, it should be noted that a key feature of the more adaptable models is the fact that they are essentially technology neutral, focusing on functionality rather than on specific technology.

Drawing on the more commonly used methodologies, it is possible to identify a number of clearly defined phases that the data analysis process undergoes when applied to fraud detection:

1. *Problem definition:* The nature of the problem and the final objectives are identified and documented.

2. *Business analysis or modeling stage:* The business and information technology environment is defined and clear rules developed to identify activity that is either acceptable or unacceptable.

3. *Data acquisition stage:* The primary data is identified and extracted from an organization's information technology infrastructure.

4. *Knowledge discovery stage:* The primary data is analyzed according to the rules developed in the modeling stage.

5. *Evaluation and validation stage:* The results of the analyses are reviewed and evaluated.

Drawing on the key features of a number of data analysis models, it is possible to construct a simple seven-stage model (Figure 4.8), specifically adapted to fraud detection utilizing data analysis techniques. In addition to the description of each stage, the following discussion also highlights the interdependencies between stages and their relative information flows

4.6.1   Stage 1: Define objectives

4.6.1.1   Overview

The objective definition stage is the starting point of the fraud analysis process. Before considering how to analyze the data, it is important to first

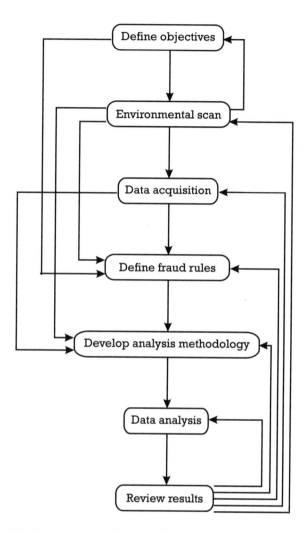

Figure 4.8 Seven-stage fraud detection model.

consider the key objectives. Merely stating that the objective is to identify fraud within a given set of data is too broad a statement. Rather, consideration should be given to the types of fraud being considered (e.g., overpayments). By being specific, it is possible to identify which systems and processes should become the focus of the analysis. In addition, we are also able to better define the fraud rules which we will rely upon during the analysis of the data.

The definition of objectives should include a clear definition of the underlying problem and should provide a clear understanding of how the results are to be measured [13]. Once clearly defined and documented, the stated objectives should be referred to throughout the analysis process to ensure that the data selected is appropriate and that the interpretation of the results is undertaken within the right context.

4.6.1.2 Inputs

The definition of objectives will very much be governed by both internal and external environmental factors which will underlie the initial reason(s) for undertaking a fraud data analysis. Such factors include regulatory requirements, customer or third party requests, corporate governance and ethics issues, industry trends, and issues.

In situations where the environmental scan has identified key issues that impact on the attainment of goals, it may be necessary to modify the objectives before proceeding on to data acquisition.

4.6.1.3 Outputs

The result of the objective definition stage is the creation of clear objectives that reflect the underlying goals and drivers behind the fraud analysis process. These objectives are used by the next stage (environmental scan) to identify relevant issues and data sources.

4.6.2 Stage 2: Environmental scan

4.6.2.1 Overview

The environmental scan seeks to identify key systems and sources of data that are pertinent to the meeting of the stated objectives. During the environmental scan, consideration is given not only to internal systems and data sources, but also to external sources of data. Consequently, the environmental scan should not be restricted to data that originates from within an organization's own IT systems, but consideration should be given to independent external sources of data that may provide corroboration.

In addition to identifying key systems and data sources, the environmental scan should also include regulatory, legal, and ethical considerations. It is no good identifying a source of data for use in the analysis stage, if access to it or its use would constitute a breach of a regulation or legal statute. Consequently, the environmental scan may result in potential risks being identified and analyzed to determine their impact on the selection of data.

Finally, the environmental scan includes an understanding of the relevant business processes and the subsequent flow of information within the organization. Understanding business processes and information flows help in the identification of core data, and is of value in the development of fraud rules. It should be noted that business processes for like functions may vary between departments within an organization. In such instances, it will be necessary to consider each department on a case-by-case basis.

4.6.2.2   Inputs

The environmental scan is very much governed by the results of the objective definition stage, and, where the fraud analysis is an ongoing activity, the review of the previous analysis results. Additionally, as has been highlighted, environmental factors and issues feed into the environmental scan stage to help determine outputs.

4.6.2.3   Outputs

The results of the environmental scan are used to identify the key information systems within the organization, which contain useable data. In addition, the results are used to define relevant fraud rules and develop an appropriate analysis methodology.

4.6.3   Stage 3: Data acquisition

4.6.3.1   Overview

The data acquisition stage takes place when the source data is extracted, cleansed, enriched, and reviewed. Extraction is the duplicating of raw data from the various sources identified by the environmental scan. Once extracted in its raw state, the data may contain fields that are irrelevant or data that is inconsistent with the other data in the same field. Data entry and system errors can result in invalid data residing within the dataset. Consequently, the raw data must be cleansed to ensure that all-irrelevant information and irregularities are removed. Where data is derived from a number of sources, it may be necessary to join a number of datasets to form one core dataset. This process of joining various data sources results in an enriched dataset that can provide more meaningful results. After enrichment has taken place the resulting dataset(s) is reviewed to ensure that it contains no irrelevant information or erroneous data, and that all necessary information is present for the purposes of meeting the stated objectives.

4.6.3.2   Inputs

In acquiring the data it is necessary to know which systems contain relevant data. As such the primary input for the data acquisition stage, comes from the environmental scan. Furthermore, in instances where the fraud analysis is an ongoing process, additional sources of data may be identified as a consequence of a review of the final analysis results. This is particularly so if the analysis results do not meet the stated objectives due to incorrect data being analyzed.

4.6.3.3   Outputs

The end structure and content of the acquired data is used in the development of appropriate fraud rules as well as the design of a suitable analysis methodology.

4.6.4   Stage 4: Define fraud rules

4.6.4.1   Overview

The definition of fraud rules relates to the identification and development of appropriate measures and indicators of fraudulent activity within an organization. The fraud rules bring together the objectives of the fraud analysis and the business processes that underlie normal business behavior. During this process, abnormal and potentially fraudulent activity is defined and relevant areas of high risk are identified and marked as *red flags*. Utilizing these known *red flags* and their associated indicators, clear rules are developed to describe the fraudulent activity. These rules, when combined, form a profile of fraudulent activity. The fraudulent activity should relate to a specific area or function within the organization, and should not necessarily be applied, as a generic indicator of fraud, across the organization as a whole. Indeed, the same type of fraud may have different profiles between departments. This is a critical factor when designing fraud rules for a number of departments.

4.6.4.2   Inputs

The development of appropriate fraud rules draws heavily on the key objectives of the fraud analysis as well as the results of the environmental scan. Where the analysis is an ongoing process, the fraud rules may need to be modified or new rules added after previous analysis results are considered.

4.6.4.3 Outputs

From the definition of fraud rules, clear fraud profiles should emerge that assist in differentiating normal activity from abnormal and potentially fraudulent activity.

4.6.5 Stage 5: Develop analysis methodology

4.6.5.1 Overview

The development of an analysis methodology involves the identification of relevant and appropriate analysis techniques to achieve the stated objectives. Drawing on any one or combination of analysis techniques, as described in this chapter, a suitable analysis methodology is developed. In undertaking the process, a number of key factors must be considered. These key factors are in essence the outputs from previous stages. The choice of which technique(s) to use is very much dependent upon the nature and structure of the source data. Additionally, the analysis objectives will factor in deciding what analysis techniques are appropriate in the given circumstances. Finally, the nature of the fraud rules defined in the preceding stage will have a bearing on how best to measure the relevant rules. Given the detailed explanation of the various techniques, it is not proposed to cover this area in any more detail.

4.6.5.2 Inputs

Analysis objectives, data structure and content, fraud rules, and environmental conditions contribute to determine the best technology and techniques to use. Additionally, in circumstances where the analysis is ongoing, the choice of technology and techniques may need to be reviewed and adapted in light of previous analysis results.

4.6.5.3 Outputs

The choice of technology, relevant fraud rules and appropriate data combine to form the basis of the analysis methodology which then facilitates the data analysis stage.

4.6.6 Stage 6: Data analysis

4.6.6.1 Overview

The data analysis stage involves the application of the chosen technology and techniques, and the application of the fraud rules in the analysis of the acquired data. This is, in data mining terms, the knowledge discovery component of the fraud analysis process. Here, the chosen technology and selected fraud rules are tested in a real world situation. While much of the data analysis stage is taken up with data processing, an important activity is the configuration of hardware and software to facilitate the accurate and timely processing of data. Here, the computer technology is mated with the chosen techniques and rules.

4.6.6.2   Inputs

The data analysis brings together through practical application, the outputs of the analysis methodology phase. Additionally, where the analysis is part of an on going process, it is possible for the results of previous analysis to be used to reanalyze the data. This is particularly the case when system error is suspected or validation of past results is needed.

4.6.6.3   Outputs

The data analysis stage delivers the results of the data analysis process.

4.6.7   Stage 7: Review results

The final stage in the fraud analysis process is the review of analysis results. The success of this stage is very much dependent upon a number of factors. These include

1. The clarity of the output;

2. The knowledge of the reviewer (both in terms of business processes and identifying fraudulent activity);

3. The reviewer's ability to interpret the output;

4. The accuracy of the source data;

5. The application of the most appropriate analysis techniques and technology.

Essentially the review stage is the culmination of all previous stages within the analysis process. As part of the review stage, the results and their subsequent interpretation, may be used to further refine the analysis methodology, thereby resulting in additional data analysis taking place.

References

[1] "Internal Auditing and Corporate Governance—Where's the Connection?," *Issues and Answers—The Institute of Internal Auditors*, http://www.theiia.org/ecm/iiapro.cfm?doc_id=344, visited Jan. 2003.

[2] Van Homrigh, D., and Garnett, M., "Forensics—New Bloodless Hounds," *Proctor—The Journal for the Queensland Law Society*, Sept. 2001.

[3] Le Grand, C. H., "Information Technology in Auditing," *Institute of Internal Auditors White Paper,* Oct. 2001.

[4] Australasian Centre for Policing Research, "The Virtual Horizon: Meeting the Law Enforcement Challenges, Developing an Australasian Law Enforcement Strategy for Dealing with Electronic Crime," Scoping Paper, Report Series No. 134.1, 2000, http://www.police.govt.nz/resources/2001/ecrimeforum/law_enforcement_challenges.pdf, visited Sept. 2002.

[5] Grabosky, P., R. G. Smith, and G. Dempsey, *Electronic Theft: Unlawful Acquisition in Cyberspace*, Cambridge, U.K.: Cambridge University Press, 2001.

[6] *Concise Oxford Dictionary*, Sixth Edition.

[7] Loebbecke, J., M. Eining, and J. Willingham, "Auditor's Experience with Material Irregularities: Frequency, Nature, and Detectability," *Auditing: A Journal of Practice and Theory*, Vol. 9, Fall 1989, pp. 1–28.

[8] Moulton, G. E., "Profile of a Fraudster," *Deloitte & Touche Conference on Corporate Fraud: Detection and Avenues for Resolution*, Toronto, Canada, 1994 (see also http://www.deloitte.ca/en/pubs/Forensic/corpfraud_conference/moulton.asp, visited Sept. 2002).

[9] Kusnierz, R., and A. Livesey, "Data Mining into the Next Century: Part 1," *IT Audit,* Vol. 2, Altamonte Springs, FL: The Institute of Internal Auditors, Sept. 1999 (see also http://www.theiia.org/itaudit/index.cfm?fuseaction=forum&fid=110, visited Jan. 2003).

[10] Kusnierz, R., and A. Livesey, "Data Mining into the Next Century: Part 2," *IT Audit,* Vol. 2, Altamonte Springs, FL: The Institute of Internal Auditors, U.S.A., Nov. 1999 (see also http://www.theiia.org/itaudit/index.cfm?fuseaction=forum&fid=111, visited Jan. 2003).

[11] "Introduction to Fraud Detection," http://dinkla.net/fraud/types.html, visited March 2002.

[12] Cahill, M., D. Lambert, J. Pinheiro and D. Sun, "Detecting Fraud in the Real World," *Handbook of Massive Data Sets*, J. Abello *et al.* (Editors), Dordrecht, The Netherlands, Kluwer Academic Publishers, 2002, pp. 911–929.

[13] "Data Mining: Using Data Mining Techniques for Fraud Detection," *SAS Institute Inc. White Paper*, 1999.

[14] Abbott, D. W., P. Matkovsky, and J. F. Elder, "An Evaluation of High-End Data Mining Tools for Fraud Detection," *IEEE Conference on Systems, Man and Cybernetics,* 1998.

[15] Lanza, R., "Digital Analysis—Real World Examples," *IT Audit, Vol. 2,* Altamonte Springs, FL: The Institute of Internal Auditors, July 1999 (see also http://www.theiia.org/itaudit/index.cfm?fuseaction=forum&fid=58, visited Jan. 2003).

[16] Nigrini, M., "Digital Analysis—A Computer Assisted Data Analysis Technology for Internal Auditors," *IT Audit, Vol. 1,* The Institute of Internal Auditors, December 1998, http://www.theiia.org/itaudit/index.cfm?fuseaction=forum&fid=95, visited Jan. 2003.

[17] Nigrini, M., "Digital Analysis—Part 2: A Review of the Audit Tests to Detect Anomalies in Data Subsets," *IT Audit, Vol. 2,* The Institute of Internal Auditors, February 1999, http://www.theiia.org/itaudit/index.cfm?fuseaction=forum&fid=96, visited Jan. 2003.

[18] Salierno, D., "Tools of the Trade—Internal Auditors Seventh Annual Software Survey," *Internal Auditor,* Aug. 2001.

[19] Levin, R. I., *Statistics for Management,* 4th ed., Upper Saddle River, NJ: Prentice Hall, 1987.

[20] Lapin, L. L., *Quantitative Methods for Business Decisions, with Cases,* 5th ed., Sydney: Harcourt Brace Jovanovich, Inc., 1991.

[21] Nigrini, M. J., "The Peculiar Patterns of First Digits," *IEEE Potentials,* April/May 1999, pp. 24–27.

[22] Coderre, D. G., *Fraud Detection: Using Data Analysis Techniques to Detect Fraud,* Vancouver, Canada: Global Audit Publications, 1999.

[23] Nigrini, M. J., "Understanding Benford's Law," http://www.acl.com/benford/Understand_Benford.pdf, visited June 2002.

[24] "Background on Digital Analysis—a look at Benford's Law," http://www.audittools.com/ekstern/atools.nsf/, visited June 2002.

[25] Coderre, D. G., "Ratio Analysis—An Understated Data Analysis Technique," *IT Audit,* The Institute of Internal Auditors, Vol. 4, April 2001, http://www.theiia.org/itaudit/index.cfm?fuseaction=forum&fid=70, visited Jan. 2003.

[26] Deshmukh, A., and J. Romine, "Assessing the Risk of Management Fraud Using Red Flags: A Fuzzy Number Based Spreadsheet Approach," *Journal of Accounting and Computers,* No. 12, article 1.

[27] Cortes, C., and D. Pregibon, "Signature-based Methods for Data Streams," *Data Mining and Knowledge Discovery,* Dordrecht, the Netherlands: Kluwer Academic Publishers, March 2001.

Appendix 4A

The equations listed in this appendix relate to sample datasets and not populations.

- Linear trend regression analysis

$$\hat{Y} = a + bX$$

where

> $\hat{Y}$ represents the calculated values of the dependent variable;
> X represents the values of the independent variable;

$$b = \frac{\sum XY - n\overline{X}\overline{Y}}{\sum X^2 - n\overline{X}^2}$$

$$a = \overline{Y} - b\overline{X}$$

Y represents the observed values of the dependent variable, $\overline{X}$ and $\overline{Y}$ are the mean values of X and Y, respectively, and n is the number of pairs of X, Y values.

- Multiple regression analysis (with two independent variables)

$$\hat{Y} = a + b_1 X_1 + b_2 X_2$$

where

> $\hat{Y}$ represents the calculated value of the dependent variable;
> X_i represent the values of the independent variables.

- Coefficient of determination

$$r^2 = 1 - \frac{\sum (Y - \hat{Y})^2}{\sum (Y - \overline{Y})^2}$$

- Standard deviation

$$S_X = \sqrt{\frac{\sum (X - \overline{X})^2}{N - 1}}$$

where N is the number of observations.

5

Case Studies

Contents

5.1 Introduction

5.2 The case of "Little Nicky" Scarfo

5.3 The case of "El Griton"

5.4 Melissa

5.5 The World Trade Center bombing (1993) and Operation Oplan Bojinka

5.6 Other cases

References

5.1 Introduction

It is useful to place the tools and techniques discussed in previous chapters into a real world context. In this chapter, we will highlight some of the forensic techniques mentioned previously in their operational deployment in some cases that have achieved prominence, including

- The case of "Little Nicky" Scarfo;

- The case of "El Griton";

- The Melissa virus;

- The World Trade Center bombing (1993) and Operation Oplan Bojinka.

Section 5.6 discusses some cases encountered personally by the authors, selected in order to highlight some key issues and technologies.

5.2 The case of "Little Nicky" Scarfo

The Philadelphia, Pennsylvania faction of La Cosa Nostra (the Mafia) has been one of the strongest in the American Cosa Nostra since its start in 1911. Nicodemo "Little Nicky" Scarfo, Sr.

223

was one of the major leaders of the Gambino crime family under John Gotti operating out of Atlantic City, New Jersey and Philadelphia. In 1989 Scarfo, Sr. was convicted of the murder of Frank "Frankie Flowers" D'Alfonso but remained in control until 1991 [1]. His son, Nicodemo S. Scarfo, Jr., carrying on the family tradition, participated in organized criminal activities including gambling and loan sharking operations.

Scarfo, Jr. kept information on his activities on a personal computer in his office, like many other modern businessmen. Like many other computer literate businessmen, he also used encryption technology to protect incriminating and sensitive information about gambling from prying eyes, including those of the government. To provide this protection Scarfo, Jr. chose commercially available Pretty Good Privacy (PGP) software from Network Associates.

On January 15, 1999, Federal Bureau of Investigation (FBI) agents executed a search warrant, one of a number that day, on Scarfo, Jr.'s New Jersey office of "Merchant Services of Essex County." During the search they forensically duplicated Scarfo's computer system and on subsequent examination discovered a single file named *Factors* that was encrypted using PGP. Investigating agents suspected that the file contained information on gambling debts which would support their loan sharking investigation but found that they were unable to crack the password using "normal investigative procedures to decrypt the codes" that would allow them to access the contents of the file [2–4].

On May 8, 1999, the FBI applied to U.S. Magistrate, Judge C. Donald Haneke, for a court order permitting them to covertly install a keystroke logging system (KLS) on Scarfo's computer in an effort to recover the PGP pass phrase that would allow decryption of the *Factors* file. Based on the documentation placed before him by the government, Judge Haneke signed an order that

- Found there was probable cause to believe there was evidence of a crime in the encrypted file;

- Permitted the FBI to covertly enter Scarfo's office, install the "keystroke logger" and capture keystrokes [5].

On May 10, 1999, FBI agents used the court order to enter the Belleville, New Jersey office of Scarfo, Jr. and planted the KLS in the computer system. It was in place for a period of 2 months, after which the court order further permitted the FBI to again covertly enter Scarfo's office to retrieve the device and the output.

When retrieved and downloaded, the output from the KLS contained only 45 pages of keystrokes, including literally hundreds of nonsense characters. The last entry on the last page of the output file was, however, the entry that included Scarfo, Jr.'s PGP pass phrase. This, as it happened, was his father's Federal Bureau of Prisons identification number, something that was of course known all along [5].

This information was then used to decrypt the *Factors* file and, as suspected, incriminating information was located.

On June 21, 2000, a federal grand jury in the District of New Jersey issued a sealed indictment charging Scarfo, Jr. and one Frank Paolercio with various illegal gambling acts.

5.2.1 The legal challenge

In June 2001, Scarfo, Jr.'s attorneys motioned to suppress information gathered through the use of the KLS as the use of a search warrant to capture keystrokes on the computer system was improper. The defense alleged that a Title III electronic communications interception warrant was more appropriate as the computer system was being used for legitimate business purposes and electronic communications. Scarfo's lawyers further motioned that the use of a surreptitiously installed computer program to monitor all the keystrokes was a violation of the Fourth Amendment of the U.S. Constitution, which specifically forbids conducting general searches [7].

In hearings before Federal Judge Nicholas Politan of Newark, New Jersey, both Scarfo's lawyers and the government made arguments. On August 7, 2001, Politan ordered the government to submit to the defense and the court, a report that revealed the nature of the technology employed by August 31, 2001 as the court "harbors serious concerns as to whether the key logger device, either intentionally or unintentionally, intercepted a communication from defendant Scarfo's desktop computer" [8]. Interception of electronic communications is dealt with in a different legal manner to data which is stored on a computer hard disk drive and if it could be determined that the KLS had captured keystrokes related to an e-mail communication or on-line chat, all data recorded by the KLS could potentially be ruled unlawfully obtained and, as a consequence, the decryption of the *Factors* file would be similarly invalidated as "fruit from a poison tree" under exclusionary evidence legal principle [9].

Despite the fact that KLSs have been publicly available for many years, the employment of such a technology in a criminal investigation was deemed so sensitive that on August 23, 2001, Justice Department attorneys sought to have the technique and technology employed suppressed as being prejudicial to National Security under the 1980 Classified Information Procedures Act (CIPA), a little-used federal law usually reserved for espionage cases. Prosecutors indicated that they wanted to file two reports:

1. An unclassified summary of the "keystroke logger" the FBI used to eavesdrop on Scarfo and learn his pass phrase;

2. A classified document that only Politan would read, which provided a detailed description of the operation of the system.

The Justice Department also requested the Judge to order that Scarfo's defense counsel be barred from releasing the unclassified summary document to the public or press [10].

Scarfo's lawyers filed an opposing motion to that request, which the court considered at a hearing on September 7, 2001.

On October 4, 2001, Judge Politan granted the Justice Department's request for CIPA protection, denying discovery of information on national security grounds [11]. The FBI subsequently provided the defense with an unclassified affidavit purporting to describe the functionality of the "key logger system." The then defense renewed its motion to suppress evidence.

On December 26, 2001, Politan handed down a decision upholding the legality of the FBI's use of the key logger system and denied the defense motion to suppress evidence obtained through the technique [6].

On February 28, 2002, the Justice Department and Scarfo entered into a plea agreement for the racketeering charges for which Scarfo was subsequently convicted [12].

5.2.2 Keystroke logging system

Owing to the computational infeasibility of cracking Scarfo's PGP encryption using conventional means, the FBI had to employ other technical means to obtain Scarfo's password. In these circumstances, employment of a keystroke monitoring or logging system seemed appropriate.

Keystroke monitoring or logging was developed in the late 1980s, and privacy experts say current U.S. law still does not appear to adequately address this technology. In 1999, the Clinton administration in fact failed to get the U.S. Congress to pass a law, known as the Cyberspace Electronic Security Act (CESA), authorizing keystroke-monitoring surveillance [13].

Keystroke monitoring has been used for a long time, for both good and nefarious purposes, in the field of computer security. Computer intruders have surreptitiously installed keystroke-logging programs on compromised computer systems for many years with the aim of capturing username and password information that would allow them to break into further systems. Their adversaries, system administrators and computer security professionals, have likewise employed keystroke monitoring to obtain evidence of intruders' activities on compromised computer systems.

The Computer Emergency Response Team-Coordination Center (CERT-CC) based at Carnegie Mellon University first issued an advisory on keystroke monitoring based on legal advice from the Department of Justice on December 7, 1992. The advisory indicates that legitimate system operators should have set up banners on systems advising users that logging into and using the system constitutes consent to monitoring and results of such monitoring may be supplied to law enforcement if criminal activity is detected [14]. This is known as *consensual monitoring* and evidence so obtained has been utilized in many intrusion prosecutions in the United States and other countries including Australia.

KLS typically comes in two forms, hardware- and software-based. Examples of commercially available systems of both types include the following:

- *Software:* Spector, KeyKey Monitor, 007 STARR, Boss Everywhere, and I-See-Ua;
- *Hardware:* KeyGhost, KeyKatcher, and Hardware Keylogger.

A good example of one of the more sophisticated software systems is Invisible Keylogger Stealth (IKS) from Amecisco [15]. In its most sophisticated version for Windows NT/2000/XP, IKS uses a high performance kernel-mode driver, which interfaces with the keyboard interface at the lowest level of the operating system. There is also a version for Windows 95/98/ME. Amecisco assures that the user will never find the driver except for possibly identifying the growing binary keystroke log file with the recorded keystrokes. For sophisticated companies and government agencies, Amecisco offers a "Custom Compile Edition" to ensure that virus scanners or a custom binary file signature detection program will not detect the KLS.

The users' technical ability, their knowledge of examining running processes and the level to which they actively monitor their system obviously has an effect on the effectiveness of software-based KLS. Use of software utilities, such as Process Explorer and Filemon for Windows NT/9x/ Linux from Sysinternals make detection of software-based KLS much more likely [16].

Further to the standard systems' utilities, there are also antikeystroke monitoring detection programs available including Anti-keylogger™ [17] and SpyCop [18].

Hardware KLS have distinct advantages over software-based systems. A hardware device, for example, Keyghost (see Figure 5.1) may be installed even when the target computer is logged out, has a password, is locked or switched off. In most cases, depending on the concealment of the device, it can be easily removed and the captured keystrokes retrieved on another computer [19].

In more sophisticated commercial hardware KLS, over 500,000 keystrokes can be stored, in most cases using strong encryption to protect the captured information. Usually, hardware devices use the same non-volatile flash memory as used in smart cards for storage. Most hardware-based devices are also operating system-independent and work equally well with any desktop PC and all PC compatible operating systems. This includes Windows 3.1, 9x, NT, 2000, Linux, OS/2, DOS, Sun Solaris, and BeOS. Most importantly most hardware KLS are impossible to detect or disable using software, such as process monitoring, anti-virus, or anti-spyware software.

Unlike software KLS, hardware KLS are able to record every keystroke, even those that may be used to modify BIOS settings before bootup. This enables capturing of BIOS passwords as well as other system passwords.

(a)

(b)

Figure 5.1 Before (a) and after (b) pictures of the installation of a standard Keyghost hardware key logger system. (*Source:* http://www.keyghost.com/. Reprinted with permission, Keyghost Ltd. 2002.)

For concealment, some hardware devices are now being built into standard manufacturers' keyboards that can be used to replace the keyboard being used by the suspect.

The ability to capture passwords to support data recovery will become an increasingly important forensic capability for law enforcement as hardware password protected hard disk drives become prevalent. The ATA-3 (AT attachment) standard first implemented a capability for IDE hard drives to be password protected. Only a few drives implement this feature, in particular IBM Thinkpads, and this will no doubt extend to other drives soon to be shipped with desktop machines requiring the same approach to access the data as laptops [20].

5.3   The case of "El Griton"

Much of the material here has been derived and reprinted with the permission of Special Agent Matt Parsons of the U.S. Naval Criminal Investigative Service.

From July 12 to December 28, 1995, computers in the United States, Korea, Taiwan, Chile, Brazil, and Mexico reported intrusions originating from a computer system belonging to Harvard University.

Investigators from the U.S. Naval Criminal Investigative Service (NCIS) Computer Crimes Division became involved in August 1995 when an intrusion was detected into a network operated by the U.S. Naval Command, Control, and Ocean Surveillance Center (NCCOSC), which contained information on aircraft design, radar technology, and satellite engineering.

A system administrator at NCCOSC in San Diego detected that certain system files had been altered and that extraneous files had appeared including a network sniffer, the sniffer's output file, and a rootkit for escalating privileges and concealing access. A sniffer program unlawfully intercepts and stores user identifications and associated passwords. With the identification and password information, the intruder had uncontrolled access to the computer system. The system administrator was subsequently able to determine that the intruder had accessed the command's network through the Internet from accounts on the Harvard Faculty of Arts and Sciences computer system (the FAS Harvard Host).

NCIS investigators, working together with the network manager of the Harvard Arts and Sciences Computer Services, in turn determined that the intruder had compromised an unknown number of legitimate accounts on the FAS Harvard host, and used these accounts to launch attacks on numerous military, government and educational computer networks.

As the intruder was accessing the FAS Harvard system through a widely used modem bank and remotely from other apparently compromised systems over the Internet, and as the intrusion was making use of legitimate account holders' identities as aliases, the NCIS investigators initially found it impossible to identify the intruder.

It was, however, possible to distinguish the intruder from other legitimate users of the FAS Harvard system and the Internet through their repetitive use of certain commands on the FAS Harvard host. This consistent behavior involved use of programs with unique names to obtain account names and associated passwords on overlapping groups of computer systems [21].

5.3.1 Surveillance on Harvard's computer network

The U.S. Attorney's Office for the District of Massachusetts sought and received U.S. Federal Court authorization to intercept electronic communications of the intruder to and from the FAS Harvard host commencing on October 23, 1995. Law enforcement had done similar electronic surveillance on computer systems in the past, but had always used consensual monitoring provisions with the implicit or explicit consent of the users of the monitored computer system. This electronic surveillance of Harvard's network was the first in the United States to be conducted pursuant to a Court order under the electronic surveillance statute, and continued through the end of the year.

Network surveillance was only conducted initially between a general access modem bank and the FAS Harvard host, and then over a segment of Harvard University's computer network through which all communications to and from the FAS host flowed. Due to the nature and extent of use of the network by the university population, the U.S. Attorney's Office was genuinely concerned about minimizing the potential extent of intrusion into the private communications of legitimate users of the Harvard computer network in contravention of the Fourth Amendment of the U.S. Constitution. To overcome these concerns, a multilayered filtration process was developed and employed to minimize the number of legitimate electronic communications that might be viewed by investigators searching for the intruder.

The intruder's communications to and from the FAS host were isolated using a specially configured monitoring computer which employed a government-developed software package called iWatch, an element of the Network Intrusion Detector (NID) intrusion detection system (IDS) developed by the U.S. Department of Energy's (DOE) Computer Security Technology Center (CSTC, now called the Cyber Solutions Tools Center) at Lawrence Livermore National Laboratory in California [22].

Monitoring of specific communications was triggered by the tell-tale:

▸ Use of the accounts that the intruder was known to have compromised;

▸ Use of certain Internet host computers that the intruder was known to be attacking from the FAS Harvard host;

▸ Use of unique files and programs that the intruder was known to utilize when engaged in his unlawful activities [21].

If a keyword or phrase was intercepted, iWatch would initially only display up to 80 characters surrounding the target word or phrase to give it a context. If, after these 80 characters were examined, it remained ambiguous whether what had been intercepted was the activity of the intruder or that of a legitimate user, investigators used a search utility to look for further indication of the intruder before actually examining the intercepted computer session.

The intruder's activities observed during the period of surveillance were similar to those positively identified earlier in the fall. In addition, the intruder was monitored discussing over the Internet, techniques for obtaining unauthorized access to computer systems. While the intruder could have used computer systems other than Harvard's as staging points, the intruder was not observed stealing files from Harvard or other computer systems connected to the Internet to which he gained unauthorized access.

5.3.2 Identification of the intruder: Julio Cesar Ardita

After attacking a system in Taiwan, during the course of one discussion on Internet relay chat (IRC) the intruder was overheard referring to himself using the alias "El Griton" (Spanish for "The Screamer"). He was also repeatedly observed accessing the FAS Harvard host from four computer systems in Buenos Aires, Argentina; one among them was Telecom Argentina. These clues eventually enabled investigators to identify the intruder.

In the middle of December 1995, the NCIS case officer, Special Agent Peter Garza contacted individuals in charge of the administration of the computer system at Telecom Argentina in Buenos Aires seeking their assistance in identifying persons who might have access to its system. The telephone company initiated a local criminal investigation because of the risk that the intruder posed to its telecommunications system. As a result of the Argentine criminal investigation, four members of the Ardita family (a father and three sons who were minors) were arrested on December 28, 1995 in Buenos Aires. Julio Cesar Ardita, then a 21-year old computer science student in Buenos Aires at the University of Argentina, and suspected of participating in the offences by Argentines, was reportedly brought before the court at a later date. The search and arrest in Buenos Aires were front-page news in the Clarin, a leading newspaper in Argentina. The Clarin reported that the people arrested were in a position to destroy the Telecom telephone company system.

After the search and the arrests in Buenos Aires, the identification of "Griton" as Julio Cesar Ardita was corroborated through recovery of previous activity of Griton on a bulletin board known as *yabbs* whose postings are accessible through the Internet.

In August 1993, Griton had invited readers of his posting on yabbs to visit his own bulletin board called "Scream!", devoted to hacking, cracking (identifying and using system vulnerabilities to crack computer security and gain unauthorized access), and phreaking (the practice of breaking into and misusing telephone systems). The telephone number posted for the bulletin board was that of the Arditas' residence in Argentina. Other postings to the bulletin board uniquely described Julio Cesar within the Ardita family.

5.3.3 Targets of Ardita's activities

A number of the host computers that Ardita gained unauthorized access to were affiliated with the U.S. government including a system at the Army Research Lab in Edgewood, Maryland; the Naval Research Laboratory in Washington; the California Institute of Technology in Pasadena, California; the U.S. DOE; and the NASA Jet Propulsion Laboratory. Victim sites included 62 U.S. government, 136 U.S. educational, and 31 U.S. commercial facilities. The U.S. Navy, NASA, and Department of Energy's National Laboratories were high on the list in terms of frequency of penetration.

The DoD systems that were compromised by Ardita resided on networks that contained sensitive and proprietary, although not classified, government information. These networks stored information related to Navy, Army and NASA research programs, and contained files relating to research on state-of-the-art satellites, radiation and energy-related engineering. There was no evidence of any of this data being stolen or compromised;

however, the level of Ardita's unauthorized access to the systems on the network potentially gave him these abilities.

While Julio Cesar Ardita's father was a retired senior officer from the Argentine military and was a consultant to the Argentine Congress, there was no obvious link between the Argentine government and the targets of intrusions to raise the suspicion of spying, unlike the "Cuckoo's Egg" case in Germany.

Despite this, the level of access gained to the systems and networks obviously put other systems on the violated networks at risk as well. Griton used the rootkit to obtain system administration level access on the systems he compromised. This access allowed him to alter, erase, or destroy files on the network.

The level of actual and potential harm was not limited to only the known, violated systems described in the affidavit filed by Special Agent Garza. Ardita was able to obtain significant additional user account names and passwords for other networks and systems that were equally at risk because of his installation of network sniffer programs.

5.3.3.1   The prosecution

Ardita admitted responsibility for the actions, but claimed he was guilty only of mischief. He was arraigned by a Grand Jury in December 1995. The U.S. Department of Justice filed criminal charges against Ardita; however, prosecution was initially frustrated by the fact that computer intrusions were not covered by international agreements for extradition or state-to-state agreements between the United States and Argentina.

On March 29, 1996, then U.S. Attorney General, Janet Reno, announced on national television that the FBI (when in reality it had been the NCIS) had successfully conducted its first "Internet wiretap," which resulted in the issuance of an arrest warrant for Julio Cesar Ardita. Reno stated that a wiretap of the Internet had allowed federal prosecutors to obtain enough evidence to charge Ardita with three felony counts related to his hacking into U.S. military computers. However, the United States extradition treaty with Argentina did not at that time provide for his extradition to the United States. Cases like this illustrate the jurisdictional issues that the Internet presents to computer crime investigators [23].

However, the Justice Department did not stop its pursuit of Ardita simply because of this minor jurisdictional issue. In December 1997, pursuant to a plea agreement with the Justice Department, Ardita agreed to waive extradition and travel voluntarily to the United States. The agreement contained a joint sentence recommendation of 3 years probation and a US$5,000 fine. The agreement also acknowledged that Ardita had been completely cooperative and truthful and taken part in a 2-week debriefing in Buenos Aires with investigators.

On May 19, 1998, Ardita pled guilty in the U.S. District Court to charges that he unlawfully intercepted electronic communications over a military computer and had damaged files on a second military computer. As agreed, Ardita received 3 years probation and a US$5,000 fine [24].

The NCIS, which did an excellent job of investigating the case and tracking down Ardita, received little credit with many reports incorrectly crediting the success of the case to the FBI. Credit for deploying iWatch on this matter leading to the successful resolution of the case must, however, go to the NCIS.

5.3.3.2 NID and iWatch

As previously stated, the Ardita case was the first time in the United States that a court-ordered "Internet wiretap" had been employed for the real-time monitoring of an unknown intruder. It was an excellent demonstration of the ability to chase and identify an international hacker on-line. Although, the United States was not the only country employing network surveillance technology in criminal investigations at that time. In early 1995, a very similar but slightly less sophisticated technique employing the publicly available *tcpdump* network monitoring program and using consensual monitoring authority was being used in Australia to monitor the activities of a computer intruder known as "The Crawler" [25].

It is useful to understand the technology underlying the system employed to monitor Ardita's criminal activities. In many ways, the system is not too dissimilar to the "Carnivore" or DCS-1000 system that created so much furore recently and is discussed in detail in Chapter 2.

As previously stated, iWatch grew out of the DOE NID program, which was a development of a system developed at the University of California, Davis (UCDavis), called Network Security Monitor (NSM). NSM, developed in 1989, was also the ancestor of the Automated Security Incident Measurement (ASIM) IDS used by the U.S. Air Force [26]. A DoD version of NID, known as the Joint Intrusion Detection System (JIDS) is also deployed.

NID is a suite of software tools that helps detect, analyze, and gather evidence of intrusive behavior occurring on an Ethernet or fibre distributed data interface (FDDI) network using the IP. The system is network based, but stand-alone and passive rather than residing on the hosts it is monitoring. NID is able to collect data, both packets headers and packet contents, and

statistics about network traffic. Operating on a network of host computers
referred to as a security domain, NID passively monitors network traffic
including those activities of an intruder.

The security domain is a collection of hosts and/or subnetworks that are
to be monitored. The domain may consist of either a subset of a network or
the entire network to which NID is directly connected. Only looking at traffic
from particular Internet services based on the TCP port allocated can further
refine the security domain [27].

Figure 5.2 shows a simple example in which NID is used to monitor a
collection of hosts. In the example, NID resides on a single host within
subnet 2, that operates in a broadcast rather than a switched mode, so the
NID sensor can see all network traffic that is being transmitted over subnet 2.

In the simple example, the security domain is defined to consist of hosts
C and D. For simplicity sake, we assume that NID is only monitoring traffic
entering the security domain. Therefore, NID can detect, analyze, and gather
evidence of intrusive behavior on all network traffic originating from hosts
A, B, or E and destined for hosts C or D. Traffic between C and D is neither
monitored nor is there any traffic specific to subnet 1. Correct placement of
the network monitoring system is crucial otherwise relevant network traffic
may be missed.

At the first tier of analysis, NID uses a tool called iDetect to look for
evidence of an intrusion by examining information packets for intrusion

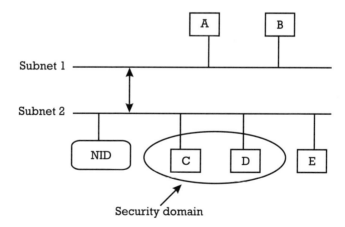

Figure 5.2 NID security domain. (*Source:* Ellen Bradley; Technical Information Department, U.S. DOE. Acknowledgment is made of the U.S. Government's right to retain
nonexclusive, royalty-free license in and to any copyright covering this material. Credit is
given to the University of California, Lawrence Livermore National Laboratory, and the DOE
under whose auspices the work was performed. Reprinted with permission.)

signatures. These signatures normally consist of a string of characters that are used for carrying out computer intrusions. Custom signatures, such as those command sequences used consistently by Ardita, can also be added to the signature database. Collected evidence is presented to an authorizer that approves the transition to iWatch, which is NID's second evidence-gathering phase.

iWatch scans network traffic passing by the NID system interface for connections that contain the same signatures found by iDetect. If iWatch provides compelling evidence, then a third tier, iScript, is used to convert the packets of data into a transcript that is suitable for use in court. Before NID software could be used in the investigation, the NCIS and FBI had to convince the authorizing judge that NID would not violate the stringent privacy standards imposed on other forms of wiretap. Accordingly, NID was modified to address the issue of civilian computer privacy. The modifications took into account the conflicting values of information protection versus privacy and made use of an evidence-gathering model that utilizes pattern analysis to detect unauthorized patterns of activity. If the pattern search triggers an apparent specific signature, explicit monitoring permission can be manually given to pursue data collection of that specific network connection [28].

Some commercial network IDS, such as Network Flight Recorder (NFR), offers capabilities similar to NID and iWatch as a native component of their functionality [29].

5.4 Melissa

The very first macro virus for Microsoft Word, WinWord.Concept (Word Prank), was discovered in the summer of 1995. It did nothing but replicate itself [30]. Since that time, many other more serious macro viruses have appeared but few have had the publicity of the Melissa macro virus. From a network security and forensic perspective, the most significant aspect of the Melissa case is the speed with which the author of the virus was tracked down.

5.4.1 A word on macro viruses

A macro virus is a piece of self-replicating code written in an application's macro language. Many applications have macro capabilities such as the automatic playback of keystrokes available in early versions of Lotus 1-2-3. More advanced macro languages are in fact cut down versions of programming languages such as Visual Basic. The distinguishing factor, which makes it possible to create a virus with a macro, is the existence of auto-execute commands within the language that allow the macro to execute in response to some event and not necessarily in response to an explicit user command. Common auto-execute events are opening a file, closing a file, and starting an application. Once a macro is running, it is free to copy itself to other documents, delete files, and create general havoc in a person's system, and these things can occur without any user interaction whatsoever.

Microsoft Word version 6 to version Microsoft Word 95 included a macro language capability known as WordBasic. From Microsoft Word 97 onwards, Visual Basic for Applications replaced WordBasic.

WordBasic was essentially the BASIC programming language with extensions to make it easy to access the contents of open documents. WordBasic was intended to allow automation of repetitive editing and formatting tasks that were not part of Microsoft Word's built-in command set. Like most other macro languages, both WordBasic and Visual Basic for Applications have the capability of creating auto-execute (AutoExe), auto-open (AutoOpen), and auto-close (AutoClose) macros, and these are the mechanisms that macro viruses use to take control of a computer and install themselves. Auto-execute macros, as their name implies, automatically run every time Microsoft Word is started. The AutoOpen and AutoClose macros run whenever you open or close the document they are attached to.

Most people employ Microsoft Word macros to initialize creation of a new document, inserting standard headers and footers, and set the default formatting including style, font types, and language. The majority of Microsoft Word users, however, still does not employ the macro capability as a matter of course.

5.4.2 The virus

Just after 7:00 A.M. eastern time on Friday March 26, 1999, a file called *Passcodes 3-26-99* was posted to the Internet newsgroup *alt.sex*. On the surface, the post seemed nothing more than a message containing a list of user accounts and passwords for pornographic Web sites. Within hours though, alarm bells began to ring.

The posting to the *alt.sex* newsgroup originated from an America on-line (AOL) e-mail account, "skyroket@aol.com". An AOL e-mail server had forwarded the message containing the virus, which was contained in an attached file named *list.zip*. The victims, who opened the *list.zip* file, which contained about 80 Web site addresses, user names, and passwords for accessing adult-entertainment Web sites, downloaded the file and opened the infected Microsoft Word document, executing the macro and serving as first stage propagators of the virus.

At approximately 2:00 P.M. eastern time on Friday March 26, 1999, CERT-CC based at Carnegie Mellon University in Pittsburgh, Pennsylvania began receiving reports of a Microsoft Word 97 and Microsoft Word 2000 macro virus that was propagating via e-mail attachments. The number and variety of reports that were received indicated that there was widespread contagion affecting a large number and variety of sites. Antivirus software vendors called this macro virus the Melissa macro or W97M_Melissa virus [31].

Fortunately Melissa's main goal was self-propagation not destruction, and two methods of infection were employed. The CERT-CC analysis of the virus indicated that user interaction in the form of opening the infected Microsoft Word document was required to execute the virus, which then infected existing Microsoft Word template files on the system, particularly the standard *normal.dot* template. The second action, far more serious, was propagating via e-mail.

When propagated via e-mail, Melissa infected messages had a subject line "Important Message From (name)," where "(name)" is someone probably known by the person receiving the message. The message also contained the infected Microsoft Word document. Content of the body of the e-mail was: "Here is that document you asked for ... don't show it to anyone else" with a winking smiley face emoticon formed by the punctuation marks ;). When the user, or in some cases their mail software, opened the attachment, if that user was running Microsoft Outlook as their e-mail application, the virus accessed the user's address book and sent copies of the original infected document to the first 50 addresses it found. Some e-mail systems were found to be configured to automatically open e-mail attachments. This significantly increased the rate of propagation. While the primary transport mechanism of Melissa was via e-mail, other methods of file transfer would obviously also support the propagation of the virus.

The e-mail propagation method had the potential to severely compromise organizational confidentiality and therefore posed a significant threat to many organizations' proprietary information. As a byproduct of its propagation, the mass e-mail replication also digested large amounts of bandwidth creating network denial of service conditions in some organizations. If you were not running Microsoft Outlook as your e-mail client, however, you were safe from e-mail-based propagation but still susceptible with respect to file transfers of infected Microsoft Word documents [32].

Many sites were aware of Melissa on Friday March 26, others over the weekend and, still others found Monday morning, March 29, 1999, to be a challenging day. By late on Friday the 26th, however, CERT-CC had already prepared and issued an excellent analysis of the virus, including measures on how to identify and contain it at the host level. According to antivirus vendor Data Fellows, many multinational firms reported infections on Friday, including both Microsoft and Intel. Microsoft even had to close down its e-mail system to prevent further spreading of the virus internally and externally [33]. Reports indicate that thousands of systems spread over hundreds of sites were affected. Media coverage was extensive, however, the virus did not carry out any intentionally malicious activities damaging systems or data.

5.4.3 Tracking the author

The FBI in collaboration with other organizations including CERT-CC, sought to track down the author of the virus. Based on the source of original posting of the infected message, skyroket@aol.com, everyone thought the author of the virus was Scott Steinmetz, a civil engineer. Steinmetz protested his innocence to AOL, which launched an investigation to track down the source of the post.

On March 29, 1999, two software engineers, Richard M. Smith, President of software tools developer Phar Lap Software Inc., and Fredrik Bjorck, a Swedish Ph.D. student at Stockholm University's Department of Computer and System Sciences, tracked down who they thought was the author of Melissa to a virus writer's Web site. They found that the Microsoft Global User Identification (GUID) contained in the original document posted to the newsgroup and containing the Melissa virus matched the GUID contained in another virus, called *PSD2000.doc*, located on the Web site, http://www.sourceofkaos.com/homes/vic/start.html, of a virus developer known as VicodinES. VicodinES also had several other aliases including, Sky Roket, John Holmes, and Johnny "One Leg" Johnson, among others, according to Smith, who provided this information to the FBI [34, 35].

VicodinES had in writings admitted that *PSD2000.doc* was based on a virus called "Shiver" that was the work of another virus developer known as *ALT-F11*. Comparison of the GUID of the "Shiver" virus with the other viruses showed that it also matched the GUID embedded in Melissa. A further virus created by ALT-F11 known as "Groovie2" was also found to have the same GUID. Examination of the other Microsoft Word macros created by VicodinES revealed that *PSD2000.doc* had a unique GUID. All other documents that VicodinES claims were his creations had a different GUID.

Unfortunately, the GUID is not a definitive method for tracing a document to its author. The GUID, described previously in Chapter 3, is stored in a Microsoft Office document only once, when that document is created. Even if a document is copied to a new computer or saved under a new name, the original GUID number does not change. Most programmers do not create programs completely from scratch but use segments of core code from other programs as the basis on which to start building the new program. It is the same for viruses.

Also the network card MAC address that was used to generate the GUID is not obtained directly from the network card's hardware. It is obtained from the software driver on the operating system. In most cases, it is derived from the actual MAC address of the hardware but in fact it can be altered in the system registry, making it possible to set up multiple computers that would generate documents with the same GUID. Therefore, although the GUID matching work done was good and assisted in the identification, it was by no means "the nail in the coffin" that some people at the time suggested.

On Tuesday March 30, 1999, obviously following the VicodinES trail, local agents from the Orlando, Florida Field Office of the FBI seized the SourceOfKaos Web server. The server, located at a local Orlando ISP, Access Orlando, was taken into custody pending a complete analysis of its contents [37].

According to Access Orlando system administrator, Dan Merillat, an agent from the FBI's New York office had contacted the ISP and asked that the SourceOfKaos server be disconnected from the Internet and preserved as evidence until a search warrant was obtained. The FBI faxed an evidence preservation order to the effect to Access Orlando [37].

FBI agents questioned the administrator of the SourceOfKaos site, Roger Sibert, about his relationship with VicodinES. Sibert indicated he had no current means of contacting VicodinES and the last time he had been in contact with them was on January 11, 1999 when he received an e-mail from the virus writer. This e-mail, which originated from an East Coast ISP, suggested that VicodinES had given up virus development [37].

The search for Melissa's author shifted to New Jersey on Monday March 29, 1999, when a lawyer for AOL, John Ryan, called Christopher Bubb, a Deputy Attorney General and head of the State's Computer Analysis and Technology Unit, and said the virus may have started in New Jersey. On that day Richard Smith also turned over to the FBI, a list of names of people who had modified the virus, including the suspect's undisguised name, David L. Smith [38].

AOL then confirmed what Richard Smith already suspected that someone had hijacked skyroket@aol.com's account. AOL "tags" newsgroup postings on its servers, including the messages on *alt.sex* with information about the account from which the post originated including information on the message itself and the software used to post the message. The tag on the post with the original infected file with details of the e-mail server was used to backtrack to a listserver in Monmouth County, New Jersey that had been used to post the original message. From there a New Jersey AOL dial up access point was identified. The real owner of the skyroket account, Steinmetz who lived in Lynnwood, Washington, was a long way from New Jersey. AOL was later able to provide investigators with calling line identification (CLI) information to determine the actual telephone that made the call, which subsequently led them to Smith's house [39].

Acting on the information from AOL tracing the culprit to New Jersey and the information from Richard Smith, a computer task force composed of federal and state agents was formed. William Megarry, an FBI special agent, later cited the joint effort as definite proof of how seriously law enforcement authorities now considered viruses, characterizing them as electronic sabotage [35].

On the evening of Thursday April 1, 1999, members of the New Jersey State Police High Technology Crime Unit, special agents of the FBI and investigators from the Monmouth County Prosecutor's Office arrived with a search warrant at the Aberdeen Township home of David L. Smith. He was not there at the time and officers searching the Ken Gardens apartment found that the central processing units from two computer systems had been removed. Police seized the remaining components of the systems, including power cables, monitors, monitor cables, floppy disks, and writeable CD-ROMs. The equipment was located on a table within the apartment indicating that it was used in conjunction with the processing units. Smith was located 3 hours later at his brother's home in nearby Eatontown, and arrested. Smith was subsequently charged with interruption of public communications, conspiracy to commit the offence, attempt to commit the offence, and third-degree theft of computer service. Alltogether, the charges potentially carried a maximum penalty of 40 years in prison and a $480,000 fine [40].

At the time of his arrest, Smith waived his Miranda rights (the various rights of a suspect when questioned, including the right to remain silent) and admitted to writing the Melissa macro virus, illegally accessing AOL for the purpose of posting the virus onto the Internet, and also destroying the personal computer he used to create and post the virus. He also admitted that he had lived for a time in Florida and named the virus after a topless dancer whom he knew there. He was subsequently released on $100,000 bail.

At a news conference on Friday, April 2, 1999, State Attorney General Peter Verniero appeared along with New Jersey Governor, Christie Whitman, and said authorities found Smith through "good old-fashioned gumshoe police work," canvassing neighborhoods and identifying other family members who led them to Smith's brother's house where he was arrested. Officials representing the interagency task force that cooperated in cracking the case also said a controversial Microsoft GUID document identification technology did not play a significant role in leading to the arrest as had been widely speculated [41].

On December 9, 1999, Smith pled guilty to charges and in his plea document agreed that he caused $80 million in damages.

On Wednesday May 1, 2002, Smith was sentenced to 20 months in federal prison and ordered to pay a $5,000 fine. The sentence included 3 years of supervised release, during which no use of the Internet, computer networks, or bulletin boards was permitted unless authorized by the court. The judge also ordered Smith to complete 100 hours of community service, which will take advantage of his computer skills in a supervised atmosphere [42].

Two days after his federal sentencing, Smith was sentenced to 10 years in state prison. He also was fined $2,500 by state Superior Court Judge Lawrence M. Lawson. Lawson upheld a plea deal allowing the state term to end when the federal term does, meaning the actual 10-year state term would end in 20 months along with the federal sentence [43].

5.5 The World Trade Center bombing (1993) and Operation Oplan Bojinka

In February 1993, a minibus packed with 500 kg of explosives was driven into the car park beneath the World Trade Center Towers in New York. The resulting explosion killed 6 people, injured over 1,000, and caused in excess of $300 million damage [44]. Immediately after the explosion a large federal government task force was formed. Suspicion immediately turned to a terrorist attack. As the investigation progressed, a number of key suspects emerged. In particular U.S. investigators sought the whereabouts of three suspected international terrorists: Ramzi Ahmed Yousef, Abdul Hakim Murad, and Wali Khan Amin Shah.

On January 6, 1995, in a seemingly unrelated incident halfway across the world, a fire broke out in a suspected terrorist safe house at Room 603

Josefa Apartment, Quirino Avenue, Malate, Manilla [45]. Inside the apartment it is believed that Yousef and Murad were preparing a range of explosive devices. After the fire Yousef fled the Phillipines, while Murad returned to the apartment to remove all evidence of their activities. While at the apartment, police arrived and arrested Murad. During a search of the apartment police recovered a range of bomb-making equipment including chemicals, bomb-making instructions, and a range of timers. In addition, police also seized a laptop computer [46].

A joint FBI and Manilla police taskforce interrogated Murad, while the laptop computer was sent away for analysis. What law enforcement officials found on the laptop both amazed and tantalized investigators. The forensic analysis of the laptop established that it belonged to Yousef and contained disturbing information pertaining to past and future terrorist activities. The information recovered from the laptop included

> Airline flight schedules;

> Detailed plans pertaining to the bombing of a Philippine Airline flight from Cebu to Narita, Japan;

> Detailed plans to blow up 11 U.S. owned commercial airliners;

> Details of project Operation Oplan Bojinka.

The details of project Operation Oplan Bojinka were particularly disturbing. Essentially, Bojinka involved the hijacking of commercial airliners flying to the United States and using them to attack key targets within the United States. It is reported that during his interrogation, Murrad indicated that the key targets included the FBI headquarters and CIA headquarters [45].

While it may seem that the analysis of the laptop yielded a wealth of information pertaining to the terrorist activities, what is perhaps more vital from a computer forensic perspective is the information that was present on the laptop that was inaccessible to investigators. For not only was damming evidence located, but more disturbing was the existence of encrypted material. In his address to the U.S. Senate Select Committee on Intelligence, then-director of the FBI, Louis J. Freeh acknowledged the growing problems for law enforcement in keeping pace with technology, particularly when used by terrorists and criminals. Indeed, in his address Freeh stated, "Law enforcement remains in unanimous agreement that the widespread use of robust nonrecovery encryption ultimately will devastate our ability to fight crime and terrorism" [47]. In support of this claim, Freeh acknowledged the role that encryption played with respect to Yousef's laptop [47]. Given the quality and value of the information retrieved from Yousef's laptop, it must be tantalizing for investigators to wonder what other detailed information was contained within the encrypted material recovered from the laptop.

Indeed, the use of encryption presents computer forensic specialists with perhaps their greatest challenge. This challenge essentially involves two key issues. First, there is the ability of computer forensic specialists to detect the presence of encrypted material. Second, there is their ability to recover the plain-text information from its encrypted state. This is made all the more difficult given the widespread availability of strong encryption systems that permit both the secure storage and transmission of data across computer networks. This has a profound effect on the computer forensic process.

In February 1995, Yousef was finally arrested in Islamabad, Pakistan, and subsequently extradited to the United States, where he eventually stood trial for the bombing of a Philippines airline, seen by many as a test run for the planned attack on 11 U.S. airlines, as well as the 1993 bombing of the World Trade Center. For his role in the World Trade Center bombing, Yousef was convicted and sentenced to 240 years in jail.

5.6   Other cases

The following case studies are derived from the authors' own experiences, and have been chosen to highlight key issues or technologies that have arisen from the computer forensic process.

5.6.1   Testing computer forensics in court

While much has been written about computer forensic techniques and technologies, the ultimate test of their suitability in supporting an investigation, arises when the technology and techniques are tested in a court of law. Indeed, no matter how sophisticated the technology underlying the computer forensic process may be, if the final output is not acceptable as evidence, then the computer forensic process could be seen to have failed.

Not surprisingly it is becoming more common in both criminal and civil matters for electronic-based evidence, derived from a computer forensic process, to be challenged by lawyers who are themselves supported by computer forensic specialists. Indeed, the use of computer forensic specialists for both prosecution and defense adds a new dimension to the underlying computer forensic process. Not only can the technology be challenged, but also the expertise and methodology of the forensic specialist can now be put under the microscope. While there are numerous instances of computer forensic examinations being challenged, the following example highlights the difficulties that a computer forensic specialist may face when giving expert evidence before a court.

5.6.1.1 Misbehaving employees

A number of employees working within the manufacturing plant of a global automaker began to disseminate by way of their company's e-mail system a wide range of pornographic material. The material was in the form of pornographic jokes embedded in the text of e-mail messages, as well as pornographic pictures and movie files attached to individual e-mail messages. Initially the e-mails were forwarded internally amongst a small group of employees, however, over time the distribution list began to grow. Eventually, the pornographic e-mails were being forwarded to entities external to the organization. Naturally, this behavior was in breach of the company's computer usage policy, and, as was to be established later, known by the employees to be a breach of such a policy.

Eventually, a number of pornographic e-mail messages made their way to external parties who promptly complained to the organization about the nature and inappropriateness of the material. Consequently, the company's human resource department commenced an internal investigation. The investigation soon focused its attention on a small number of employees. During the course of the investigation, the company sought the services of a computer forensic specialist, who was engaged to take forensic images, after hours, of the hard disk drives of each of the employee's computer workstations. From the resulting forensic analysis a wealth of pornographic material and e-mail communications was recovered that supported the earlier complaints. As the investigation proceeded, a wealth of electronic evidence was gathered. This evidence included system logon information, e-mail messages recovered from the exchange server, as well as the results of the computer forensic analysis. After preparing their case, the company formally interviewed the employees. During the course of one interview, the particular employee maintained that while material may have been recovered from his computer system, the fact that everyone in the office had access to this computer and that everyone knew his ID and password made it difficult for the company to say that it was definitely him. Indeed, the employee alleged that his ID and password were recorded on a small sticky note attached to his computer monitor. Something that surprised the HR manager, who attended the night of the forensic data capture and who did not recall seeing such a note. Not surprisingly, after being interviewed and eventually taken back to his desk to retrieve his personal items, the employee pointed out to the HR manager a sticky note attached to his monitor containing his password and user ID. Following the interviews, all employees were dismissed on the grounds that they had breached the company's IT policy.

Appealing their dismissal before the Australian Industrial Relations Commission, the employees sought to test various aspects of the computer forensic evidence. In particular, the underlying methodology and technology used in the forensic analysis was tested. What follows is a brief summary of the key issues raised during the cross examination of the computer forensic specialist:

1. Counsel for the employees raised the issue of identity, in particular the ability of the computer forensic specialist to be able to say who was using the computer at the time of the alleged misbehavior. This being particularly relevant in one case where the employee is alleged to have had his userid and password recorded on a sticky note attached to his computer's monitor.

   The computer forensic specialist gave evidence that he was able to establish the logical identity of the user logged in to the network, but was unable to say who was actually physically sitting at the keyboard. With regard to the sticky note, the computer forensic expert gave evidence that the night he duplicated the computer system, there was no such note present. When asked how he could be so sure, the computer forensic specialist advised that it was part of his normal methodology to document and record the work area prior to commencing his examination, and in addition, a key goal of his preliminary reconnaissance of the work area was to look for any passwords that may be written down. None were found.

   As an interesting aside to the issue of the sticky note, the HR manager, on returning to the employees work area, removed the note and sent it away for forensic analysis. The results came back indicating that the handwriting on the note did not match the handwriting of the employee concerned—evidence that was later used to discredit the employees claims.

2. Counsel for the employees speculated whether the pornographic material recovered from the hard disk drives was in fact material

that had been resident on the computers hard disk drives prior to the employees being assigned their particular computers.

In response to this line of questioning, the full history of each computer was presented in evidence. Due to careful and meticulous records, maintained by the company, it could be shown that the computers were unique (i.e., new) to each employee, with no recorded difficulties or changes having taken place.

3. When evidence was given that the taking of a forensic image had duplicated the data from the employees' computers, counsel for the employees sought to ascertain how the imaging process worked and sought evidence that the forensic image obtained was in fact a true and accurate copy of the original.

   In response the computer forensic specialist detailed, under oath, how the imaging process for the particular forensic software operated and gave evidence as to how the inbuilt verification process could be used to ensure the authenticity of the duplicated data.

4. Exploring the forensic analysis process used in the examination of imaged data, counsel for the employees speculated that the results obtained were not derived from the duplicated data, but rather derived from previous, unrelated, forensic analysis conducted by the computer forensic specialist in other cases.

   In explanation, the computer forensic specialist detailed that the methodology employed in the forensic analysis process prevented such an occurrence for the following reasons:

   ‣ Each forensic image was stored as a series of discrete files.

   ‣ The image files were created on hard disk drives that had been independently sanitized (i.e., overwritten by a series of reads and writes of 1 and 0) and then newly formatted.

   ‣ That each image file was immediately backed up onto permanent read only (CD-ROM) media.

   ‣ The analysis was performed on either the discrete image files or an image restored to a freshly sanitized hard disk drive.

5. Throughout the course of cross examination, counsel for the employees sought to test the computer forensic specialist's knowledge with regard to the operations of the various forensic software applications used. In particular, counsel sought clarification and confirmation that the output of the forensic software was both accurate and reliable.

In response to this, the computer forensic specialist had to have a detailed understanding and knowledge of the operations of each piece of forensic software. In addition, the specialist had to give evidence as to the reliability of the software used (something that came with extensive use as well as independent testing).

While this example may not be overly technically exciting, it does highlight the very essence of computer forensics, that is, the end result may be challenged in a court.

Consequently, the success of any computer forensic evidence not only hinges on the strength of the technology employed, but also on the methodology employed during the forensic process as well as the expertise of the computer forensic specialist.

5.6.2   The case of the tender document

With the competitive nature of today's business environment, it should come as no surprise to learn that a number of individuals seek to cheat the system by exploiting either their position or taking advantage of an opportunity that may present itself without warning. In either situation, the exploitation will invariably leave an electronic trail that can, and usually is, their ultimate undoing. Whether it is e-mails or user created documents and spreadsheets, the resulting electronic trail provides investigators with a wealth of direct and indirect evidence. Getting at this trail is ultimately the responsibility of the computer forensic specialist, who through the use of advanced techniques must be able to delve deep into computer hard disk drives and network servers in an effort to identify and extract potentially incriminating evidence.

In one particular case that highlights the complexity of the computer forensic process that may need to be applied, an employee of a government organization was investigated for collusion with a tendering party during the tendering of a government contract. During the tendering process, a number of rival bidders submitted to the particular government department their respective tender documents. As each tender was submitted, it crossed the desk of the particular employee who was responsible for processing each document. Unknown to the head of the department, the employee had links with a rival bidder who had yet to submit their final tender proposal. Utilizing his own knowledge of the competing bidders, the employee began to disseminate details of the other bidders' tenders to his associates. Miraculously, hours before the deadline for all tender proposals to be submitted, the final bidder submitted their proposal. Not surprisingly, it exceeded all other bidders both in terms of quality of service, price, and timing.

Shortly after the tender process had closed, rumors began to reach the department head that the employee responsible for processing tender documents had links with one of the bidders. The department head subsequently notified his superiors who immediately commenced an investigation. As part of the investigation, the government department utilized the services of a computer forensic specialist.

One day during the working week, the investigators, in company with the computer forensic specialist, arrived at the government offices. Their intention was to search the employee's workspace and conduct an examination of both the employee's computer system and the department's e-mail servers. The computer forensic specialist took a forensically sound image of the employee's hard disk drive and conducted an examination of the e-mail server. All e-mail communications pertinent to the employee were identified and extracted to CD-ROM for future reference by the investigators. In addition, a number of data and e-mail server backup tapes were seized for the period relevant to the tender process.

Returning to a secure computer forensic laboratory, the painstaking process of analyzing the information began. The data and e-mail backup tapes yielded no additional information and were subsequently eliminated from the investigation. Attention then turned to the hard disk drive image of the employee's computer. A search of current files failed to find any information that linked the employee with any of the tendering parties. Even a search for compressed and encrypted files yielded no value. A search of deleted information also failed to identify deleted documents that could provide the vital nexus between the employee and the tendering party. A search of unallocated and residue space, however, turned up references to one particular tendering party. These references appeared to be a part of a much larger document.

While it was possible for the computer forensic expert to manually extract the relevant data from the unallocated space, the computer forensic specialist sought to recover complete instances of the document through the use of a technique known as signature analysis or recognition and this is discussed later. After applying this process, the computer forensic specialist was able to retrieve a number of instances of the tender document relating to the tendering party to whom it was alleged that he, the employee, had a relationship with. In fact not only was one complete copy recovered, but rather a number of copies of the tender document, in Microsoft Word format, were recovered in various stages of completion. This indicated that the rival tender document had at some point been prepared on the employee's computer. Because the entire document had been recovered, reference to the metadata contained within the header of each document provided information as to when the document was prepared and even when it was printed. Not surprisingly, when all the documents were placed in a logical sequence and the relevant properties examined, it became clear that the document was prepared prior to the closing date, with most activity taking place on weekends or after normal work hours, a profile that fitted the employees known work habits.

5.6.2.1. Signature analysis and recognition

Therefore, how is it that standard undelete techniques were unable to recover the deleted files, and yet signature analysis was able to recover the missing information. Standard undelete software utilizes the logical structure of the filing system to identify files that have been deleted, and to piece together the content of the file by rebuilding its structure by referencing data blocks that are not allocated to any current file. This approach is made possible by the fact that in most filing systems, when a file is deleted, its logical information, such as file name, creation date, and access date remains intact with the exception of some minor change that tells the operating system that the file has been deleted. In addition, the various data blocks or clusters assigned to the file generally retain their information even though they are marked as being free for future use.

On the other hand, signature analysis does not rely upon the logical structure associated with the mechanism by which files are stored on a hard disk drive. Rather, signature analysis relies on identifying potential files or file fragments by way of a unique signature. A significant number of files stored on a computer system comprise a relatively simplified structure: the file header, the file body, and the file footer. The file header contains information specific to the particular type of file. For example, in the case of a graphics file such as JPEG, the file header contains information regarding the horizontal and vertical resolution, horizontal and vertical pixel count, and units used for resolution. In addition, the file header contains a unique string, *magic number* or signature (Hex string: FF D8 FF E0). Figure 5.3 depicts the header of a common JPEG file in which the signature can be seen at offset 0. In addition to the signature, another identifier, ''JFIF'' can be found at offset 6.

In the case of Microsoft Word documents created with MS Office, as referred to in the earlier case study, the signature commonly used is D0 CF 11 E0 A1 B1 1A E1 00 00 00 00. This signature can be seen at offset 0 in Figure 5.4.

Offset	0	1	2	3	4	5	6	7	8	9	A	B	C	D	E	F	
00000000	FF	D8	FF	E0	00	10	4A	46	49	46	00	01	01	01	01	2C	ÿØÿà..JFIF.....,
00000010	01	2C	00	00	FF	DB	00	43	00	08	06	06	07	06	05	08	.,..ÿÛ.C........
00000020	07	07	07	09	09	08	0A	0C	14	0D	0C	0B	0B	0C	19	12	....
00000030	13	0F	14	1D	1A	1F	1E	1D	1A	1C	1C	20	24	2E	27	20	............$.'
00000040	22	2C	23	1C	1C	28	37	29	2C	30	31	34	34	34	1F	27	".#..(7),01444.'
00000050	39	3D	38	32	3C	2E	33	34	32	FF	DB	00	43	01	09	09	9=82<.342ÿÛ.C...
00000060	09	0C	0B	0C	18	0D	0D	18	32	21	1C	21	32	32	32	32	........2!.!2222
00000070	32	32	32	32	32	32	32	32	32	32	32	32	32	32	32	32	2222222222222222
00000080	32	32	32	32	32	32	32	32	32	32	32	32	32	32	32	32	2222222222222222
00000090	32	32	32	32	32	32	32	32	32	32	32	32	32	32	FF	C0	22222222222222ÿÀ
000000A0	00	11	08	01	93	02	81	03	01	22	00	02	11	01	03	11	...."...."....
000000B0	01	FF	C4	00	1F	00	00	01	05	01	01	01	01	01	01	00	.ÿÄ.............
000000C0	00	00	00	00	00	00	00	01	02	03	04	05	06	07	08	09	....
000000D0	0A	0B	FF	C4	00	B5	10	00	02	01	03	03	02	04	03	05	..ÿÄ.µ..........
000000E0	05	04	04	00	00	01	7D	01	02	03	00	04	11	05	12	21	......}........!
000000F0	31	41	06	13	51	61	07	22	71	14	32	81	91	A1	08	23	1A..Qa."q.2•'i.#
00000100	42	B1	C1	15	52	D1	F0	24	33	62	72	82	09	0A	16	17	B±Á.RÑð$3br,....
00000110	18	19	1A	25	26	27	28	29	2A	34	35	36	37	38	39	3A	...%&'()*456789:
00000120	43	44	45	46	47	48	49	4A	53	54	55	56	57	58	59	5A	CDEFGHIJSTUVWXYZ
00000130	63	64	65	66	67	68	69	6A	73	74	75	76	77	78	79	7A	cdefghijstuvwxyz
00000140	83	84	85	86	87	88	89	8A	92	93	94	95	96	97	98	99	ƒ„…†‡ˆ‰Š'""•—™
00000150	9A	A2	A3	A4	A5	A6	A7	A8	A9	AA	B2	B3	B4	B5	B6	B7	š¢£¤¥¦§¨©ª²³´µ¶·
00000160	B8	B9	BA	C2	C3	C4	C5	C6	C7	C8	C9	CA	D2	D3	D4	D5	¸¹ºÂÃÄÅÆÇÈÉÊÒÓÔÕ
00000170	D6	D7	D8	D9	DA	E1	E2	E3	E4	E5	E6	E7	E8	E9	EA	F1	Ö×ØÙÚáâãäåæçèéêñ

Figure 5.3 JPEG file header.

Offset	0	1	2	3	4	5	6	7	8	9	A	B	C	D	E	F	
00000000	D0	CF	11	E0	A1	B1	1A	E1	00	00	00	00	00	00	00	00	ÐÏ.à¡±.á........
00000010	00	00	00	00	00	00	00	00	3E	00	03	00	FE	FF	09	00	........>...þÿ..
00000020	06	00	00	00	00	00	00	00	00	00	00	00	03	00	00	00	....
00000030	4C	01	00	00	00	00	00	00	00	10	00	00	4E	01	00	00	L...........N...
00000040	01	00	00	00	FE	FF	FF	FF	00	00	00	00	49	01	00	00	....þÿÿÿ....I...
00000050	4A	01	00	00	4B	01	00	00	FF	FF	FF	FF	FF	FF	FF	FF	J..K...ÿÿÿÿÿÿÿÿ
00000060	FF	FF	FF	FF	FF	FF	FF	FF	FF	FF	FF	FF	FF	FF	FF	FF	ÿÿÿÿÿÿÿÿÿÿÿÿÿÿÿÿ
00000070	FF	FF	FF	FF	FF	FF	FF	FF	FF	FF	FF	FF	FF	FF	FF	FF	ÿÿÿÿÿÿÿÿÿÿÿÿÿÿÿÿ
00000080	FF	FF	FF	FF	FF	FF	FF	FF	FF	FF	FF	FF	FF	FF	FF	FF	ÿÿÿÿÿÿÿÿÿÿÿÿÿÿÿÿ
00000090	FF	FF	FF	FF	FF	FF	FF	FF	FF	FF	FF	FF	FF	FF	FF	FF	ÿÿÿÿÿÿÿÿÿÿÿÿÿÿÿÿ
000000A0	FF	FF	FF	FF	FF	FF	FF	FF	FF	FF	FF	FF	FF	FF	FF	FF	ÿÿÿÿÿÿÿÿÿÿÿÿÿÿÿÿ
000000B0	FF	FF	FF	FF	FF	FF	FF	FF	FF	FF	FF	FF	FF	FF	FF	FF	ÿÿÿÿÿÿÿÿÿÿÿÿÿÿÿÿ
000000C0	FF	FF	FF	FF	FF	FF	FF	FF	FF	FF	FF	FF	FF	FF	FF	FF	ÿÿÿÿÿÿÿÿÿÿÿÿÿÿÿÿ
000000D0	FF	FF	FF	FF	FF	FF	FF	FF	FF	FF	FF	FF	FF	FF	FF	FF	ÿÿÿÿÿÿÿÿÿÿÿÿÿÿÿÿ
000000E0	FF	FF	FF	FF	FF	FF	FF	FF	FF	FF	FF	FF	FF	FF	FF	FF	ÿÿÿÿÿÿÿÿÿÿÿÿÿÿÿÿ
000000F0	FF	FF	FF	FF	FF	FF	FF	FF	FF	FF	FF	FF	FF	FF	FF	FF	ÿÿÿÿÿÿÿÿÿÿÿÿÿÿÿÿ
00000100	FF	FF	FF	FF	FF	FF	FF	FF	FF	FF	FF	FF	FF	FF	FF	FF	ÿÿÿÿÿÿÿÿÿÿÿÿÿÿÿÿ
00000110	FF	FF	FF	FF	FF	FF	FF	FF	FF	FF	FF	FF	FF	FF	FF	FF	ÿÿÿÿÿÿÿÿÿÿÿÿÿÿÿÿ
00000120	FF	FF	FF	FF	FF	FF	FF	FF	FF	FF	FF	FF	FF	FF	FF	FF	ÿÿÿÿÿÿÿÿÿÿÿÿÿÿÿÿ
00000130	FF	FF	FF	FF	FF	FF	FF	FF	FF	FF	FF	FF	FF	FF	FF	FF	ÿÿÿÿÿÿÿÿÿÿÿÿÿÿÿÿ
00000140	FF	FF	FF	FF	FF	FF	FF	FF	FF	FF	FF	FF	FF	FF	FF	FF	ÿÿÿÿÿÿÿÿÿÿÿÿÿÿÿÿ
00000150	FF	FF	FF	FF	FF	FF	FF	FF	FF	FF	FF	FF	FF	FF	FF	FF	ÿÿÿÿÿÿÿÿÿÿÿÿÿÿÿÿ
00000160	FF	FF	FF	FF	FF	FF	FF	FF	FF	FF	FF	FF	FF	FF	FF	FF	ÿÿÿÿÿÿÿÿÿÿÿÿÿÿÿÿ
00000170	FF	FF	FF	FF	FF	FF	FF	FF	FF	FF	FF	FF	FF	FF	FF	FF	ÿÿÿÿÿÿÿÿÿÿÿÿÿÿÿÿ

Figure 5.4 MS Office document header.

The signature analysis approach ignores any logical filing structure and scans a block of data for any instance of the chosen signature, whereas standard undelete programs rely on the existing logical structure of the filing system to recover the deleted information. The block of data may be an image file (uncompressed), a swap file, or unallocated space extracted into a discrete file. When an instance of the signature is found, a block of data of a user defined size, is extracted into a discrete file. It is this discrete file that forms the recovered document.

Signature analysis is a relatively powerful technique that allows the computer forensic specialist to recover seemingly unrecoverable files. Despite its power, signature analysis does have some significant drawbacks, which include

> ▸ Signature analysis does not discriminate between current and deleted files. This is due to the fact that the process disregards the logical filing system structure.

> ▸ If the file is large and fragmented, the extraction of a block of data does not account for fragmented data, and as such may result in a high level of corruption for the recovered files.

> ▸ The relevant file properties (e.g., creation date, access date, and file name) are not recovered with the file data. This is particularly a problem if there is no metadata stored within the recovered file header.

> ▸ The usability of any extracted data is reliant upon the alignment applied during the extraction process. In the case of an image file the alignment data is extracted in 512-byte blocks. In the case of a swap file, the data is not necessarily stored in a 512-byte alignment; consequently, a misalignment can result in a corrupted file.

5.6.2.2 File signatures used to verify file types

Another common use of file signatures in computer forensics is as a means of verifying the accuracy of each resident file's extension (Table 5.1). Essentially, all resident files are scanned and their file signatures are matched against known signatures for their particular file type. If a file has been incorrectly labeled as a data file (e.g., extension equates to *dat*), but contains an MS Office signature, then a red flag is set for the computer forensic specialist to follow up. As a result of this process, files that have been deliberately disguised by means of changes to their file extensions can be correctly identified.

Table 5.1 Examples of File Signatures

File Type	Extension	Hexadecimal Signature
Bitmap graphics file	bmp	42 4D 00 00 00 00
Cursor	cur	00 00 02 00 01 00 20 20
Excel 2	xls	09 00 04 00
Excel 3–4	xls	09 00 06 00 00 00 10 00
GIF graphics file	gif	47 49 46 38 37 61
JPEG graphics file	jpg	FF D8 FF E0
Lotus 1-2-3	wk	00 00 02 00
MS Access	mdb	00 01 00 00 53 74 61
MS Office	doc/xls/mdb	D0 CF 11 E0 A1 B1 1A E1 00 00 00 00
Microsoft Word 1	doc	9B A5
Microsoft Word 2	doc	DB A5
Microsoft Word 4–5	doc	31 BE 00 00
Netscape 3 mail	snm	23 20 4E 65 74 73 63 61 70 65
Outlook mail file	pst	21 42 44 4E
Outlook personal address file	pab	21 42 44 4E
Paintbrush graphics file	pcx	0A 05
PowerPoint 3.0	ppt	ED DE
Printer spool file	emf	01 00 00 00 58 00 00 00
WAV sound file	wav	52 49 46 46 00 00 00 00 57 41 56 45
Microsoft Windows metafile	wmf	D7 CD C6 9A 00 00
Word Perfect document file	doc	FF 57 50 43

References

[1] Machi, M., "Philadelphia, PA," http://www.americanmafia.com/Cities/Philadelphia.html, visited May 2002.

[2] Sammut, H., "The Untouchable Keyboard," http://www.alert.com.mt/webresources_articles_detail.asp?i=63, visited Feb. 2001.

[3] Schwartz, J., "F.B.I. Use of New Technology to Gather Evidence Challenged," *The New York Times*, July 30, 2001.

[4] Anastasia, G., "Big Brother and the Bookie: How the Feds Turned Top-Secret Spy Technology Against the Son of a Mafia Don, and Made a Low-Level Wiseguy into a Poster Boy for the Fourth Amendment," http://www.motherjones.com/magazine/JF02/mafia.html, Mother Jones, Jan./Feb. 2002.

[5] Rasch, M., "Break the Scarfo Silence," *Business Week Online*, http://www.businessweek.com/technology/content/sep2001/tc2001094_186.htm, Sept. 4 2001.

[6] United States v. Scarfo, Criminal No. 00–404 (D.N.J.), Nicholas H. Politan, District Judge, United States District Court, District Of New Jersey, "Court Order Denying Motion to Suppress Evidence, December 26, 2001," http://www.epic.org/crypto/scarfo/opinion.html, visited May 2002.

[7]   United States v. Scarfo, Criminal No. 00–404 (D.N.J.), Vicent C. Scoca, Esquire, Norris E. Gelman, Esquire, "Defense Motion to Suppress Evidence Seized by the Government Through the Use of a Keystroke Recorder, June 2001," http://www2.epic.org/crypto/scarfo/def_supp_mot.pdf, visited May 2002.

[8]   United States v. Scarfo, Criminal No. 00–404 (D.N.J.), Nicholas H. Politan, District Judge, United States District Court, District of New Jersey, "Court Order Requiring Government Submission of Report "Detailing How the Key Logger Device Functions", August 7, 2001," http://www2.epic.org/crypto/scarfo/order_8_7_01.pdf, visited May 2002.

[9]   Norton, J. E., "The Exclusionary Rule Reconsidered: Restoring The Status Quo Ante," Wake Forest Law Review, April 4, 1998, http://www.law.wfu.edu/lawreview/V33/docs/33-2-2.pdf, visited May 2002.

[10]   United States v. Scarfo, Criminal No. 00–404 (D.N.J.), Ronald D. Wigler, Assistant U.S. Attorney, "Government's Request for Modification of Court Order Pursuant to Classified Information Procedures Act, Aug. 23, 2001," http://www.epic.org/crypto/scarfo/gov_cipa_motion.pdf, visited May 2002.

[11]   United States v. Scarfo, Criminal No. 00–404 (D.N.J.), Nicholas H. Politan, District Judge, United States District Court, District Of New Jersey, "Protective Court Order Granting Motion Denying Discovery of Classified Information, Oct. 2, 2001," http://www.epic.org/crypto/scarfo/gov_ex_parte_mot.pdf, visited May 2002.

[12]   Anastasia, G., "Scarfo's High-Tech Case Ends with Plea," The Philadelphia Inquirer, March 1, 2002, http://www.philly.com/mld/inquirer/news/local/2769774.htm, visited May 2002.

[13]   Electronic Privacy Information Center, "Court Hears Arguments on Use of Secret Keystroke Monitor." EPIC Alert Volume 8.14, July 31, 2001, http://www.epic.org/alert/EPIC_Alert_8.14.html, visited May 2002.

[14]   Computer Emergency Response Team-Coordination Center, "CERT[r] Advisory CA-1992-19 Keystroke Logging Banner," Dec. 7, 1992, http://www.cert.org/advisories/CA-1992-19.html, visited May 2002.

[15]   Amecisco, Inc., "Keylogger.com Hardware and Software Products," http://www.amecisco.com/products.htm, visited May 2002.

[16]   Sysinternals Freeware, Utilities—Windows NT/2K, http://www.sysinternals.com/ntw2k/utilities.shtml, visited May 2002.

[17]   Raytown Corporation, "Flagship Software Application—Anti-Keylogger[TM]," http://www.anti-keyloggers.com/products.html, visited May 2002.

[18]   Spycop, http://www.spycop.com/index.html, visited May 2002.

[19]   Keyghost Ltd., http://www.keyghost.com/products.htm, visited April 2002.

[20]   Joe, "Joe's Unlock Thinkpad[TM]," http://www.ja.olm.net/unlock, visited May 2002.

[21] Parsons, M., "Behind the Scenes Account," *The Department of the Navy Information Technology Magazine*, March 29, 1996, http://www.chips.navy.mil/archives/96_jul/file3.htm, visited May 2002.

[22] DOE-CIAC's Network Intrusion Detector (NID) Distribution Site, http://ciac.llnl.gov/cstc/nid/nid.html, visited April 2002.

[23] "Federal Cybersleuthers Armed with First Ever Computer Wiretap Order Net International Hacker Charged with Illegally Entering Harvard and U.S. Military Computers," *U.S. Department of Justice Press Release*, March 29, 1996, http://www.usdoj.gov/opa/pr/1996/March96/146.txt, visited May 2002.

[24] "Argentine Computer Hacker Pleads Guilty," *Miami Herald*, May 20, 1998, http://www.cmcnyls.edu/bulletins/AGCHPGPF.HTM, visited April 2002.

[25] Collie, B., "Data Monitoring on Victim Computer Networks," *Proc. 3rd Int. Law Enforcement Conference on Computer Evidence*, Feb. 14, 1996.

[26] Heberlein, T. L., "A Network Security Monitor," *Proc. 1990 IEEE Symp. on Res. in Security and Privacy*, Oakland, CA, May 1990.

[27] "NID Introduction," http://ciac.llnl.gov/cstc/nid/intro.html, visited April 2002.

[28] Mansur, D., "Making Information Safe," *Science and Technology Review*, Jan./Feb. 1998, http://www.llnl.gov/str/Mansur.html, visited April 2002.

[29] NFR Intrusion Management System, NFR Security, http://www.nfr.com/products/NID/, visited April 2002.

[30] U.S. Department of Energy Computer Incident Advisory Capability (CIAC), "WinWord Macro Viruses," *CIAC Information Bulletin G-10A*, Feb. 8, 1996, http://ciac.llnl.gov/ciac/bulletins/g-10a.shtml, visited April 2002.

[31] CERT[(r)] Advisory CA-1999-04, "Melissa Macro Virus," CERT-CC, March 27, 1999, http://www.cert.org/advisories/CA-1999-04.html, visited May 2002.

[32] McMillan, R., "Lessons Learned from Loving Melissa," Australian Computer Emergency Response Team, July 5, 2000, http://www.auscert.org.au/Information/Auscert_info/Papers/loving-melissa.html, visited May 2002.

[33] F-Secure Virus Descriptions, "Melissa," http://www.f-secure.com/v-descs/melissa.shtml, visited May 2002.

[34] Lemos, R., "Melissa Creator May Be Uncovered," *ZDNet News*, March 29, 1999, http://zdnet.com.com/2100-11-514170.html?legacy=zdnn, visited April 2002.

[35] "Officials: AOL Info Cracked Virus Case," *ZDNet News*, April 1, 1999, http://zdnet.com.com/2100-11-514222.html?legacy=zdnn, visited April 2002.

[36] Reiter, L., and J. Louderback, "Melissa Trail Leads to 'ex' Virus Writer," *ZDTV*, http://zdnet.com.com/2100-11-514175.html?legacy=zdnn, visited April 2002.

[37] Reiter, L., and J. Louderback, "FBI Hunting for Virus Writer," *ZDTV*, March 31, 1999, http://zdnet.com.com/2100-11-514175.html?legacy=zdnn, visited April 2002.

[38] Takahashi, D., and D. Starkman, "It's Getting Harder to Hide in Cyberspace," *The Wall Street Journal Online*, April 4, 1999, http://zdnet.com.com/2100-11-514239.html?legacy=zdnn, visited April 2002.

[39] Shankland, S., "Melissa Suspect Arrested in New Jersey," *CNET News.com*, April 2, 1999, http://news.com.com/2100-1023-223857.html?tag=rn, visited April 2002.

[40] Silverstrini, E., "Prosecutor: 'Melissa' Creator Confessed," *USA Today Tech Report*, August 25, 1999, http://www.usatoday.com/life/cyber/tech/ctf941.htm, visited May 2002.

[41] "Man Charged With Unleashing 'Melissa' Computer Virus," *CNN*, April 2, 1999, http://www.cnn.com/TECH/computing/9904/02/melissa.arrest.03/, visited May 2002.

[42] "Creator of Melissa Computer Virus Sentenced to 20 Months in Federal Prison," *U.S. Department of Justice Press Release*, May 1, 2000, http://www.usdoj.gov/criminal/cybercrime/melissaSent.htm, visited May 2002.

[43] "Melissa Author Sentenced and Fined Again," *Sophos Virus Info*, May 14, 2002, http://www.sophos.com/virusinfo/articles/melissa3.html, visited May 2002.

[44] "Chilling Reminder of 1993 World Trade Centre Attack", *Australian Broadcasting Corporation (ABC) News Online*, Sept. 2001, http://www.abc.com.au.

[45] "PNP Exec Links US Attacks to 1995 Manilla Terrorist Plot," *ABS-CBN News*, Sept. 13, 2001, http://www.abs-cbnnews.com.

[46] "Plane Terror Suspects Convicted on All Counts," *CNN Interactive*, Sept. 5, 2001, http://www.cnn.com/US/9609/05/terror.trial/.

[47] Congressional Statement, FBI. "Statement for the Record of Louis J. Freeh, Director FBI on Threats to U.S. National Security, Before the Senate Select Committee on Intelligence, Washington DC," FBI Press Room, Jan. 28, 1998.

CHAPTER

6

Contents

6.1 Intrusion detection, computer forensics, and information warfare

6.2 Intrusion detection systems

6.3 Analyzing computer intrusions

6.4 Network security

6.5 Intrusion forensics

6.6 Future directions for IDS and intrusion forensics

References

Intrusion Detection and Intrusion Forensics

6.1 Intrusion detection, computer forensics, and information warfare

Intrusion detection (ID) takes over where preventative security fails. It is designed to identify, and in some cases limit the occurrence and effect of intrusions into computer systems accessible via a wider computer network, typically the Internet. Such intrusions are difficult to prevent by traditional access control techniques as they typically circumvent access control by exploiting flaws in the implementation or design of the systems being attacked or intruded upon and possibly of the systems being used to mount the intrusion. The Computer Emergency Response Team Co-ordination Center (CERT-CC) at Carnegie Mellon University provides the following definition for intrusion [1]:

> Any intentional event where an intruder gains access that compromises the confidentiality, integrity, or availability of computers, networks, or the data residing on them.

Chapter 1 presented definitions for the terms *computer forensics* and *intrusion forensics* (IF), and subsequent chapters have dealt largely with computer forensics of the traditional sort, that is, computer forensics as it relates to the imaging

257

and analysis of individual computer systems in a manner which meets evidentiary requirements. In this chapter, we focus on IF and its relationship to ID.

ID relies in the case of host-based ID (see later) upon event information similar to that used in computer forensics. The two are nonetheless otherwise quite different. ID relates very specifically to the detection of computer intrusions namely, activities which are unauthorized or unintended by those properly managing the computer or network and which may be harmful in the ways described in the above definition. In particular, ID differs from traditional computer forensics in three important dimensions:

> *Its domain:* Computer forensics deals with any activity and the computer evidence which serves to confirm or deny the occurrence or nature of the activity.

> *Its time frame:* Computer forensics is typically concerned with post hoc investigation.

> *The type of event information scrutinized:* Computer forensics typically makes use of noncomputer related information as well as computer-related information in order to arrive at a conclusion.

ID uses standard computer logs and computer audit trails, gathered as a matter of routine by host computers, and/or information gathered at communication routers and switches, in order to detect and identify intrusions into a computer system. Successful detection of intrusions is based either upon recognition of a known exploitation of a known vulnerability or upon recognition of unusual or anomalous behavior patterns or a combination of the two. The former is referred to as signature or misuse ID, the latter as anomalous behavior (or simply anomaly) ID. It is clear that anomalous behavior per se cannot in general be equated with intrusive behavior and this is indeed at the heart of some of the challenges facing the development of successful anomaly intrusion detection systems (IDS). The differences between anomaly-based and signature-based ID are examined in detail in Section 6.2 later.

Computer forensics on the other hand is concerned with the analysis of any information stored by, transmitted by or derived from a computer system in order to reason post hoc about the validity of hypotheses which attempt to explain the circumstances or cause of an activity under investigation. Computer forensics therefore, covers a much broader scope of activities than does ID, the scope of the latter being limited to reasoning about activities or detecting activities relating to computer system abuse.

Before proceeding to explore the detailed nature of IDS, it is useful to present definitions for the following terms as they are used throughout this chapter:

1. *Intrusion:* This refers to "any intentional event where an intruder gains access that compromises the confidentiality, integrity, or availability of computers, networks, or the data residing on them" [1].

2. *Intrusion detection:* This typically refers simply to what is achieved by an IDS—an integrated software package, be it signature- or anomaly-based, without human intervention.

3. *Intrusion forensics:*

   a. The recovery and analysis of information from a computer or computer system or computer network suspected of having been compromised or accessed in an unauthorized fashion; information which includes host-based data and will typically also include communications traffic and payload data;

   b. Analysis of information from other sources, for example call records, PDA flash memory contents, and business organizational structure;

   c. *Purpose:* To allow investigators to reason about the validity of hypotheses attempting to explain the circumstances and cause of the activity under investigation, and possibly provide evidence to support litigation either criminal or civil.

4. *Network forensics (NF):*

   a. The recovery and analysis of information from one or more computer networks suspected of having been compromised or accessed in an unauthorized fashion, information which includes communications traffic and payload data (and may also include host-based connection information);

   b. *Purpose:* To allow investigators to reason about the validity of hypotheses attempting to explain the circumstances and cause of the activity under investigation, and possibly provide evidence to support litigation either criminal or civil.

5. *Incident Response (IR)/incident handling:* This is related to IF and NF, but the emphasis in IR is more on how best to protect a computer or computer network against possible damage rather than on

elucidation of the precise nature of the activity as in IF and NF. IF is concerned primarily with achieving an outcome which identifies the circumstances and agents behind an intrusion (in some cases leading to prosecution), whereas IR is typically concerned primarily with safeguarding a computer or computer network against damage which may include deciding whether or not to shut the system down in the face of an attack.

The CERT-CC defines IR/incident handling as follows: "actions taken to protect and restore the normal operating condition of computers and the information stored in them when an adverse event occurs; involves contingency planning and contingency response."

6. *Incident/security incident:* When using the term *incident* in what follows, we are referring to what is more properly known as a *security incident* and for which the CERT-CC [2] provides the following general definitions by way of illustration:

Any real or suspected adverse event in relation to the security of computer systems or computer networks

or

The act of violating an explicit or implied security policy.

We note that regarding CERT-CC's published statistics, the term incident means something rather different, there it means an incident-type, so that, for instance, all reports related to the Melissa virus are in that context counted as one incident not tens of thousands. We also note that the above definitions include both *insider* and *outsider* activity.

If anomaly-based, an IDS uses a typically statistical profile of activity to decide whether the occurrence of a particular computer event or event pattern is normal or anomalous. If normal, then the activity is considered to be harmless and thus legitimate, if anomalous then it is potentially unauthorized and harmful.

If signature-based, the IDS attempts to match a sequence of observed events with a known pattern of events which is characteristic of an attack of some sort, such as a buffer overflow attack and password guessing. If there is no match to be found with any of the known attack event patterns (signatures), then the activity under scrutiny is considered to be harmless and thus legitimate. A solely signature-based IDS cannot recognize a new or previously unknown type of attack; an anomaly-based IDS on the other hand cannot categorically identify a sequence of events as an attack.

In both cases, the IDS reaches a conclusion based upon computer data that is more informative than what is allowed by the legal definition of what constitutes computer evidence. This is because the latter is constrained by formal rules of law that might require the exclusion of information that might nonetheless be relevant and informative. By contrast, an IDS can exploit any and all such informations including knowledge of the target operating system and architecture. That is, any relevant computer system information is grist for the IDS mill, and is used by the IDS as a basis for its decision-making. As a result, whether signature- or anomaly-based, the operation of an IDS is based upon three working assumptions that are typically technically sound but which do not necessarily or even typically stand up in court. These assumptions are

1. That disruptive or malicious user behavior can be distinguished from innocent actions (in the case of anomaly IDS, this includes the assumption that user profiles of *normal activity* are uncompromised).

2. That it is possible from a knowledge of system behavior, that is from a knowledge of the state of a system and its previous states to identify the disruptive or malicious user behavior referred to in the first assumption.

3. That the event logs upon which IDS decisions are based are tamper-free, that is they are a true record of system events and state.

All three assumptions can present a problem to the routine use of ID records as computer forensic evidence. The first assumption can be a particular problem. For a start, the distinction between disruptive or malicious user behavior and innocent actions is made typically on the basis of the *effect* of the behavior rather than the intent of the user and as such this in itself may present a potential difficulty for the courts. Secondly, to draw an analogy with the acceptance of DNA evidence after 1984, the missing elements in the case of using IDS data seem to be instrumental calibration and a protocol for testing. Calibration establishes the known level of false positives and negatives tolerated by the instrument. A test protocol shows that in a specific test, the IDS's test result is measuring what it claims to be measuring. At present, although some IDS perform a type of calibration by operating in learning mode until a satisfactory low level of false results is reached, this is not an objective model of background system behavior. Likewise, many anomaly-based IDS continually update their model of what is normal. What is missing is an objective platform-independent model.

Having said all that, the information used in a computer forensic investigation will likewise potentially include system logs and audit trails namely, the exact same or at least similar data as is used by host-based IDS (HIDS). In the case of computer forensic investigation however, there is the additional requirement (reminiscent of the now rescinded Section 69 of the 1984 Police and Criminal Evidence Act of the United Kingdom) that the above three assumptions be explicitly justified on a case-by-case basis. While ID is not subjected to the same burden of evidentiary requirements that constrains computer forensics, there are many computer forensic cases in which information gathered by an IDS is useful and feeds into the forensic process but the usefulness of such information as evidence in court may be uncertain. It is worth noting in this context that, while there is an increasing emphasis on the evaluation of IDS with regard to their effectiveness in identifying intrusions, IDS evaluation has not extended to a systematic evaluation of their success either individually or collectively as collectors of evidentiary material. Indeed, this will be difficult until there is a body of case history involving IDS evidence given the other variables affecting the legal outcome of any specific court case. Sommer [3] has addressed the issue and has proposed a set of desirable properties of IDS to be targeted by IDS designers. These properties include the following, inter alia:

‣ A focus on the admissibility of gathered evidence and its presentation in the court;

‣ The use of multiple corroborating streams of evidence;

‣ Transparency of the working of the IDS to allow explanation in the court;

‣ Retention of raw logs for possible use in the court when needing to demonstrate the validity of IDS conclusions.

An interesting irony noted in the NATO report *Intrusion Detection: Generics and State-of-the-Art* [4] is that early ID prevents or mitigates an attack or intrusion before it achieves its full impact, but this in turn works against obtaining the detailed information that may be necessary for evidence purposes. There is a parallel here with property offences—the sounding of a perimeter motion-sense burglar alarm may prevent a visitor from attempting a break-in, in which case the ID has been effective, but since detection occurred at an early stage, no crime has been committed and no evidence of an intended break-in exists.

There is a clearly identifiable area of overlap between ID and computer forensics that relates to the forensic investigation of attacks on a computer system. IF is the investigation of activities which use or access, or which threaten to use or access, a computer or computer system or computer network in a manner that is illegal or unintended by the proper administrator of the computer and its network. It differs from ID, which typically refers simply to what is achieved by an IDS—an integrated software package, be it signature- or anomaly-based, without human intervention. IF does not necessarily imply a need to satisfy the evidentiary requirements of a court of law although it may in some circumstances lead to prosecution. In that case evidentiary requirements will be relevant and the computer and other evidence upon which the case is based will need to meet those requirements.

There is another domain in which IF comes into play. Information and its management lie at the heart not only of commerce but also of national infrastructures and national security. As a result, national defense policies now incorporate information warfare strategies for the purposes of both offence and defense, something explored in more detail in Chapter 3 under Section 3.8.1. Information warfare (or information operations as it is also known) used for offence is subject to the security and ID techniques of one's opponent, while defensive information warfare is reliant upon defensive security and detection measures. Computer forensic techniques come into their own in this case both for post hoc analysis of captured information and captured computer systems as well as for the investigation of intrusions. As a result, the concepts of computer security, ID, and computer forensics are inextricably linked when it comes to information warfare. Their intersection in the context of defensive information warfare and in the wider context of protecting computer systems at large is captured in the term IF. (For a comprehensive tour de force on the subject of information warfare—its history, its practice and the implications it holds for the security of IT systems—see Denning [5].)

The remainder of this chapter is organized as follows. In Section 6.2, we examine how IDS has developed and is developing, its importance to computer systems of the future rivaling the impact in the 1990s of firewall technology. Section 6.3 provides an account of computer intrusion analysis with a particular focus on event log analysis and alert correlation techniques, including time-lining. Section 6.4 examines the nature of modern network security and its reliance on the concept of "defense in depth" and the crucial roles that network and host monitoring and logs, and vulnerability analysis play. Section 6.5 returns to the topic of IF, discussing the relationship between IR and intrusion analysis and examines two intrusion situations. Section 6.6 concludes with an examination of future research and development in the areas of ID and IF.

6.2 Intrusion detection systems

6.2.1 The evolution of IDS

The raison d'etre for IDS is the realization that firewalls and access control on their own do not provide an adequate defense against attack. As discussed later in the chapter, this inadequacy has resulted in widespread adoption of "defense in depth" strategies which achieve the security required by an organization through a graduated reliance, firstly upon prevention (based typically upon identity based access control), secondly upon detection (for instance, through IDS) and then upon reaction (e.g., shutting down a network connection where this is appropriate). Prevention is intended to filter out the majority of potential attacks; detection then identifies those relatively few attacks that have not been prevented while reaction applies in relatively fewer cases still.

The reality is that modern software is typically designed and implemented without security in mind [6] and this is exacerbated by continuing increases in system complexity and the resulting overall number of bugs and security vulnerabilities. As a consequence, preventative security even with the deployment of firewall and filtering technology remains an unrealistic pipedream. For instance, even with the problem of buffer overflow now well understood, it is still a common programming error [7] which continues to leave systems vulnerable in a way which makes preventative security at the system level very difficult. A large part of the overall insecurity of the Internet arises because the Internet was designed to be an open system with sharing, not security, as its prime objective. This has been exacerbated by the use of underlying network protocols formulated to support sharing rather than security at a time when the security of systems was not a priority [8].

IDS has evolved significantly over the past two decades since its inception in the early 1980s. For an account of this evolution see [9]. The simple IDS of the early days was based either upon the use of simple rule–based logic to detect very specific patterns of intrusive behavior or upon historical activity profiles to confirm legitimate behavior. In contrast, we now have IDS which use data-mining and machine-learning techniques for the dynamic compilation of new attack signatures and which allow for quite general expressions of what may constitute intrusive behavior (see MADAM ID in Section 6.2.2). Other modern IDS may use a mixture of sophisticated statistical and forecasting techniques to predict what is legitimate activity. In short, we now have a variety of quite sophisticated approaches to the ID problem which represents a considerable advance on the early systems.

IDS is still, however, commonly characterized according to a two-fold taxonomy involving detection method on the one hand and placement on the other, and this taxonomy has stood the test of time notwithstanding the important advances indicated immediately above. Detection method relates to signature-based versus anomaly-based IDS, while placement relates essentially to host-based (HIDS) versus network-based IDS (NIDS).

In the case of signature-based IDS, the IDS identifies known intrusive behavior. Other behavior is by default not reported, that is, these systems provide a default outcome of *permit* (or *legal*). Such IDS rely on statically or dynamically compiled libraries of attack signatures or attack signature types which are matched, either post hoc or in real-time, to a candidate activity trace. Anomaly-based systems on the other hand identify deviations from normal behavior. They use a model of normal behavior and report any activity which does not conform with the normal behavior, thus providing a default outcome of *deny* (or *illegal*). Signature-based IDS constrain the range of attacks that can possibly be detected in return for an acceptable error rate in detection, while anomaly-based IDS cover the entire attack space, at the cost of increased error rates. The latter is due to the fundamental problem that an anomaly is not necessarily an attack, something alluded to earlier. It is indeed often not an attack, and this leads to the major failing of many such systems, that is, the problem of a high false positive or *false alert* rate. In addition, it can be difficult to identify exactly why an activity is anomalous and whether such an activity is truly threatening. Alternatively, if the IDS is unable to do that, it is then up to the security administrator to do so. This leaves many administrators at a loss as to the correct procedure to follow when an anomaly detector gives an alert, and leads to administrators ignoring or simply switching off the anomaly IDS. Signature-based IDS too can suffer from this problem of false alerts—though not nearly to the same extent as do anomaly-based systems—mainly due to incomplete signatures.

It is partly as a result of this problem of false alerts that we have seen a recent focus on the evaluation of IDS effectiveness. The higher the proportion of genuine intrusions detected (true positives) and the lower the number of false alerts generated (false positives), the more successful or effective is the IDS. Unfortunately, the generally poor performance of IDS in this regard has been a major obstacle to their overall success and deployment. The U.S. Defense Advanced Research Projects Agency (DARPA) and the U.S. Air Force Research Laboratory (AFRL) have been major sponsors of IDS research for many years and have recently focused attention on evaluation of IDS as a means of targeting this. The off-line evaluations performed at Lawrence Livermore Laboratories in 1998 [10] and 1999 [11] are the most comprehensive evaluations of IDS performance to date. The evaluations measured the performance of various IDS in the face of a combination of detailed attack simulations and a composite of real and synthetic attack data. A lucid account of the methodology is provided in [12]. The work has been criticized [9] for a number of reasons including failure to validate the background data and the data analysis methodology used. Nonetheless, an important and unmistakable conclusion in both evaluations was that, tested systems fell considerably short of DARPA expectations. The off-line evaluation methodology used is applicable only to passive IDS, and cannot be applied to those IDS which interact with their environment either by query or by modifying the network configuration. Durst et al. [13] have reported separately on the more limited but real-time evaluation of four IDS carried out by the AFRL in 1998, three of the IDS evaluated being DARPA funded developments, while the fourth was a "government off-the-shelf" system. This evaluation while more limited, addressed the issue of accommodating real-time interactions between the IDS and other system components. Researchers at the MIT Lincoln Laboratory, which had performed the 1998 and 1999 off-line evaluations and developed the off-line evaluation tools used in those evaluations, have since then reported on some significant extensions to those evaluations including the development of the LARIAT real-time testbed for IDS [12] which they report is more easily configurable than the environment used in [13] and which likewise accommodates real-time interaction between IDS and other components. Other noteworthy research in the related area of IDS test data generation is taking place at Carnegie Mellon University [14]. A recent paper by NFR's Marcus Ranum on IDS performance entitled "Experiences Benchmarking Intrusion Detection Systems" details some of the pitfalls to be avoided in measuring IDS performance and how to focus on measuring those aspects of performance which are meaningful [15].

Signature-based systems can suffer from performance problems of a different sort when high bandwidth networks are involved and there are many signatures to be checked against. As a result, IDS performance in the sense of computational performance (rather than in the sense of effectiveness as discussed earlier) has likewise become an important issue and new products are being developed to perform load-balancing for IDS on high-speed networks, to lessen the requirement for one IDS to look at all the traffic on a given network segment [16].

IDS differ also according to the domain in which they operate. HIDS use host logs and host event records to provide a record of current activity which can then be analyzed with either signature- or anomaly-based logic. NIDS on the other hand promiscuously capture packet headers and packet content (payload) to identify activity at the network level, typically using packet header and packet payload information as their working data which can then be processed using either signature- or anomaly-based logic, typically the former. There is a recent and growing interest in application-based IDS and it is clear that such IDS can in certain circumstances present some advantages arising from application specificity. It is properly regarded as a specialized form of HIDS.

The above evolution of IDS has taken place at a time when there is an increase in system complexity and thus the number of bugs and security vulnerabilities overall. This has led to the recognition that IDS need to interoperate with one another and possibly also with the other components of network security architecture in order to provide comprehensive coverage of potentially intrusive behavior. In addition, as alluded to earlier, individual IDS have their own areas of specialization and effectiveness. As a result, and with the broad range of systems and networks in use today, many environments use multiple IDS which brings with it associated challenges with regard both to the management of heterogenous IDS and other components and the analysis of the data gathered by those IDS and other components.

6.2.2   IDS in practice

IDS have historically been categorized as network-, host-, anomaly- or misuse- (signature-) based. This simple categorization is, however, no longer adequate. IDS can also be distributed or centralized, can be passive or reactive, can be application-specific or general-purpose, can focus on real-time or after-the-event analysis. The five IDS described later are not intended to be exemplifiers of these or other various categories, but are presented as an indication of how major trends have developed.

The NIST Special Publication on IDS by Rebecca Bace and Peter Mell [17] lists the following output produced by IDS, a list useful to keep in mind in reading through these next sections:

Almost all IDS will output a small summary line about each detected attack. This summary line typically contains the information fields shown below.

1.   Time/date;
2.   Sensor IP address;
3.   Vendor specific attack name;
4.   Standard attack name (if one exists);
5.   Source and destination IP address;

6. Source and destination port numbers;

7. Network protocol used by attack.

Other more general information is also often provided; information such as a textual description of the attack, identification of the software attacked, information that identifies the patches required to fix the vulnerability, and advisories regarding the attack.

6.2.2.1 A lightweight network intrusion detection system

The free utility Snort (http://www.snort.org) is a packet sniffer and logger that can act as a *lightweight* IDS. It has been favorably compared to similar NIDS commercial products. A sniffer is software that exploits the promiscuous mode of operation of a local area network (LAN) adaptor in order to capture all packets on the network thereby—for good or bad—having access to all packet information. Such sniffer capability provides the basis for NIDS, which can then protect networks by monitoring network activity. On the other hand, if installed surreptitiously by a hacker, a sniffer can capture and provide the hacker with, for instance, account and password information. See [18] for an account of sniffer technology.

Snort is a misuse detector identifying attacks by analyzing packet contents or sequences of packets. Hence, Snort requires a library of known attack signatures that needs to be updated, but its increasing popularity means that signatures are rapidly available, and additional rules can be written for local requirements.

Three main components make up Snort's architecture: the packet decoder, the detection engine and the alerting/logging subsystem. In the decoder, Snort prioritizes speed, flagging packet data for immediate analysis by the detection engine, where signatures are embodied in rule chains. A chain header contains common attributes used in the entire ruleset for an attack (e.g., same source and destination port) while chain option lists attached to each chain header present different optional rule subsets dependent on that type of attack [19]. Hence, if a match is found on the chain header, a bundle of attack-specific rules will be invoked using, for example, the following options:

1. Pattern matching on packet contents;

2. Check IP header's time-to-live (TTL) field;

3. Match on packet payload size;

4. Offset payload search start point.

Snort's logging feature permits packets to be logged in both readable and tcpdump binary format. Alerts generated by the detection engine via rule invocation are sent to nominated consoles.

6.2.2.2 A distributed anomaly-based intrusion detection system

In contrast to Snort, SRI's Event Monitoring Enabling Responses to Anomalous Live Disturbances (EMERALD) [20] provides a rare example of distributed IDS offering both signature- and anomaly-based detection, and also real-time response. A research product, EMERALD's objectives were aimed to build on lessons learned from previous SRI research in IDS, notably the Intrusion Detection Expert System (IDES) and Network IDES (NIDES) projects. Specific problems to be overcome included scalability in a distributed environment, and the integration of audit data from different sources and at different levels of abstraction. Particularly relevant for anomaly detection was the observation from NIDES that profiling functionality, such as clients and applications, was more successful than user profiling.

EMERALD's principal focus is its resource objects, attached to targets, routers or gateways, and services such as FTP or HTTP, which are frequently the subject of malicious attacks. Its architecture consists of three main components—profiler engines, signature engines, and resolver—and is designed to permit communication with external data sources and alternative analysis platforms. Independently configurable monitors watch over the resource objects, working in conjunction when a coordinated attack is suspected. Monitors can operate both passively (log observation or packet sniffing) and actively, probing for extra evidence. The monitors can either act as standalone IDS, analyzing local activity in real-time, but they may also communicate the results to higher level monitors which correlate these for an overall picture, for example, of a possible coordinated attack across a domain.

Whether freestanding or working in conjunction, each EMERALD monitor can act both as a signature detector and as a statistical anomaly detector working with usage profiles derived from event logs and/or third party security product outputs. The brains of each EMERALD monitor is in its resolver, accepting alerts from other communicating monitors and reconciling these to invoke an appropriate counter response, for example, more detailed monitoring. However, all monitors also contain both a signature analysis subsystem and an anomaly detection subsystem to feed the resolver. The usage profiles for anomaly detection are based on four types of variables (categorical, continuous, traffic intensity, and event distribution) extracted from the target object's event history. With the signature subsystem, security administrators can develop a detection ruleset customized for each resource object, for example, for buffer overflow or SYNflood symptoms.

The EMERALD project offers a library of downloadable resource objects for analyzing various specific service and network elements, and has in more recent times concentrated on enhancing distributed monitoring capabilities. These include *multiperspective analysis*—analysis of the same target from different perspectives—and *commonality detection*, where local results may fall short of triggering a local response, but the combination of these with overall monitoring results may indicate a global response.

6.2.2.3 A hybrid network and host intrusion detection system

Unlike Snort and EMERALD, "Dragon" from Enterasys Networks [21] is a commercial product. Developed for Linux or UNIX systems, it is a centralized IDS offering signature analysis on a host-based system and a network monitor, and focused on forensic evidence collection for later analysis. The three main components of "Dragon" are a network IDS, Dragon Sensor, that monitors traffic for signs of attack (as for Snort); a HIDS, Dragon Squire, which tracks system file usage and firewall activity; and a reporting system, Dragon Server, which collates, summarizes and presents information from network and host monitoring, as well as securely communicating with and managing Dragon NIDS and HIDS.

Dragon Sensor is a packet sniffer and monitor equipped with a misuse signature library, which can also be customized. Its response repertoire extends to alerts, packet dropping, session termination, and detailed logging for postanalysis. Hence, its strength lies in its capacity for collecting sufficient evidence to determine what kind of attack was mounted, and whether this was successful or not. Multiple sensors can extend cover to a large network, although they do not interoperate.

The host IDS Dragon Squire can also be run either freestanding or with sensors. It monitors system files for tampering, performs system log analysis and can also monitor firewall and router events, reportedly with minimal impact on system performance. System file change indicators include MD5 hash values, file permission alterations, deletions, and file trunca-tions. The HIDS is also equipped with a signature library for checking for suspect events like file transfers, reboots, and failed log-ins. Suspicious application or service activity can also post messages for signature matching. Dragon Server's role is NIDS and HIDS management, event correlation, alert customizing and report presentation, although unlike EMERALD it reportedly does no automatic correlation.

As a *lightweight* IDS, "Dragon" requires substantial participation from its administrators. However, it has useful additional features such as out-of-order or fragmented packet reassembly to frustrate IDS-avoidance techniques and is scalable. Successfully laboratory-tested with some 100 sensors, the server is claimed to manage 50 HIDS as well. Its focus is not so much on frustrating or preventing attacks as on to establish whether and in what manner an attack has taken place. Dragon server's ability to manage operations securely makes the product a useful adjunct in forensic computing, as does its feature of replaying stored attacks for postmortem analysis.

6.2.2.4 A pattern-matching network intrusion detection system

The research prototype graph-based IDS (GrIDS) resists categorization as either signature- or anomaly-based. The anomalies that GrIDS looks for are at a higher level than characteristic packet sequences or statistical profiles, and either network- or host-based. In the sense that large-scale network behavior in a hierarchical organization can have characteristic patterns, it is anomaly-based, and GrIDS aims to detect suspicious variations from the norm. However, it is based on the principle that a particular kind of attack will present as distinctive variations from this norm, that is, these attacks also have signatures.

The GrIDS project set out to address some of the more sophisticated attacks emerging in recent times. As examples, [22] mentions

> ‣ Multistep coordinated attacks, where stages of the attack are spread across several sessions, for example, simultaneous sweeps from several sources;

> ‣ Diversion attacks, where one user makes a highly visible attack to obscure the real attack from a collaborator;

> ‣ Worm attacks, where the pattern of infection can only be detected by looking at several hosts.

GrIDS is a meta-IDS, integrating output from other IDS and network monitors to maintain and interpret network and host activity. It views the organization as a hierarchy, and its working models, known as *activity graphs*, are reduced at each level in the hierarchy, to aggregations of activity at the next level down. The aggregation mechanism was designed to be scalable and the prototype offers dynamic reconfiguration to assist deployment in large organizations.

The basic idea of GrIDS is that activities among hosts will be related if they occur closely in time. To illustrate, [22] describes a simple graph-building example of worm tracking: successfully infecting the first host, it spreads to two others and the connection is reported to a GrIDS module, which creates a timestamped graph. If no further activity is detected, the graph lapses, but if further spread occurs within a time-limit, the graph is updated. Eventually, the size and/or spread of the activity graph will trigger a threshold value identifying it as suspected worm. However, graphs are more complex than this implies. Particular kinds of activity are registered as distinct graph spaces, each holding several graphs, which GrIDS attempts to consolidate into a recognizable pattern. Graph nodes represent hosts, and the edges are traffic. These edges can be furnished with attributes representing information passed up from lower level graphs or from other IDS products. Users can define the attribute contents, for example, to represent domain-specific security policies. Edge attributes are then used by the graph space's autonomous ruleset to correlate activities across the graphs in the space. New information is presented to each ruleset to see whether it is applicable in that particular space. If so, it will be added to a graph, cause a new graph to be started, or consolidate several graphs into one.

Within an organization, each department appears as a separate GrIDS module tracking activity within the department. Interdepartmental traffic is passed up to the next level to be aggregated as a reduced graph, where the nodes are entire departments, resulting in a much lower processing load. Thus, at each level, GrIDS can detect suspicious activity at lower levels from the size and depth of the reduced graph, depending on how strongly the edge attributes permit it to reach a conclusion. Therefore, GrIDS depends on the quality of information supplied from other monitors or IDS as well as on how deftly the attributes have been crafted. Its benefit is in its ability to gain leverage off other IDS products that do not interoperate, and also to exploit the hierarchical structure of a typical organization.

6.2.2.5 MADAM ID—a data mining approach to intrusion detection

Java Agents for Meta-Learning (JAM) is an agent-based data mining system designed to be highly scalable and extensible. It is based on the concept of metalearning which is the discovery of knowledge by combining higher-level concepts extracted from a number of data sets. JAM has been successfully applied to intrusion detection and credit card fraud detection.

We provide here an overview of the JAM-based MADAM IDS framework. More detailed information about JAM and MADAM ID can be found in [23–25]. The 1998 DARPA intrusion detection evaluation found MADAM ID to be one of the best performing IDS evaluated. The name MADAM ID stands for *Mining Audit Data for Automated Models for Intrusion Detection*.

MADAM ID uses data mining techniques to process system audit records in order to develop (extract) rules from that data which intrinsically define an ID model or ruleset, that is, a model which is expressed in terms of rules that characterize misuse signatures. The rules are automatically derived from patterns in the audit logs discovered through the data mining techniques employed. This addresses some of the weaknesses of current IDS. For example, signature-based IDS typically require that the signatures (rules) be manually derived and added into an IDS; formulating the signatures is a laborious and skilled task, successful automation of this task provides significant benefits of economy and accuracy. Similar considerations apply with regard to anomaly-based IDS, which rely upon the intuition and knowledge of the system designer to select the important statistical measures of audit data needed to characterize anomalous activity.

MADAM ID identifies frequent patterns in the connection records derived from the audit logs and then expresses those patterns as association rules and frequent episodes. Association rules are used to express correlations and are expressed in the form "XY, confidence, significance," where confidence is a measure of the strength of the association between X and Y and significance is a measure of the number of occurrences of X in the data (see also Chapter 7). For example, "trn rec.humor, 0.3, 0.1" means: if the user invokes trn, 30% of the time he/she is reading the newsgroup rec.humor and that trn makes up 10% of activities. The frequent episodes algorithm is used to characterize sequential patterns in audit logs, such as a sequence of events that occurs frequently within some time frame. A detailed description of the association rules and frequent episodes algorithms can be found in [25]. A classification program, for example, RIPPER is then used to learn or produce the detection model, also known as the base classifier, by training on test data that has already been manually classified.

It has been shown that short sequences of system calls are very consistent in normal invocation of commands, with significant variation being displayed if an exploit such as a buffer overflow attack is attempted [23]. Moderate success was reported with the use of the JAM rule induction agent, RIPPER, to generate a set of association rules from *sendmail* system call-data. A sliding window with a fixed period of time was used and if more than a certain level of abnormal system calls occurred in that time, then the trace being analyzed was considered to represent an intrusion. An important requirement was that the rule induction agent needs a training data set that includes the complete range of normal behaviors, otherwise normal behaviors unseen by the learning agent will be classified as abnormal.

The frequent episodes algorithm was tested using network traffic data in the form of *tcpdump* data that was first preprocessed to extract the meaningful data, producing a series of connection-level records suitable for data mining. Each connection record contained data such as the start time, duration, and statistics of the connection along with the hosts, ports, protocol used and a flag (used to indicate connection or termination errors). Temporal statistical features were added to the connection records to add a measure of frequency of connections of this type in the past n seconds and the average duration and bytes for connections of this type in the past n seconds. RIPPER was then used to build the classifiers for detecting network intrusions. It was reported that this approach proved highly successful for detecting some types of network intrusions.

The effectiveness of an IDS can be increased by combining a number of ID models. This can be achieved by the use of metalearning agents that use the base classifiers generated by several IDs models to determine if an intrusion is in progress. MADAM ID provides the framework for this type of distributed IDS.

A particularly important advantage of this architecture is that performance of the IDS can be improved by using a number of lightweight subsystems distributed over several hosts, rather than relying on a single system. A further improvement in performance can be gained by using metalearning agents to process a large data set in parallel to find different anomalies in the same data set (e.g., one agent may examine network traffic for denial of service (DOS) attacks, while another may examine the traffic for port scans). An additional advantage of the distributed metalearning agent architecture is that the vulnerability of the IDS is reduced, as the attacker is faced with the need to subvert several subsystems in order to avoid detection, rather than a single monolithic system.

6.2.3 IDS interoperability and correlation

The core issue in the increasingly important objective of IDS interoperability is the ability to allow data from different, heterogeneous IDS to be pooled. Some proposals for achieving IDS interoperability specify the standardized semantics of that communication, while others provide standardized data formats and protocols.

The common intrusion detection format (CIDF) was one of the first serious efforts to attempt to address the challenge of IDS interoperability. It did so by specifying the language used by IDS to communicate, and by defining the various roles of the IDS components needing to communicate. These roles are *Event Generators, Event Analyzers, Event Databases,* and *Event Response Units.* Event Generators produce data streams for Event Analyzers to interrogate in order to detect intrusions. Event Analyzers can then use Event Databases for the storage of events originating from generators, as well as of events generated by Analyzers. Event Response Units can be used to perform operations to react to intrusions, such as killing processes or dropping connections. Event Analyzers may be used in multiple levels, with lower level Analyzers acting as Event Generators to those above. CIDF also allows the systems to complement or reinforce events notified by other generators. A distinguishing feature of CIDF was the specification of S-Expressions, which describe events in a Lisp-like syntax.

The Intrusion Detection Working Group (IDWG) has emerged as a result of the work done by CIDF. However, it took a slightly different approach than did CIDF: Instead of specifying interfaces, it specifies protocols and formats. The development so far has focused on two objectives, a data exchange format and a protocol for communication. The intrusion detection message exchange format (IDMEF) builds on the experience of CIDF in S-Expressions, but given the increased deployment of XML for specifying protocols across the Internet, uses XML for expressing message formats. The IDMEF data type definition (DTD) is currently in version 1.0, and caters to a broad range of applications in an extensible and easy-to-use manner. Specified at the same time as IDMEF, a protocol for communication was proposed, called the Intrusion Alert Protocol (IAP). This has recently been superseded by the Intrusion Detection eXchange Protocol (IDXP) based on the Blocks Extensible Exchange Protocol (BEEP).

IDS interoperability and the facility for data aggregation, which that provides in turn enables alert correlation which is the discovery and identification of relationships between alerts. The objective of alert correlation is to relate successive alerts from the one sensor, in order to identify multistep attack scenarios and to relate alerts emanating from different IDS in order to identify attack scenarios that cannot be identified by a standard individual IDS. If the appropriate IDS control and management interface is in place, this may then also allow dynamic reconfiguration of individual hosts and sensors and their IDS for early attack detection, with the opportunity for preemptive action against the source of attacks, such as disconnection (which may alert the intruder) or filtering (which is less likely to do so). Correlation between alerts may occur either in real-time when alerts are generated, or may be performed off-line sometime afterwards and may occur at different levels. A simple form of correlation is aggregation and selection of aggregate groups from within the alert database. An example of this could be a query on a database for all alerts within a given period that had related to a particular target host.

An aspect of IDS that is sometimes overlooked is their potential to produce *alert storms,* overloading a sensor and resulting in possible denial of service, analogous to *event storms* produced by network management systems (NMS). NMS alleviate this problem with correlation and aggregation of data, and this can also be of use in the IDS domain. Correlation and aggregation is most effective on a centralized server where all the alerts possible can be used for correlation, and where aggregation can be performed on the greatest amount of alerts possible.

There have been a number of significant recent developments in this area of IDS interoperability and management, some ad hoc, others with a clear research purpose in mind. One of the most significant has been the development of Snort (described previously) which is the most widely used open source NIDS currently available and which has several useful tools to aid its interface and usability. Importantly, for this and future work, this includes availability of XML and IDMEF output plug-ins. Another major advantage that Snort has over other free NIDS is its capability to log to a database, something used by many Snort monitors as an effective way to store Snort alerts. Both Demarc and Analysis Console for Intrusion Detection (ACID) use this alert store to provide quite advanced analysis tools for Snort, comparable to those offered in commercial systems.

6.3   *Analyzing computer intrusions*

In investigating a possible computer intrusion, investigators will sift through large amounts of log data from the targeted system and/or network of computers, possibly having in mind also the evidential nature of such information for subsequent court proceedings. A simple analysis may examine event records contained wholly within the one event log, while more sophisticated analyses will examine event records across several event logs. The intention of such analysis is typically to confirm or deny the occurrence of suspicious activities that is, activities which constitute successful or unsuccessful attempts to make illegal use of a system, or which may facilitate or are intended to facilitate the illegal use of a system. For example, in response to a simple compromise of a user account, the analysis may consist of as little as a simple search of an event log to locate any record which indicates recent access to the password file—legal or illegal—by any user. This in itself involves no across-event analysis, that is, no interevent correlation. Alternatively, in the investigation of large-scale network intrusion, the analysis may be far more sophisticated and involve analysis of event records across a range of computer or network event logs, as well as external information such as transaction logs from utilities, such as phone companies and communication services.

In any case, the analyst aims essentially to determine the who, what, when, and where of suspicious or illegal activities:

1. *Who:* This relates to attribution: who is responsible for the activity (this can be difficult to determine as the intention of the attacker is often to disguise his identity by masquerading as a legitimate user).

2. *What:* This relates to impact determination and motive: what has the attacker done, for example, files accessed and the mode of access (read, write, or execute), the network traffic generated, the network nodes targeted, and the user accounts attacked (again, an attacker will often attempt to disguise what they have done or possibly even delete entire logs or individual log records in order to hide their activities).

3. *When:* This relates to the time of each event in order that an accurate picture can be drawn of the sequence of activities employed by the attacker.

4. *Where:* This relates to identifying the location or identity of the computer from where the attacker operated (once again, an attacker will often attempt to falsify his apparent computer address, that is, his/her IP address, and operate through intermediary computers—giving rise to the terms *connection laundering* and *stepping stone*—in order to frustrate this).

The why and the how too are important, these will shed light on how best to mitigate against such activity in future. These relate to

‣ *Why:* Is the intrusion to do with hacking, political motives, revenge, financial fraud, other criminal activities?

‣ *How:* What were the tools employed, the vulnerabilities exploited? Can the scenario be reconstructed?

As discussed in Chapter 2, one of the crucial items of information that appears in event log records is the time at which a logged event took place, the so-called time attribute of the event—the when. This is important as a computer intrusion or attack consists typically of a specific sequence of events—an attack scenario—and knowing the sequence or order in which the events occurred is important in identifying the nature of the attack and hence the who, what, and where are referred to above. The general objective of event sequencing and event correlation is to identify the presence of temporal and other relationships between events in order to assess whether an attack has occurred or is occurring by matching a set of related events against known attack scenarios.

6.3.1   Event log analysis

Most operating systems have some sort of facility for recording event data. For example, UNIX systems can record log-ins and log-outs, user command histories, root access events, and ftp logs. Some operating systems can potentially collect large streams of security data. For example, Microsoft Windows 2000 has the capability to collect log-on activity, the files that users access, the programs that users run, and the operations that administrators perform. Unfortunately, skilled attackers have many ways of avoiding detection, namely disabling the collection of, or deleting or hiding this log evidence, producing large quantities of noise events to purposely fill up the log data files prior to penetration, and modifying the computer operating system. However, in many cases, the attacker may not cover all of his/her tracks and remnants of log activity may still be found in the system.

While Microsoft Windows 2000/XP has a comprehensive functionality for logging security-related events, this needs to be explicitly enabled to be effective. There exist many different types of events included in three default logs: application logs, system logs, and security logs. Examples of event types are log-on failure, opening an object, deleting an object and so on. However, some events have cryptic descriptions and many are relatively unimportant in the context of security needs. Furthermore, this capture of security-related events provides very little in the way of real-time monitoring or notification of suspicious activities, analyzing of logs, fusing, and correlating of networked computer logs.

There exist some commercial tools that perform some basic log analysis functions. For example, LANguard Security Event Log Monitor, LANSELM, from GFI [26] is a network-wide event log monitor that retrieves event logs from networked NT/2000 servers and workstations and alerts the administrator of possible intrusions. LANSELM provides the following functionality:

1. Real-time monitoring and notification (via e-mail);

2. Archiving all security events from different machines in a single database;

3. Central archiving of events for reporting and backup;

4. Clarification of some event descriptions to concise explanations and suggestions for action;

5. Removing noise events that make up a large ratio of all security events.

The system security administrator can define categorization rules in LANSELM that detect certain events, for example, events that occur at certain hours, events that arise on high security computers, network log-ons to workstations, and the deletion of logs.

However, no commercial software is yet available to undertake more in-depth analysis of event logs, such as fusing or correlating event logs from multiple computers. The ability to identify a set of interesting log events could be a useful function. A pattern of repeated yet unusual events could be extracted from the event logs by using data mining tools.

For example, using event data from a middle manager's workstation Linux *wtmp* log file (see Table 6.1) together with a concept hierarchy to generalize the log event data attributes, we obtain an association rule of the form (see Chapter 7 for more details about concept hierarchies and association rules):

$$(\textbf{StaffType} = \text{contractor}) \wedge (\textbf{Console} = \text{ttyp})$$
$$\implies (\textbf{DayOfWeek} = \text{weekend}) \wedge (\textbf{TimeOfDay} = \text{morning})$$
$$\wedge (\textbf{Duration} = \text{fewMinutes})$$

which indicates that a contractor, who is logging-in onto the manager's workstation via the ttyp console port on weekend mornings, is logged-on for a few minutes only. This may raise some suspicion as we could infer that the contractor is illegally accessing the manager's files on his/her local drive(s) outside work hours, or that the contractor is using the manager's workstation for impersonating the manager and employing the manager's

Table 6.1 Time Slice from a Linux *wtmp* Log File, Listing Past and Current Logins (User Names and Domain Names Have Been Modified To Preserve Privacy)

User	Console Source	Day/Date	Time	Duration
gwaihir	ttyp1 tosca2.braves.com.au	Wed Apr 18	14:17–23:09	08:51
legolas	ttyp1 tosca3.braves.com.au	Wed Apr 18	11:27–12:21	00:54
legolas	tosca3:0	Wed Apr 18	11:27–12:56	01:29
saruman	ftp 187.219.47.157	Wed Apr 18	10:42–10:58	00:15
shelob	ftp 187.219.47.170	Tue Apr 17	11:33–11:34	00:01
shelob	ttyp2 tosca4.braves.com.au	Tue Apr 17	10:32–10:34	00:01
shelob	ttyp2 187.219.47.170	Tue Apr 17	09:25–09:30	00:05
saruman	ttyp1 verdi.braves.com.au	Tue Apr 17	10:24–10:44	00:20
saruman	ftp 187.219.47.157	Mon Apr 16	16:11–17:12	01:01
saruman	ftp 187.219.47.157	Mon Apr 16	11:11–11:41	00:30
saruman	ftp 187.219.47.157	Mon Apr 16	10:05–10:11	00:06
contractor	ttyp1 tosca1.braves.com.au	Sun Apr 15	09:56–10:07	00:11
sauron	ttyp1 cpe-61-9-19.isp1.net	Wed Apr 11	08:19–08:32	00:13
elron	ttyp1 cai-56k-089.isp2.net	Wed Apr 11	07:14–07:49	00:34
sauron	ttyp1 cpe-61-9-18.isp1.net	Wed Apr 11	03:03–03:08	00:04
sauron	ttyp2 cpe-61-9-17.isp1.net	Tue Apr 10	23:44–00:10	00:26
celeborn	ttyp1 213.134.110.85	Tue Apr 10	23:04–01:15	02:11
boromir	ftp 1cust212-22.isp3.net	Tue Apr 10	21:17–21:19	00:02
boromir	ttyp2 1cust212-18.isp3.net	Tue Apr 10	21:11–23:06	01:54
boromir	ttyp1 1cust212-09.isp3.net	Tue Apr 10	21:07–22:41	01:34
saruman	ftp 187.219.47.157	Tue Apr 10	15:00–15:15	00:15

access rights to log into other servers on the company's LAN. More complex rules could also be derived, such as

$$(\textbf{ObjectType} = \text{files}) \wedge (\textbf{AccessType} = \text{write}) \wedge (\textbf{StaffType} = \text{contractor})$$
$$\Longrightarrow (\textbf{TimeOfDayRange} = [T1, T2]) \wedge (\textbf{LoginStatus} = \text{failure})$$

indicating that certain files were written to by the contractor staff person between times T1 and T2 after one or more unsuccessful log-ins.

6.3.2 Time-lining

Time is a critical component in any criminal investigation. Time-lining is concerned with ordering events of interest in time, with the intention of obtaining an overview of the sequence of events and an insight into cause and effect relationships and motive. The technique is sufficiently important and applicable across forensics in general and computer forensics in particular that we address it in the book in several different chapters (in Chapters 2 through 4 and in this chapter).

The prosecution of hackers generally involves piecing together a large number of items of evidence that have different time attributes. Such evidence includes telephone call records, people movements, sounds, forensic analysis of weapons or accessories, and witness testimonies. A major problem for the prosecution and defense is organizing, time-sequencing, correlating, and presenting all this information. When a criminal investigation involves a computer, or a network of computers, the complexity of the time-sequence and correlation phases increases significantly due to potential explosion in the amount of additional time-related information or evidence contributing to the investigation. The additional information can come from computers and servers distributed over different geographical sites; such as from local and networked peripheral devices, from network transit nodes (e.g., routers), from back-up media, and from network data. Such information can be sourced from computers running different operating systems (which may interpret time in different ways) in different countries covering multiple time-zones. Evidence with a time attribute can, for example, include the following:

1. Log-on and log-off events;

2. Internet browsing site location and content and chat room activities;

3. Application usage (file transfer and software usage);

4. E-mail communication events;

5. File usage events (e.g., creating, modifying, and accessing documents);

6. Router log events.

Techniques that simplify the process of accurately identifying, correlating and establishing the sequence of evidence timestamps and aid the investigator visualize these timestamps are therefore required. Time-lining is one simple technique. A time line is a one-dimensional time-ordered linear display of the evidence timestamps. It marshals time-related events into a chronological sequence where one or more evidence type can be simultaneously displayed on the time line.

Time-lining is useful, for example, for

‣ Identifying time-related patterns (e.g., periodic events, timestamp fingerprints such as operating systems file Modification-Access-Creation (MAC) timestamps);

- Cross correlating different time sequences;

- Correlating these sequences with other physical evidence.

We note in passing that time-lining is helpful in broadly identifying the timing sequence of evidence and its relationship to other evidence but, as a tool by itself, will not automatically undertake any correlation or extract any cause-effect dependencies. This, currently, is still left to the investigator to complete.

Time-lining forensic tools are available in Encase (Guidance Software) and CFIT® (DSTO). The CFIT® Timelining tool (see Chapter 2 for a brief description of the CFIT® system) displays all time-related events (e.g., file timestamps and log events) onto a one-dimensional time-line, together with event type information. Selecting appropriate event icons on the time-line provides information relating to each event type and its contents. The investigator can thus quickly identify unusual patterns and zoom in on time ranges of interest.

6.3.2.1 Event time frames

While a time-line is a convenient way of displaying time-related evidence, a time-line has no inherent understanding of time offsets such as those created by differences in time-zones nor of the time difference between *real* or *wall-clock* time and what the computer thinks it is. As a result, errors in the interpretation of the time line could occur, thereby making the ordering of events by time difficult, if not impossible. To overcome such errors, time referencing and adjustment features need to be incorporated into the forensic analysis procedure that allows the investigator to deal with different time references that might be found both within and across the various case data sets.

Leaving aside for the moment, considerations of *informal* time (e.g., as in the e-mail message "meeting at 3:00 P.M. tomorrow"), there are essentially two separate issues affecting time information or timestamps:

- That of time zones or relative time;

- That of clock synchronization.

If time-lining is being carried out within a single time-zone, then the first is not an issue but clock synchronization is. That is, in comparing the event timestamps of event records from different computers or different sources, there is the need to ensure that the clocks of the two computers or sources are synchronized with each other and with *real* or *wall-clock* time and were so at the time the events were recorded. This applies even in cases where there is only a single computer system being investigated or seized, since there is still the need to relate computer time (i.e., the timestamps of the events recorded on that computer system) to *real* or *wall-clock* time. It is in this situation that the action of noting computer time at the time that media, files or system information are being seized or imaged (Chapter 2) assumes importance. This applies equally to the start of an on-line investigation. Noting the time in this way will provide the basis for subsequently allowing the necessary adjustments to be made to the various timestamps when carrying out the time-lining to ensure that these adjusted times are comparable to each other and with *real* or *wall-clock* time.

When there are several time zones involved, there is then the additional need to ensure that the timestamps are adjusted to a common time zone for purposes of comparability unless they have already been so adjusted. Typically this means adjusting timestamps to what used to be known as Greenwich Mean Time (GMT), now known as Universal Coordinated Time (UTC) (see later). To achieve this, the time-lining and analysis logic needs to have access to time-zone information for all timestamps (or to know that all timestamps are UTC + 0) and to incorporate that knowledge when comparing and processing timestamp information. In addition, there is *informal* time which appears in free text (e.g., as in an e-mail message "meeting at 3:00 P.M. tomorrow") rather than as part of system-recorded timestamps, and this too must be assessed and adjusted to the appropriate time scale, probably UTC. To summarize, case data may be sourced from different machines with possibly different operating systems, from different time zones, and may be subject to time zone variations also. For example, an e-mail has multiple time event information—file timestamps, send timestamps, receive timestamps, transit route timestamps, and possibly time-related content events (e.g., "meeting at 3:00 P.M. tomorrow"). Each one of the related timestamps may have to be treated differently.

The above discussion focuses on time as determined by geographical location on the Earth's surface and time readings whether from computers or clocks which provide a local time frame. Each geographical location is located in some defined time zone and subject to local time zone variations (e.g., daylight savings, special events, such as the 2000 Sydney Olympic Games). This is what people generally refer to as the *real, civil* or *wall-clock* time.

We have in addition another issue regarding time and synchronization. Noting the time displayed by a computer at the start of a computer forensic investigation is intended to negate any potential problems that would otherwise arise from the computer time being out of synchronization with real-time or wall time. Nonetheless, a well-administered computer system will ensure that the *system clock* is regularly synchronized with an external source. Most computers have a clock chip on the motherboard, which is powered by a battery and uses low-power CMOS technology for storage, so that the clock still keeps time when the computer is powered off. This is what is often referred to as the *BIOS clock*. This clock can usually be set and read through the BIOS set up just after power up. Most computer systems also have what may be called a *system clock*. This is normally set to be the same as the *BIOS clock* at computer boot time, but over a period of time may not necessarily be the same as the *BIOS clock* (due to clock drift and deliberate changes). Neither the *BIOS clock* nor the *system clock* can provide accurate measures of time over the long term. However, the Network Time Protocol (NTP) [27] enables Internet clients to synchronize regularly with UTC which is the commonly used worldwide time reference and is the basis for the worldwide system of civil time replacing the previously used GMT. UTC time is kept by time laboratories around the world and is determined in terms of an atomic transition of the element cesium under specific conditions. Both Microsoft Windows 2000/XP and UNIX-like systems maintain time zone information so that synchronization with UTC is accomplished "under the hood" and the appropriate time zone adjustment made when displaying or recording local time. Microsoft Windows 2000/XP provides a Date/Time icon under the Control Panel to set the time zone while UNIX-like systems provide the *-u* switch to the date command to display UTC time and maintain time zone information in a system shell variable. NIST operates an Internet Time Service from Boulder, Colorado, using various timing protocols including NTP and using multiple servers around the United states. It also provides a related time setting service—Automated Computer Time Service—(ACTS) for computers connected to the Internet via modem that can provide accuracy of setting to within 10 ms [28]. For a comprehensive account of synchronization of clocks, time and Internet timeservers and some of the above topics see [29].

Readers are referred to an article by Steve Romig entitled "Correlating Log File Entries" [30] which addresses some of the detailed steps required in time-lining and the manual analysis of log file entries in order to achieve event synchronization. Romig comments also on *event lag*, which occurs irrespective of synchronized time. Event lag occurs as a result of the fact that events are not atomic and the related logged event timestamps may be generated at different points in the lifetime of an event depending upon the nature of the event and the operating system environment. As a result, time-lining or the correlation of events based upon their timestamps needs to take into account not only time-zone information and synchronization but also that the ordering of events across logs may be inexact due to this phenomenon.

For the purpose of forensic analysis, the CFIT® forensic clock definition software (see Chapter 2) uses a set of clock times, or time references, to associate time with different case data and to achieve uniform time synchronization across all time-related events. For a given case data event, CFIT® computes its equivalent UTC date–time value, thereby ensuring that the event has a uniform (UTC) date–time reference.

With some forensic tools in CFIT®, an investigator is also able to override a time reference for a particular event contained within previously referenced case data. For example, in the hard disk analyzer forensic tool, it is possible to associate one time reference instance with a hard disk partition, and another time reference instance with the various time events of a particular file found within that partition, if that file were found to be an e-mail.

6.4 Network security

6.4.1 Defense in depth

The traditional view of computer security is that it needs to provide confidentiality, integrity, and availability—typically abbreviated to CIA. The sea change that has occurred over the past decade in computer usage namely, from using and securing individual computers to using and securing the Internet—has not changed these goals; it has, however, changed the means by which the goals can be achieved. This subsection and the next explore how security technologies have evolved in the age of the Internet to continue to meet the goals of CIA in this changed environment.

The most difficult security threat to counter in the Internet environment is the existence of security flaws in the wide range of software that is employed. These flaws arise as a result of errors in the software engineering processes involved in software development, design or implementation or both. Analysis of security flaws shows us the kinds of security flaws that exist in our systems and which parts of systems and the system lifecycle are vulnerable in this regard. Security flaw analysis is useful in informing security assessors and security analysts of the vulnerabilities of systems.

A landmark paper "A Taxonomy of Computer Program Security Flaws" [31] published in 1994 provides a comprehensive review of actual software security flaws and identifies the kinds of errors in the software development process that was their genesis. The article includes a list of seven categories of operating system security flaws:

1. Incomplete parameter validation;
2. Inconsistent parameter validation;
3. Implicit sharing of privileged/confidential data;
4. Asynchronous validation/inadequate serialization;
5. Inadequate identification/authentication/authorization;
6. Violable prohibition/limit;
7. Exploitable logic error.

A similar analysis of security faults in the context of the UNIX operating system from Purdue University [32] resulted in a simple, two-tier flaw taxonomy consisting of

1. *Coding faults:*

    a. Synchronization errors;

    b. Condition validation errors.

2. *Emergent faults:*

    a. Configuration errors;

    b. Environment faults.

The occurrence of security flaws as a direct result of errors in the software development process is a particular cause for concern in an era in which there is an increasing reliance upon open systems, often accessible worldwide through the web, and upon the integrity and security of Commercial-Off-The-Shelf (COTS) products. As a result, a new concept, that of defense in depth, has arisen. It is not possible to secure an organizational intranet and yet have it remain connected to the Internet for the reasons outlined above and previous security strategies based upon the idea of risk avoidance have had to be replaced by risk management. While the former would be preferable it is unfortunately incompatible with an open and universally connected Internet.

Defense in depth relies upon mutually supportive layers of security operating at network and host levels to avoid a single point of failure of system security. It incorporates conventional host-centric access control based security measures as merely one of the several layers of defense that are employed to provide the required security. It recognizes the reality that host-centric access controls based upon user identity, while crucial in securing a local host, are not a sufficient foundation upon which to base security in a networked or distributed system. Instead defense in depth relies upon multiple layers of defense, their number and nature depending upon the hosts or network segments being protected. These layers may, inter alia, include the following:

1. Access control at routers;

2. Traffic blocking at firewalls;

3. Monitoring of general network traffic by network intrusion detection systems;

4. Monitoring of activity by host and application IDS;

5. Host-centric identity-based access control;

6. Secure protocols for the confidentiality, integrity, and authentication of communications;

7. Encryption.

One result of this is that there is now potentially a plethora of event logs, one or more associated with each layer, which will or can maintain a record of host or network activity. These logs are kept by network routers, by firewalls, by IDS, by application servers including IIS and Netscape and Apache Web servers, by application proxies, by application clients including Web browsers, and by the common host system platforms, such as UNIX and Windows. Some of this logging is blind, that is, the information logged is general in nature and identifies things such as time of day, identity of user, identity of files or objects being accessed, type of access or network traffic that is taking place, and IP (Internet address) of computer originating the traffic in the case of a network log. Some of the logging is, however, more specific: logs kept by applications, for example, are necessarily more detailed, they provide a record of the specific application server activity being invoked while the log output provided by IDS is attack-specific namely, an alert generated and logged by an IDS will identify that a particular attack has occurred and will indicate the vulnerability being exploited and possibly also the steps comprising the attack. This plethora of log records comprises a rich vein of information to be explored during investigation of an intrusion. The event records stored in these logs may have interrelationships with one another and some of these will in turn be of potential interest. For example, related event records may identify the following:

▸ The same user successively accessing or attempting to access different files;

‣ Unsuccessful log-on attempts originating from the same network IP address targeting bona fide user accounts in quick succession;

‣ An e-mail chain, that is, an e-mail from e-mail account E1 to E2 to E3.

The core of IF is concerned with the mining and analysis of these logs. In many cases other information may also have been gathered at the time of initial intrusion response, in which case that information gathered by the careful execution of special commands or utilities to report on some aspects of system state (e.g., identity of users logged on and the state of network connections) is also grist for the mill. The objective is in all cases to unravel the who, what, when, and where of an intrusion. Knowledge of attack types and vulnerabilities is a vital ingredient in doing this and Section 6.4.3 deals with these concepts. Section 6.4.2 addresses the nature of the logs available in a typical networked environment and which are an essential part of successful intrusion investigations.

6.4.2 Monitoring of computer networks and systems

Communication of information between computers connected by the Internet is based squarely on the following three concepts:

1. Packets;

2. Packet switching;

3. Communication protocols.[1]

Before a message is transmitted across the Internet, it is fragmented into a number of packets, whose order within the message is identified within the packet headers and the packets are then transmitted from the originating or source computer to the destination computer. The Internet provides multiple links between its various components such as routers, hosts, and subnetworks and so the different packets may travel to their destination along different paths (packet switching), as network routers along the way adapt dynamically to choose a completion path which best avoids congestion or failed links. The packets will therefore arrive generally out of order as a consequence of this packet switching. Once all the packets constituting a particular message have arrived at the destination, they are reconstructed into a sequence representing the original message.

---

1. The two protocols of interest here are IP and TCP.

This begs a number of questions, in particular:

» How do the sending and receiving computers know that their communication is proceeding satisfactorily—that packets are not being lost, that packets have arrived where they are meant to arrive, and when to acknowledge receipt of a packet?

» When a packet is launched into the Internet directed to a specific destination, how is that destination computer identified so that intermediate hosts known as routers can know which next router to send the packet to for their next hop?

The answers to these and similar questions lie in the protocols mentioned earlier. Computers on the Internet are identified by IP addresses and each such address (with some exceptions which we shall not dwell upon) uniquely identifies a particular computer on the Internet. IP addresses are hierarchic so that forwarding of packets is efficiently directed. The IP address appears in the IP header which is the information added to the front of the packet by the IP layer of the communications software (often referred to as the *protocol stack*) of the sending computer (see Figure 6.1). This header is a minimum of 20 bytes in size; it can be longer in cases where optional extra information is included. The header includes the IP address of the destination computer as well as that of the sending computer. So the IP protocol manages the communication of information between IP addresses, that is, between computers.

In addition to the IP header there is typically a previous header added to the front of the data packet, that is added, for example, either by the User Datagram Protocol (UDP) layer of the protocol stack or by the TCP layer of the protocol stack. The TCP protocol is intended to provide two additional features that the IP protocol on its own does not:

1. *Communication between processes, for example, between a client on one computer and a server on another as in the case of a Web browser and a Web server.* The logic of TCP expects incoming packets to include a port number in the TCP packet header and this port number in turn identifies the local application process for which the packet is intended. (The header also includes the port number of the sending process so that a reply can subsequently be made to that sending process.) A port can be used by only one process at a time and this therefore achieves process-specific communication. The several processes executing on the same computer that are each waiting for

TCP packets from the Internet will be waiting on different port numbers. The common servers have preset port numbers which are publicized and upon which they wait for the purpose of receiving client requests from other computers. For example, Web servers wait upon port hexadecimal 80, domain name service (DNS) servers wait upon port hexadecimal 53.

2. *Reliable transmission is achieved by inserting packet sequence and acknowledgement information into the TCP header.* The TCP logic at the sending and receiving computers uses that information for detection of missing packets, re-assembly of packets into messages and for managing retransmissions. UDP and TCP are both transport protocols (process to process protocols) but UDP is simpler and does not provide reliable transmission, as does TCP.

While UDP and TCP over IP provide the information needed for transport across the Internet, once a packet with its various headers reaches the router of a corporate or in-house intranet, such as an Ethernet LAN, yet another protocol layer is executed to add yet another packet header (commonly referred to as a *frame header*) appropriate to that intranet hardware. In this case, an Ethernet header is added which contains local LAN addressing information amongst other things.

The resulting packet structures are shown in Figures 6.1 and 6.2.

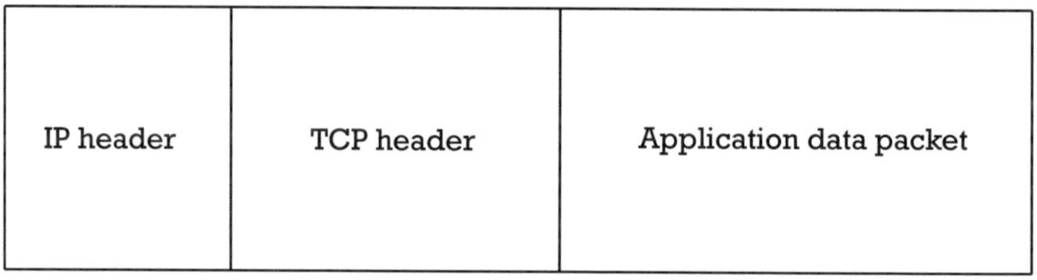

IP header	TCP header	Application data packet

Figure 6.1   TCP/IP packet.

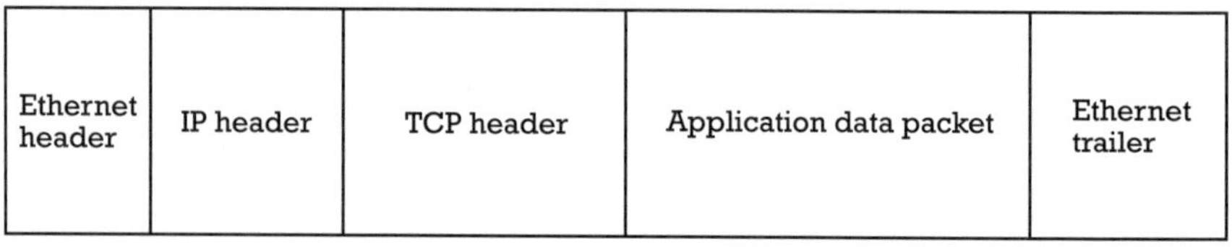

Ethernet header	IP header	TCP header	Application data packet	Ethernet trailer

Figure 6.2   TCP/IP/Ethernet packet.

Figure 6.3 shows a typical network configuration of a corporate intranet connected to the Internet or other wide area network (WAN) and the various layers of defense, including firewalls and IDS, required to secure that intranet according to defense in depth concepts. The components shown in the figure are discussed in the following paragraphs. Special Publication 800–41 of U.S. NIST provides a comprehensive description of the nature and deployment of firewalls [33], and how firewalls and IDS together provide a defense in depth network security architecture. Figure 6.3 is from that report.

The boundary router packet filter (or *screening router*) is a firewall component with a dual purpose. It connects the Internet and the corporate network and acts as both a router and a packet filter. It forwards network traffic at the packet level but does so only for packets that conform to

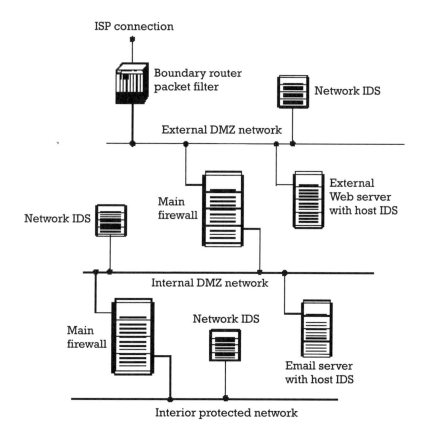

Figure 6.3 Internet to corporate intranet. (*Source:* [33]. This document may be used voluntarily by nongovernmental organizations. It is not subject to copyright.)

the rules with which the firewall has been configured—the firewall ruleset. Inward bound traffic arriving from the Internet is forwarded to the appropriate internal node or subnet while outward bound traffic is forwarded one hop to the next Internet router node. In either case the packets are forwarded only if they conform to rules that express the attributes to which packets must conform in order to pass through the router in that direction. The rules are expressed in terms of simple packet attributes, such as

- *Source and destination IP addresses* (e.g., disallow incoming packets with a supposed source IP address which is equal to an inside host address—this is symptomatic of a number of attacks and is clearly fraudulent; IP header information can be forged to achieve this);

- *The type of traffic, or service being accessed based on the protocol or port number* (e.g., allow Web server traffic, disallow SNMP traffic to deny incoming network control packets, disallow telnet traffic);

- *The particular network interface card on which the packet arrived and the one to which it is destined in the case of more sophisticated filters operating on router hosts with several interfaces.*

This packet filtering is in effect, access control at the packet level—access into the corporate network is permitted only to packets whose headers conform to the rules, and likewise for access out of the network. The rules will typically detect and screen out packets constituting some of the simpler DoS attacks, for example, screening out packets without an ACK field and packets which have the same source and destination address. The rules will also detect and prevent other simply detected attacks such as IP impersonation attempts whereby an incoming packet purports to have a source address of one of the corporate subnets. While packet filter rules are regularly enhanced by the addition of new rules that expands the scope of attacks they can detect and prevent, they operate at the packet level and so are vulnerable to attacks whose nature is evident only after packet reassembly.

While the boundary router/packet filter provides simple ruleset-based protection against rogue traffic and protects externally accessible servers (e.g., Web server and external DNS), the main firewall undertakes one or more of a variety of more detailed protection safeguards:

- It may incorporate a virtual private network (VPN) server to encrypt traffic between the firewall and dial-in telecommuters or between the firewall and other sites on the Internet.

‣ It can restrict connections from the servers to internal systems, to guard against the possibility of the servers being compromised and then attacking internal systems, the subnet thus formed between the boundary router and the main firewall is termed an external DeMilitarized Zone (DMZ) subnet.

‣ It can accept inbound traffic, determine which application is being targeted, and then hand-off the traffic to the appropriate proxy server, for example, an e-mail proxy server (in the DMZ, but not shown in Figure 6.3). The proxy server typically will perform filtering or logging operations on the traffic and then forward it to internal systems. A proxy server can also accept outbound traffic directly from internal systems, filter or log the traffic, and then pass it to the firewall for outbound delivery.

Internally accessible servers such as the e-mail server can then be located between the main firewall and an internal firewall on the subnet formed between the main firewall and the internal firewall—an internal DMZ subnet—thus protecting the servers from internal attack.

Application proxies or application proxy agents are often used to provide an additional level of protection. An application proxy agent is a software application that acts effectively as a front-end or guard to the regular server situated on a different host and on whose behalf it acts, hence the term proxy. An application proxy runs on a firewall or on a dedicated proxy server and has the role of filtering a protocol and routing it to the regular server. It may provide application level authentication and further protection against malicious activity or malicious software which in the case of e-mail may take the form of e-mail attachments. Separating out such additional preprocessor functionality allows use of a regular unmodified server that is protected by the proxy. Figure 6.4 adapted from [33] shows the network topology in this case.

IDS are then used to augment the above firewall strategy for the reasons mentioned earlier: In summary, firewalls enforce security policy by forwarding legitimate network traffic or requests—or not as the case may be if the traffic is deemed illegitimate. In contrast to firewalls, IDS in most cases carry out no enforcement but monitor and analyze traffic in a more sophisticated fashion in order to identify potentially harmful activity that needs to be investigated further before action can be taken. (There are active IDS which do operate in real-time in order to preempt activity deemed on analysis to be harmful, however, there are performance and other issues here and these IDS are still in the minority). Both HIDS and NIDS assume

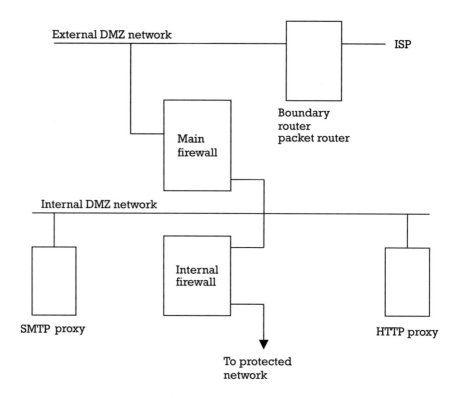

Figure 6.4 Application proxy servers. (*Source:* adapted from [33]. This document may be used voluntarily by nongovernmental organizations. It is not subject to copyright.)

importance and a large corporate intranet may need to accommodate a number of each, one NIDS per LAN segment and a HIDS on critical server hosts as appears in Figure 6.3.

The network components described earlier and shown in Figure 6.3 all have specific functions to carry out in terms of allowing or disallowing or monitoring activities. Depending upon how the components are deployed or configured, this leads potentially to event logs that can provide a comprehensive record of activity across an entire network. In addition, and as described in this and earlier chapters, there are logs also maintained by the various host operating systems (e.g., UNIX, Microsoft Windows) deployed on the end systems. Table 6.2 lists some of these network and system components and the kinds of information they record. Readers are referred to Tina Bird's "Log Analysis Resources" which provides a comprehensive account of logging and log analysis tools [34].

Table 6.2(a) Example of Logged Information Available for Analysis

Routers, packet screening filters, firewalls
▸ Packet source and destination IP addresses and port numbers, protocol; denied/allowed connections; configuration (e.g., ruleset) changes; traffic statistics; outages; reboots
Application proxies (e.g., ftp-proxy [35])
▸ All regular user actions: user commands, transfer statistics, user rejection on authentication, ftp configuration problems
NIDS (e.g., SNORT [19])
▸ Unique alert ID, the type of the alert; the origin of the alert, textual description of the alert, date and time the alert occurred, packets and their attributes, alert industry cross reference (CVE[1], Bugtraq, arachnids, McAffee)
HIDS (e.g., Emerald [20, 36])
▸ Key attributes of the attack including the rule (by name) fired by the attack, the severity of the attack (Debug, Informative, Warning, and Severe_Warning, Attack), the hostname of the machine attacked, and the number of times the attack occurred; the name of the sensor that produced the alert; the IP address of the host on which the observer is run, start time (mandatory) and end time (optional) of the attack; the name of the operation that is being performed (if BSM[2] is being used, this represents the system call name or high-level audit event name provided by the BSM audit trail of the key record used to distinguish the attack); the outcome field reports the audit return value on a given operation, the identity of the attacker (if at all possible, this represents the username of the individual responsible for the attack, for network-related attacks, this represents the remote IP address of the attacking host)
▸ And optionally: an alert-dependent enumeration of supportive information, where applicable provides additional information regarding the arguments used to invoke an operation, with respect to BSM analysis, the exec_args parameter with respect to process executions, where applicable, additional information regarding resources (usually files) that are manipulated during the malicious activity, and the owner of the object, and recommended countermeasure directives for responding to intrusive activities

[1]Common Vulnerabilities and Exposures.
[2]Sun Solaris Basic Security Module.

6.4.3 Attack types, attacks, and system vulnerabilities

The terms *intrusion* and *incident*, defined at the start of this chapter, are closely related to each other and also to the concept of an attack. In this section, we examine attack types and attacks, and the system vulnerabilities that they target before moving on to Section 6.5, to focus on IF.

Attacks and intrusions or intrusion attempts consist typically of several steps undertaken by the intruder(s), either manually or via execution of software toolkits. An attack, by definition, targets a particular computer or set of computers. This is typically done by identifying a computer either by specifying a MAC address in the case of attacking a target computer within LAN, or in the case of attacks launched across the Internet by specifying an

Table 6.2(b)   Example of Logged Information Available for Analysis

---

Application servers—Web servers
Web servers typically use the common log format [37]: Remote hostname or IP address, remote logname of the user, username as which the user has authenticated himself, date and timestamp, the (http) request itself, http status code returned to client, bytes of content transferred to client; other fields may also be logged (e.g., browser type)

Application servers—DNS[1] (e.g., BIND[2])
The BIND logging mechanism is very powerful and can be configured very flexibly with a wide variety of options; there are a range of predefined message categories which can be selectively channeled to files or the syslog server or discarded. Categories of message include:
▸ default, configuration file processing, query message, type of message (e.g., DNS query, reverse DNS query), statistics, panic, update, zone transfers, actual packets sent and received, operating system problems

Application servers—e-mail [38]
For example, UNIX sendmail (supported by the *syslogd* (8) facility, messages logged under the "*LOG_MAIL*" facility):
▸ *On receipt*: From, message size in bytes, priority, number of envelope recipients for this message (after aliasing and forwarding), message id of the message (from the header), relay—the machine from which it was received
▸ *On delivery*: To, the controlling user (the user whose credentials were used use for delivery), delay (between the time this message was received and the time it was delivered), mailer (name of the mailer used to deliver to this recipient), relay (name of the host actually accepting or rejecting), stat (delivery status)

Application servers—dialup server [39]
Dialup servers provide standard call detail (CDR) records which include the following per-call information:
▸ Calling and called numbers, call origination/connect time, the time the call was disconnected, the disconnect reason
Most dialup servers provide additional session information, for example, number of packets transferred, number of bytes transferred

---

[1]Domain Name Service.
[2]Berkeley Internet Name Domain.

IP address. In addition, an attack will usually target some specific vulnerability of the computer host being attacked, and hence rely upon host-specific information for its success. Apart from some DoS attacks that are broadcast across either intranets or the Internet promiscuously, targeting all and sundry, most attacks consist broadly of three phases as shown in Figure 6.5.

These three phases are

1. *The reconnaissance phase:* This is an information gathering phase that identifies a vulnerable host or hosts and the nature of the software executing on those host(s).

Table 6.2(c) Example of Logged Information Available for Analysis

---

Operating systems

UNIX-like systems log a variety of information, such as

‣ Command history files which list the most recent commands executed by a user, the set of currently open files, the set of current processes

‣ The most recent login time for each user in the system, information such as the terminal line, login time, logins, and logouts since reboot; accounting information (username, command, CPU time used, timestamp of the process, status, records everyone that has executed an "su" on the system

The most commonly used server and applications logging facility in UNIX-like systems is "syslog" which allows a wide range of log information to be recorded selectively. Commonly recorded information includes

‣ Mail system, line printer system, authentication system (or programs that ask for user names and passwords, such as *login, su, getty,* and *ftpd*), system daemons, news subsystem, UUCP[1] subsystem

Microsoft Windows NT/2000/XP:

‣ There are several "syslog" look-alike packages for Microsoft Windows (e.g., NTsyslog [40], WinSyslog [41], see also [42])

‣ Microsoft Windows NT/2000/XP all have powerful event logging systems of their own. For instance [43], Microsoft Windows 2000 auditing supports six different types of log of which the last three are present only if the appropriate services are installed: application, system, security, directory service, file replication, and DNS server. The security log can be configured to record events in the following categories:

  Remote logon, account management, local logon, directory service access, object (including file) access, policy change, privilege use, process tracking, and system security events.

---

[1]UNIX-to-UNIX copy.

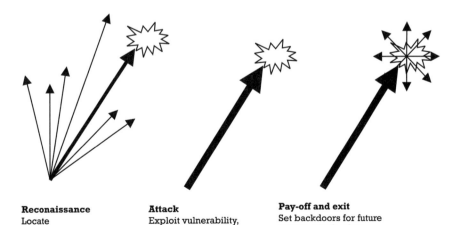

Reconaissance
Locate
vulnerability

Attack
Exploit vulnerability,
gain access

Pay-off and exit
Set backdoors for future
access, attack other
systems

Figure 6.5 A typical attack scenario.

2. *The attack phase:* This is the attack proper at the potentially vulnerable host. This phase may itself split into several subphases:

   a. Establishing a toe-hold by finding an accessible account (perhaps already identified in the reconnaissance phase) and then logging into it;

   b. Running scans on the local host and network configuration from within to identify further vulnerabilities;

   c. Attempting to access previously inaccessible files (e.g., password files);

   d. Attempting to access more privileged accounts which will in turn provide access to previously inaccessible files (e.g., achieving access to the root account in UNIX systems which effectively allows access to any object or resource in the system).

3. *The "pay-off and exit" phase:* This phase will provide some of the following:

   a. Access to confidential information (compromise of confidentiality);

   b. Corruption of information (compromise of integrity);

   c. Destruction of information (compromise of availability);

   d. Use of the current host and account as a base or *stepping stone* for extending the attack(s) to other connected hosts;

   e. Setup of attacker accounts for use in future attacks;

   f. Leaving backdoors for use in future attacks;

   g. Installation of trojaned system utilities to prevent their proper detection of the attack;

   h. Elimination of log file entries further to hide the attack.

(Trojaned utilities are utilities which have been compromised in such a way that while they seem to function correctly, they hide any evidence of the intruder. For example, a trojaned "ps" command utility will list all processes as one would expect but it will not list the processes launched by the intruder).

The first phase typically means identifying the nature and version of the operating system (e.g., Linux and Microsoft Windows) or that of the application servers (e.g., IIS, Netscape, and DNS Berkeley Internet Name

Domain (BIND) implementation) executing on that host. The steps, both the reconnaissance and attack proper, may be conducted as follows:

▸ Manually by the attacker, who keys in sequences of commands at the attack computer, commands that locate and identify other computers and then perform some form of remote execution on a targeted computer;

▸ Via preexisting executable software invoked by the attacker.

The latter is the most frequently conducted step and the software involved may either take the form of previously written programs which have been compiled or need to be compiled for the particular platform for which they are intended or take the form of scripts which are directly and readily interpreted by any platform. To that extent, script attacks are correspondingly more threatening. Depending on the sophistication of the attack, it will require the execution of software on the attacking computer (the attack computer) being used to launch the attack as well as the subsequent execution of software on the host being attacked. In some cases, an attack will involve the use of intermediate *stepping stone* computers between the attack computer and the target host to facilitate the attack on the target in which case the attack may also involve the execution of software on those intermediate hosts to erase evidence of the intrusion and to leave backdoors for subsequent re-entry.

The intention of the reconnaissance phase of the attack is to identify security vulnerabilities, which may allow the attacker to access services (e.g., internal corporate computer services intended for internal use only) or files or other objects on the target computer which are not intended to be accessible to the attacker. These attacks include what Cisco classifies [44] as

▸ Network;

▸ Operating system;

▸ Application attacks.

While Graham [45] identifies three categories of attack:

1. Reconnaissance;

2. Exploits;

3. DoS attacks.

In the context of "programs run by people to gain unauthorized control over a computer," NIST in its May 1999 *ITL Bulletin* [46] lists the following types of computer attack: remote penetration, local penetration, remote denial of service, local denial of service, network scanners, vulnerability scanners, password crackers, and sniffers.

The National Institute of Standards and Technology in conjunction with the SANS Institute issues a list of the 20 most common Internet security vulnerabilities [7] which categorizes vulnerabilities according to

- General;

- Microsoft Windows;

- UNIX vulnerabilities.

The list provides a comprehensive account of each vulnerability and provides a detailed description of how to correct the vulnerability, linking each of them via its unique common vulnerability and exposures (CVE) number to the Internet categorization and analysis of threats (ICAT) indexing system [47] maintained by NIST. The ICAT indexing service is maintained by the Computer Security Division at NIST and covers all known CVEs that hackers can exploit. ICAT provides a short description of each vulnerability, a list of the characteristics of each vulnerability (e.g., associated attack range and damage potential), a list of the vulnerable software names and version numbers, and links to vulnerability advisory and patch information. The Top 20 list includes only those vulnerabilities published within the last year in order to remain focused on relevant problems. The list is updated daily.

The Top 20 Web site also includes an appendix which lists "common vulnerable ports." Noteworthy is the continuing threat posed by exploitation of buffer overflow—a generic flaw which was most famously first exploited by the Morris worm [48]. Attacks which target this flaw fall into the class "application" attacks and are becoming increasingly common, testimony to the failure of attempts to improve the quality of software through software engineering techniques which are intended to emphasize correctness and software development process as a means of securing IT infrastructure. Perhaps most interesting is the basic nature of the dangers presented by the first four classes of vulnerabilities noted in the Top 20 as on May 2, 2002. These vulnerabilities either individually or collectively can completely undermine system security, they comprise the following:

- Vulnerabilities introduced by software misinstallation and misconfiguration;

‣ Vulnerabilities introduced by accounts with no passwords or weak passwords;

‣ Vulnerabilities introduced through nonexistent or incomplete backups;

‣ Vulnerabilities introduced from having open a large number of ports.

Vulnerability analysis or vulnerability assessment tools—VA tools— have assumed importance over recent years. While the use of vulnerability scanners (as they are also known) was initially focused largely within the hacker community as a means of identifying vulnerabilities during the first, reconnaissance, phase of an attack, they have today become an indispensable part of the security armory of system administrators, so that they can identify the vulnerabilities of their systems. The irony here is that a good knowledge of the nature and level of system security assists both system administrator and attacker alike. Port scanners are a special kind of VA tool which allow users to scan hosts on local or remote networks in order to report back to the user on the kinds of communication ports which are open on each host and related information, for example, the type of operating system executing on each host. The best known and most widely used such tool is NMap [49] which is described as follows:

NMap ("Network Mapper") is an open source utility for network exploration or security auditing. It was designed to rapidly scan large networks, although it works fine against single hosts. Nmap uses raw IP packets in novel ways to determine what hosts are available on the network, what services (ports) they are offering, what operating system (and OS version) they are running, what type of packet filters/firewalls are in use, and dozens of other characteristics. Nmap runs on most types of computers, and both console and graphical versions are available. Nmap is free software, available with full source code under the terms of the GNU GPL.

Vulnerability scanners may be passive or active according to their mode of operation [50]:

‣ Passive vulnerability scanners undertake passive vulnerability assessment and audit the security of a host or hosts on a system against a specified security policy by inspecting their security settings (file system, processes, and ports).

▸ Active vulnerability scanners undertake active penetration testing and launch attacks against the host or hosts thus actively probing for vulnerabilities and allowing system administrators to evaluate the susceptibility of their systems to attack.

Two of the earliest, best known and most widely used vulnerability scanning tools were Computerized Oracle and Password System (COPS) and Security Administrator Tool for Analyzing Networks (SATAN). COPS is a host-based UNIX security auditing tool and was developed at Purdue University by Spafford and Farmer. Farmer, in conjunction with Venema later went on to create SATAN that used the ideas from COPS in a networked environment. The ideas in COPS and SATAN laid the basis for what has turned out to be a miniindustry of such tools. Two recent free open source scanners now in use and directly descended from SATAN are SARA and SAINT. The emphasis on VA tools since the development of COPS has swung completely towards active, network scanners. Forristal and Shipley published a comparison of (mainly commercial) active scanners in 2001 in Network Computing Magazine [51] which identified the free, open source Nessus Security Scanner [52] and Internet Scanner [53] from Internet Security Systems (ISS) as the best of the eight evaluated. The ability of such packages to provide remote scanning for security vulnerabilities means that these packages and their successors are an integral part of the security tools necessary for achieving defense in depth. In July 2001, *SC Magazine* [50] published an account of an evaluation of a number of IDS and VA tool products (all commercial). In the latter category they concluded that CyberCop Scanner from Network Associates [54] deserved the "best buy" award on account of its presentation of results, IDS testing capability and its scripting language. Custom Attack Simulation Language (CASL), SecureScan NX from VIGILANTe.com Inc. [55] also received an accolade.

Noordergraaf [56] lists four kinds of attack tools most commonly used by hackers:

▸ Port scanners;

▸ Vulnerability scanners;

▸ Rootkits;

▸ Sniffers.

We have discussed all of these tools except rootkits. Rootkits achieve most of the objectives of the second and third phase of an attack, the attack proper and pay-off and exit phases above. They comprise a package of scripts and executable programs whose overall objective it is to hide the presence of the attack by ensuring that system utilities return information from which evidence of the attack is absent. They also install hidden back-door access points in the system for future use.

Loadable Kernel Module rootkits exploit the LKM feature of some UNIX systems to attack them at kernel level [57, 58]. LKM is intended to be a feature that provides a system with the flexibility needed in those cases where dynamic loading of kernel functionality is a requirement. An LKM rootkit is able to access kernel functions and tables and sanitize them to hide the presence of the rootkit and the intrusion by hiding processes, files, and connections at the kernel level. The utilities in this case do not need to be trojaned in order to achieve the intended effect of misinformation. The best defense against rootkits is to disable the LKM feature. However, if this, for some operational reason is not desirable, then it is important for system administrators to run detection measures that will identify the presence or absence of rootkits in a system. This will make use of a tool such as *kstat* [59] or similar which circumvents possibly corrupted kernel functionality by checking memory directly and essentially allows a system administrator to detect changes to the kernel system call tables and to see if these are legal (i.e., intended), or whether they are surreptitious changes emanating from a rootkit. Foundstone Inc. have developed a Linux kernel module called Carbonite which provides *lsof* and *ps* at the kernel level thus easing the burden of the Linux sysadmin or investigator charged with identifying open files and active processes during an investigation [60].

6.5   Intrusion forensics

6.5.1   Incident response and investigation

The circumstances of an intrusion incident will determine the decision as to whether or not to pull the plug on a system. It may in some circumstances be advantageous to actively and covertly monitor an intrusion, to collect whatever information one can as an intrusion continues; honeypots [61] are an extreme example of such situations. There are other situations where it is deemed appropriate for whatever reason to pull the plug as soon as an intrusion is suspected. Most intrusion investigation situations, however, fall somewhere between these two extremes and typically include an explicit information gathering phase which is just one part of what has commonly become known as Intrusion Response or IR. Other aspects of IR relate to defense of the victim system from further, unnecessary, damage and the defense of other connected systems. From a forensics viewpoint, the focus of interest is on the value of the information secured from the victim system, that is, information secured from volatile and persistent memory as discussed in Chapters 2 and 3 or gathered by probing the live victim system during this information gathering phase. Otherwise the forensic investigation of an intrusion follows the familiar pattern of the Secure, Analyze and Present (SAP) model presented earlier in the book. There is still the need to assure the integrity of the system and the evidence or information being gathered from it, but there is the additional need to gather as much state information as possible in order to identify as much as possible, the current state of the system under investigation with particular regard to logged-on users and network connections.

What does this information gathering phase of an intrusion investigation involve? First and foremost, and as noted in Chapter 2, we emphasize the importance of preparation before an investigation; the investigator needs to become as familiar as possible with the situation and prepare a plan of action, for example to identify which specific logs will be available, which are to be investigated, what tools are needed and available to analyze those logs and so on.

The book *Incident Response* by Mandia and Prosise [58] is a cornucopia of techniques and tools for use in IR situations. As reflected there and in other publications, while there are many platform dependant features whose detail needs to be understood and exploited by the incident response team (IRT), there are two overriding principles which guide the nature of the response and the tools used during this information-gathering activity. These principles are common to all platforms, Microsoft Windows and UNIX alike, and relate to what is known as the order of volatility, that is the volatility hierarchy of the information being gathered (volatile through to persistent), and to process integrity, which in turn relates to the integrity of the information yielded:

1. *Order of volatility principle [62]:* Information must be gathered in a timely manner and in an order that maintains its integrity and consistency:

    a. Current processes, current users, open files, open network connections—volatile information;

    b. User attributes, security/privilege information—persistent information but subject to change, if relevant needs to be secured at IR time;

      c. Event log information—persistent information, but needs to be secured;

      d. Persistent store (as covered in Chapters 2 and 3)—needs to be secured.

2. *Process integrity principle:* Assurance of the integrity of the tools and techniques used in order to ensure continuity of evidence:

      a. Integrity of the commands and utilities used;

      b. Integrity of the logs;

      c. Integrity of all images made including all state information;

      d. Integrity of the communication channels used for communicating any of the above.

The focus on integrity is even more crucial than it is in the case of forensic investigation of standalone systems for the simple reason that intrusion investigation involves investigation of a system which is already known or at least suspected to have been compromised. As with the investigation of standalone systems, there is a crucial reliance upon tools and utilities which are known to be uncompromised, loaded and executed from removable storage, preferably *read-only* or at least MD5-hashed. Furthermore, for the same reasons, the tools must be statically linked and not reliant upon *dlls* or other dynamic host libraries that may be compromised given that the host is suspected of being compromised.

Reference [63] provides a list of links to IR tools intended to gather information from the four information sources listed earlier. That list also includes tools (such as file integrity tools) and toolsets (such as CD-based IR toolkits) which are intended to assure the integrity objectives of an IR. The importance of these latter tools was discussed in Chapter 2. *Incident Response* by Mandia and Prosise provides a comprehensive account of the tools and procedures used in investigating intrusion incidents for both Microsoft Windows and UNIX systems and once again provides a comprehensive account of both information gathering tools and also tools to assure integrity. IR in the case of UNIX systems in particular is addressed also by Dittrich [64,65] while "NT/2000 Security Tool Kit on a Budget" [66] focuses on Microsoft Windows systems. The Incident Response Collection Report (IRCR) toolkit [67] to which one of us (Rodney McKemmish) has contributed is "a collection of tools that gathers and/or analyzes forensic data on a Microsoft Windows system. Like the Coroner's toolkit (TCT) [68] most of the tools are oriented towards data collection rather than analysis."

Both IRCR and TCT are intrusive in the sense that they are installed on the hard drive and use *dll*s from host dynamic libraries. Before installing or using these or other products mentioned in this book in an operational environment, they should be independently evaluated for their suitability in terms of operational functionality and intrusiveness.

Clearly, IR is related to IF, but the emphasis in IR is on how best to protect a computer or computer network against possible damage rather than on elucidation of the precise nature of the activity as in IF. IF is concerned primarily with achieving an outcome which identifies the circumstances and agents behind an intrusion (in some cases leading to prosecution), whereas IR is typically concerned primarily with safeguarding a computer or computer network against damage which may include deciding whether or not to shut the system down in the face of an attack. In Section 6.6, we discuss a new style of powerful forensic tool, network forensic analysis tools (NFATs) which find application in both IR and IF. There are in addition a number of reconstruction or replay tools available and these too find use in both IR and IF.

Some IDS also provide reconstruction or replay facilities. Ethereal [69] is one of the best known such tools and is described as follows:

> Ethereal is a free network protocol analyzer for UNIX and Windows. It allows you to examine data from a live network or from a capture file on disk. You can interactively browse the capture data, viewing summary and detail information for each packet. Ethereal has several powerful features, including a rich display filter language and the ability to view the reconstructed stream of a TCP session.

6.5.2 Analysis of an attack

Chapter 5 provides a detailed examination of several real-life intrusions and the IR and intrusion investigation measures that were brought to bear in those cases. Section 6.5.3 describes an intrusion and its attempted prosecution and how the nature of the intrusion evidence gathered and the processes employed to do so are critical to the admissibility of evidence and successful prosecution in court.

Here, we illustrate the points of Section 6.5.1 by summarizing an attack analysis report published by Spitzner [70]. The reader is referred to Spitzner's report for an intuitive and extremely detailed account of the actual attack, which included several phases of intrusive activity, from gaining a toehold to attempting to stage denial of service attacks against further systems. The report presents the complete detail of the actual alerts and log entries, the commands used by Spitzner to gather volatile information and the raw data gathered by those commands. The attack incident is interesting also for the reason that the attack was launched (unsuspectingly) against and scrutinized in the context of a honeypot [61].

6.5.2.1 First warning of the attack

The attack was first noticed from an alert generated by the Snort IDS. The generation of the alert was notified via SWATCH [71], a widely used real-time log monitoring tool. The alert identified the attack as a *noop* attack which was noteworthy as it is typically used by attackers launching a buffer overflow attack of some sort. The attack was on port 53, which is the DNS port. Almost immediately afterwards, there appeared system log entries showing a successful log-on followed by an "su" command to provide the logged-on user with superuser privileges thus providing the attacker with the privileges needed to access all resources on the system, protected or otherwise. This provided clear evidence that the system had been compromised.

6.5.2.2 The reconnaissance

Following this, a search for earlier Snort alerts relating to DNS-queries (from any source IP address) revealed that there were a number of DNS-version-query requests made to the attacked site on the previous day, and that they did in fact emanate from the same source IP address as appeared in the *noop* alert above. Further investigation of actual packets logged by Snort showed that the reconnaissance phase had confirmed to the attacker that the DNS server in use was in fact recursive, a prerequisite to this particular DNS server being vulnerable to the buffer overflow attack that followed.

6.5.2.3 The attack toehold

The actual toehold consisted of the attack-computer using a recursive query to exploit the DNS vulnerability and thus open an unexpected TCP connection to the DNS server. This was immediately followed by the launch of the buffer overflow attack—a query with the carefully crafted *noop* characters followed by a /bin/sh command at the end of the packets. The resulting execution of the /bin/sh at the victim computer then resulted in the attacker executing a remote root shell. This then completely compromised the victim system.

6.5.2.4   The attack—setting up attacker accounts for use in future attacks

The attack then proceeded to set up a number of accounts for future use, including an account that provided superuser status. This was to allow future attacks to proceed more easily via telnet (a remote log-in/terminal protocol), that is, without having to go through the attack toehold phase again.

6.5.2.5   The attack—leaving backdoors for use in future attacks

On a later occasion, the attacker entered the system via the attacker accounts above and downloaded a trojaned login program which allows any user with the TERM setting of vt9111 to access the system via telnet. The attacker completed this phase of his activity by deleting various log file entries to hide the attack.

6.5.3   A case study—security in cyberspace

The seminal article by Sommer [72] on admissibility of computer evidence in cases of computer intrusion arose out of a series of intrusions by the "Datastream Cowboy" and one other person into the network at the Rome Air Development Center, Griffiss Air Force Base, New York, (Rome Labs) in 1994.

On March 28, 1994, Rome Labs computer system administrators discovered that their network had been penetrated and compromised by a sniffer program. A report entitled "Security in Cyberspace" to the U.S. Senate Permanent Subcommittee on Investigations dated June 5, 1996 [73], provides an account of the IR measures that then took place. These involved first the CERT of the Defense Information Systems Agency (DISA), then the Air Force Office of Special Investigations (AFOSI) and finally the Air Force Information Warfare Center (AFIWC) in San Antonio, Texas. The report provides an account of the investigation, laborious step by laborious step, and notes that the intrusion involved installation of seven sniffer programs, the compromise of 30 Rome Labs systems and over 100 accounts as well as providing a launching platform for further attacks on sites elsewhere on the Internet.

A summarized account of the intrusions appears in [72] in which it is noted that ultimately all of the careful piecing together of host, network, and other information is of no use unless it allows investigators at the end of the day to

> make the linkage (between the person in the dock and the assumed computer identity "Datastream Cowboy") and to do so to criminal standards of proof: beyond a reasonable doubt.

The intrusions launched by the "Datastream Cowboy" commenced with the use of "phone phreaking" via a Bogota, Colombia phone facility to connect to a U.S. ISP which was then used as the base account from which to launch attacks using hacking tools. The somewhat depressing aspect of this case for forensic investigators is that the best lead, came not from forensic analysis but rather from information provided by a Web surfer who had at some previous time had e-mail contact with a hacker named "Datastream Cowboy". Furthermore and very much to the point, this information included both the geographic location (the United Kingdom) and the phone number of the "Datastream Cowboy". As it happens, although there had been a great deal of painstakingly collected computer evidence, Sommer notes that the person charged "pleaded guilty to a restricted range of charges so that the evidence was never properly tested." He further notes that "good though the structure of the investigation was, almost every individual stream of digital evidence could be challenged." In addition to hard disk information from the PC belonging to the "Datastream Cowboy," there were six other streams of digital evidence of relevance:

1. *Material on the hard disk:* There were possible inadmissibility due essentially to refusal by the prosecution to release confidential materials.

2. *Logs of phone activity:* There were some possible inconsistencies in the logs and the prosecution's argument that the logs demonstrated phone phreaking.

3. *Activity at the Bogota telephone exchange:* No evidence of this phase of activity was provided.

4. *Evidence from the U.S.-based Cyberspace ISP:* There were several reasons why this evidence might be challenged:

   a. A recent hard disk crash at the ISP;

   b. A break in the *chain of custody* (e-mailed logs and not original logs, a significant time had elapsed since the events, there were no tamper-proof controls);

   c. Violation of the hearsay rule (the person presenting the evidence was a different person to the one who gathered the evidence);

   d. Inadmissibility (the defense was not allowed access to the monitoring tools used for reasons of confidentiality).

5. *Network monitoring tools:* Possible inadmissibility of such evidence due to the prosecution not providing the defense access to the tool source code, or network topology, and problems again with hearsay.

6. *Evidence from the target computers:* Concerns with regard to evidence collected from a compromised computer—given the absence of specified measures to safeguard the recording processes, evidence gathered from the compromised computer might well be regarded as unreliable.

6.6   Future directions for IDS and intrusion forensics

There are four central technical challenges facing IDS researchers currently:

1. The performance of IDS in terms of their effectiveness in detecting new attacks;

2. The performance of IDS in terms of their capacity for ID in real-time;

3. Interoperability and inter-IDS correlation;

4. New user interfaces and new tools relating to visualization of network activity and attacks and attack threats.

As discussed earlier in the chapter, we have seen a recent focus by DARPA on the evaluation of IDS effectiveness. However, this is a moving target as attacks are becoming more sophisticated and are designed to defeat those simpler IDS with limited stateful analysis. As a consequence, this challenge cannot be cleanly separated from the other three. IDS load balancing and load management techniques are expected to attract increasing attention as a means of coping with gigabyte networks [16]. IDS correlation and integration have become increasingly important as a means of coping with the heterogeneity of network hosts and volume of alerts. There are many researchers and organizations addressing this important issue, some recent and encouraging research projects in this area are reported in [74–79].

New tools are needed and are emerging. These tools reflect a growing awareness that intrusion investigation has come a long way since the simple examination of stateless sniffer logs. Attacks and intrusions have become more sophisticated not just in terms of keeping up with new releases of system software but also in their scope and complexity. There has been an increase in the proportion of noninsider attacks, in their complexity, their novelty (see, for example, homographic attacks on Web servers [80]), their sophistication and in their scope (i.e., the range of systems and attack types, for example, routers, host platforms, application servers, and IDS attacks including IDS denial of service attacks). The resulting increased complexity of attack and intrusion investigation has led to the recent emergence of powerful NFATs which integrate a range of forensic capabilities including protocol replay and other facilities together with powerful visualization capabilities of network activity and possible attacks.

While session reconstruction tools such as Ethereal have been around for some time, these new tools of which SilentRunner [81] is a prime example, provide integrated functionality which allows the investigator to view and replay attacks and sessions using built-in protocol analyzers and three-dimensional visualization features. The latter is assisted in the case of SilentRunner by sophisticated pointing and input devices to navigate the three-dimensional visualizations. A recent article in *Information Security* [82] reviews SilentRunner and two other NFATs. SilentRunner for instance includes the facility of templates that allows it to understand a wide variety of logs, so that it will accept IDS and firewall logs for analysis.

An important aspect of computer logs and computer forensics per se relates to the integrity of computer traces and computer logs [83]. There are at least two separate developments possible within this research, one focusing on integrity of audit records per se, and the other focusing on back up of audit records in real time to a trusted repository. The latter has already received some attention within the overall context of back up techniques in general. A point emerging at RAID'98 [72] is that multiple streams of evidence (e.g., traces) are typically required in a court of law. The same paper points out

> while there are many intrusion detection tools to assist the system security administrator, there is little in the way of tools to assist the *chain of evidence* forensics requirements on the law enforcement side.

An area of future research and development which targets the overlap between IDS and IF focuses on active IDS, that is IDS that have a built-in response component. Automated intrusion response is intended to allow appropriate response measures to be launched when an intrusion or attack is detected. Clearly, as attacks become more sophisticated and active IDS follow suit, there will be a convergence between IDS and the analytic techniques employed by the forensic investigator. Once IDS move from passive to active duties they make the kinds of judgments currently made by forensic investigators. The Intruder Detection and Isolation Protocol (IDIP) project funded by DARPA has developed a technology that demonstrates the following capabilities [84]:

> (1) Cooperative tracing of intrusions across network boundaries and blocking of intrusions at boundary controllers near attack sources; (2) use of device-independent tracing and blocking directives; and (3) centralized reporting and coordination of intrusion responses.

The applicability of automated intrusion response to support the defense in depth philosophy and its clear application in defensive information warfare situations means that this area will continue to develop. At the time of publication of reference [84], the IDIP project had integrated a total of 15 COTS and research components, highlighting the importance of the related area of integration and correlation between components discussed earlier.

References

[1]   Computer Emergency Response Team-Coordination Center (CERT-CC), Carnegie Mellon University, "Responding to Intrusions," http://www.cert.org/security-improvement/modules/m06.html, visited Aug. 2002.

[2]   Computer Security Incident Response Team (CSIRT), CERT-CC, SEI, Carnegie Mellon University, "Frequently Asked Questions (FAQ)," http://www.cert.org/csirts/csirt_faq.html, visited Aug. 2002.

[3]   Sommer, P., British Computer Society Legal Affairs Committee, "Intrusion Detection Systems as Evidence," *Criminal Courts Review*, http://www.bcs.org.uk/lac/ids.htm, March 2000, visited July 2002.

[4]   North Atlantic Treaty Organization, "Intrusion Detection: Generics and State-of-the-art," *RTO Technical Report 49*, Jan. 2002.

[5]   Denning, D. E., *Information Warfare and Security*, Reading, MA: Addison-Wesley, 1999.

[6]   Garfinkel, S., and G. Spafford, *Practical UNIX Security*, Sebastepol, CA: O'Reilly, 1991.

[7]   The SANS Institute, "The Twenty Most Critical Internet Security Vulnerabilities (Updated). The Experts Consensus," http://www.sans.org/top20.htm, Version 2.504, May 2, 2002, visited July 2002.

[8]   Longstaff, T. A., et al., "Security of the Internet," in *The Froehlich/Kent Encyclopedia of Telecommunications*, Vol. 15, New York: Marcel Dekker, 1997, pp. 231–255.

[9]   McHugh, J., "Intrusion and Intrusion Detection," *International Journal of Information Security*, Vol. 1, No. 1, 2001, p.14.

[10] Lippmann, R. P., et al., "Evaluating Intrusion Detection Systems: The 1998 DARPA Off-line Intrusion Detection Evaluation," in *Proceedings of the 2000 DARPA Information Survivability Conference and Exposition (DISCEX)*, Vol. 2, IEEE Press, Jan. 2000.

[11] Lippmann, R. P., et al., "The 1999 DARPA Off-Line Intrusion Detection Evaluation," *Computer Networks*, Vol. 34, 2000, pp. 579–595.

[12] Haines, J. W., L. M. Rossey, and R. P. Lippmann, "Extending the DARPA Off-line Intrusion Detection Evaluations," in *DARPA Information Survivability Conference & Exposition 2001 (DISCEX'01)*, http://www.ll.mit.edu/IST/ideval/pubs/2001/discex01_paper.pdf, visited Aug. 2002.

[13] Durst, R., et al., "Testing and Evaluating Computer Intrusion Detection Systems," *Communications of the ACM*, Vol. 42, No. 7, July 1999.

[14] Maxion, R., "Cinnamon: Synthetic Data Generation," http://www-2.cs.cmu.edu/~maxion/invictus/cinnamon.html, Carnegie Mellon University, visited Aug. 2002.

[15] Ranum, M., NFR, "Experiences Benchmarking Intrusion Detection Systems," http://www.nfr.com/publications/white-papers/Benchmarking-IDS-NFR.pdf, visited Aug. 2002.

[16] Edwards, S., Top Layer Networks, "Vulnerabilities of Network Intrusion Detection Systems: Realizing and Overcoming the Risks. The Case for Flow Mirroring," http://www.toplayer.com/pdf/IDSB_White_Papera.pdf, May 1, 2002, visited June 2002.

[17] Bace, R., and P. Mell, "Intrusion Detection Systems," *NIST Special Publication*, http://csrc.nist.gov/publications/nistpubs/800-31/sp800-31.pdf, visited June 2002.

[18] Graham, R., "Sniffing (Network Wiretap, Sniffer) FAQ," http://secinf.net/info/misc/sniffingfaq.html, visited Aug. 2002.

[19] Roesch, M., "Snort—Lightweight Intrusion Detection for Networks," *USENIX LISA '99 Conference*, Nov. 1999.

[20] Neumann, P. G., and P. A. Porras, "Experience with EMERALD to DATE," *1st USENIX Workshop on Intrusion Detection and Network Monitoring*, Santa Clara, CA, April 11–12, 1999, pp. 73–80.

[21] DRAGON, Enterasys Networks, http://www.enterasys.com/ids/, visited Aug. 2002.

[22] Cheung, S., et al., "The Design of GrIDS: A Graph-based Intrusion Detection System," *Technical Report CSE-99-2*, U.C. Davis Computer Science Department, 1999.

[23]   Lee, W., and S. Stolfo, "Data Mining Approaches for Intrusion Detection,"
       Proc. 1998 7th USENIX Security Symposium, http://www.cs.columbia.edu/~sal/
       hpapers/USENIX/usenix.html, visited Aug. 2002.

[24]   Stolfo, S., et al., "JAM: Java Agents for Meta-Learning over Distributed
       Databases," *Proceedings of KDD-97 and AAAI97 Workshop on AI Methods in Fraud
       and Risk Management,* 1997, http://www.cs.columbia.edu/~sal/hpapers/kdd97-
       jamarch.ps.gz, visited Aug. 2002.

[25]   Lee, W., and S. Stolfo, "A Framework for Constructing Features and Models
       for Intrusion Detection Systems," *ACM Transactions on Information and System
       Security,* Vol. 3, No. 4, Nov. 2000, pp. 227–261.

[26]   GFI, "LANguard Security Event Log Monitor," http://www.gfi.com/, visited
       Feb. 2002.

[27]   Mills, D., "Improved Algorithms for Synchronizing Computer Network
       Clocks," *IEEE Transactions Networks,* June 1995, pp. 245–254.

[28]   Lombardi, M., "Computer Time Synchronization," NIST Time and Frequency
       Division, http://www.boulder.nist.gov/timefreq/service/pdf/computertime.pdf,
       visited Aug. 2002.

[29]   Coulouris, G., J. Dollimore, and T. Kindberg, *Distributed Systems—Concepts and
       Design,* 3rd ed., Boston, MA: Addison-Wesley, 2001.

[30]   Romig, S., "Correlating Log File Entries," http://www.usenix.org/publications/
       login/2000-11/pdfs/log.pdf, visited Aug. 2002.

[31]   Landwehr, C. E., et al., "A Taxonomy of Computer Program Security Flaws,"
       ACM Computing Surveys, Vol. 26, No. 3, Sept. 1994, pp. 211–254.

[32]   Aslam, T., I. Krsul, and E. H. Spafford, "Use of a Taxonomy of Security Faults,"
       Technical Report TR-96-051, Sept. 4 1996, COAST Laboratory, Department of
       Computer Sciences, Purdue University.

[33]   Wack, J., K. Cutler, and J. Pole, "Guidelines on Firewalls and Firewall Policy,"
       Special Publication 800-41 of the US National Institute of Standards and Technology,
       http://csrc.nist.gov/publications/nistpubs/800-41/sp800-41.pdf, Jan. 2002,
       visited Aug. 2002.

[34]   Bird, T., "Log Analysis Resources," www.counterpane.com/log-analysis.html,
       last updated June 4, 2002, visited June 2002.

[35]   The FTP-Proxy White Paper, The SuSE Proxy-Suite Core Development Team,
       http://www.suse.de/en/support/whitepapers/proxy_suite/ftp_proxy/
       index.html.

[36]   EMERALD *eXpert-BSM* Evaluation Edition, User's Guide, Version 1.5, SRI
       International, http://www.sdl.sri.com/projects/emerald/releases/eXpert-BSM/
       eXpert_BSM_User_Manual_1.5.pdf, Release Date: April 2002, visited June
       2002.

[37] The Common Logfile Format, W3C World Wide Consortium, http://www.w3.org/Daemon/User/Config/Logging.html#common-logfile-format, visited Aug. 2002.

[38] Allman, E., Sendmail Inc., "Sendmail Installation and Operation Guide," http://www.sendmail.org/~ca/email/doc8.9/op.html, visited June 2002.

[39] Call Detail Records, Cisco Systems, http://www.cisco.com/univercd/cc/td/doc/product/wanbu/das/das_1_4/das14/das14apd.htm#27711, visited June 2002.

[40] NTsyslog Windows NT Syslog Service, SaberNet.net, http://ntsyslog.sourceforge.net/, visited Aug. 2002.

[41] Winsyslog, Adiscon GmbH, http://www.winsyslog.com/, visited Aug. 2002.

[42] Logging via Syslog, INT Media Group Incorporated, http://www.practicallynetworked.com/support/syslog.htm, visited Aug. 2002.

[43] Internet Security Systems Inc., *Microsoft Windows 2000 Security Technical Reference*, Redmond, WA: Microsoft Press, 2000.

[44] Cisco Systems, "Raising the Bar: Extending Cisco Intrusion Detection with a Host-Based Solution," http://www.cisco.com/offer/tdm_home/pdfs/vpn/Raising_the_Bar_White_Paper.pdf.

[45] Graham, R., "FAQ: Network Intrusion Detection Systems," http://www.ticm.com/kb/faq/idsfaq.html, visited June 2002.

[46] NIST, "Computer Attacks: What They Are and How to Defend Against Them," *ITL Bulletin*, http://www.itl.nist.gov/lab/bulletns/may99.htm, May 1999, visited June 2002.

[47] ICAT Metabase, Computer Security Division, IT Laboratory, NIST, http://icat.nist.gov/icat.cfm, visited June 2002.

[48] Spafford, E. H., "The Internet Worm: Crisis and Aftermath," *Communications of the ACM*, Vol. 32, 1989, pp. 678–688.

[49] Insecure.org, NMap ("Network Mapper"), http://www.insecure.org/nmap, visited Nov. 2002.

[50] Walder, B., and J. Parkhouse, "Unearthing the Invaders," http://www.scmagazine.com/scmagazine/2001_07/, July 2001, visited Aug. 2002.

[51] Forristal, J., and G. Shipley, "Vulnerability Assessment Scanners." *Network Computing Magazine*, Jan. 8, 2001, http://www.nwc.com/1201/1201f1b1.html, visited June 2002.

[52] Nessus Security Scanner, http://www.nessus.org/intro.html, visited June 2002.

[53] Internet Scanner, Internet Security Systems (ISS), http://www.iss.net/products_services/enterprise_protection/vulnerability_assessment/scanner_internet.php, visited June 2002.

[54] Cyber Cop Scanner, Network Associates, http://www.pgp.com/products/cybercop-scanner/default.asp, visited June 2002.

[55] SecureScan, Vigilante, http://www.vigilante.com, visited June 2002.

[56] Noordergraaf, A., Enterprise Server Products Sun Blueprints Online, "How Hackers Do It: Tricks, Tools and Techniques," http://www.sun.com/blueprints, May 2002, visited June 2002.

[57] Miller, T., "Detecting Loadable Kernel Modules (LKM)," http://members. prestige.net/tmiller12/papers/lkm.htm, visited June 2002.

[58] Mandia, K., and C. Prosise, *Incident Response*, McGraw-Hill, 2001.

[59] kstat, http://www.s0ftpj.org/tools/kstat.tgz, visited Sept. 2002.

[60] Foundstone Inc., Carbonite, http://www.foundstone.com/knowledge/proddesc/carbonite.html, visited Sept. 2002.

[61] The Honeypot Project, *Know Your Enemy*, Boston: Addison-Wesley, 2002.

[62] Farmer, D., and W. Venema, "Computer Forensic Analysis," slide presentation for IBM T.J. Watson Labs, 8/6/1999, http://www.fish.com/forensics/intro.pdf. As reported by John Tan, " Forensic Readiness," http://63.251.138.38/atstake/acrobat/atstake_forensic_readiness.pdf, July 17, 2001.

[63] McLeod, J., "Windows NT/2000 Incident Response Tools," http://www. incident-response.org/windows.htm, visited Sept. 2002.

[64] Dittrich, D., "Incident Response Procedures," http://staff.washington.edu/dittrich/talks/blackhat/blackhat/incident-response.html, visited Sept. 2002.

[65] Dittrich, D., "Basic Steps in Forensic Analysis of UNIX Systems," http://staff.washington.edu/dittrich/misc/forensics/, visited Sept. 2002.

[66] "NT/2000 Security Tool Kit on a Budget," http://rr.sans.org/win/budget.php, visited Sept. 2002.

[67] Incident Response Collection Report, http://www.incident-response.org/IRCR.htm, visited Sept. 2002.

[68] Farmer, D., and W. Venema, "Bring Out Your Dead" *Dr. Dobb's Journal*, Jan. 2001.

[69] Ethereal, http://www.ethereal.com, visited Sept. 2002.

[70] Spitzner, L., "Know Your Enemy: A Forensic Analysis," http://www.enteract. com/~lspitz/forensics, Last Modified May 23, 2000, visited Sept. 2002.

[71] SWATCH: The Simple WATCHer, http://www.oit.ucsb.edu/~eta/swatch/, visited Sept. 2002.

[72] Sommer, P., "Intrusion Detection as Evidence," RAID'98, University of Louvain-la-Neuve, Belgium, Sept. 1998.

[73] "Security in Cyberspace," Staff Statement, U.S. Senate, Permanent Sub-committee on Investigations (minority staff) Hearings on June 5, 1996, http://www.fas.org/irp/congress/1996_hr/s960605b.htm, visited Sept. 2002.

[74] Cuppens, F., and A. Miege, "Alert Correlation in a Cooperative Intrusion Detection Framework," *Proc. 2002 IEEE Symposium on Security and Privacy (S&P.02)*, Berkeley, CA, May 12–15, 2002, pp. 187–202.

[75] Debar, H., and A. Wespi, "Aggregation and correlation of intrusion detection alerts," *Proc. 4th Int. Symposium on Recent Advances in Intrusion Detection (RAID 2001)*, Vol. 2212 of LNCS, Davis, CA: Springer-Verlag, October 2001, pp. 85–103.

[76] Doyle, J., et al., "Agile Monitoring for Cyber Defense," *Proc. DARPA Information Survivability Conference and Exposition (DISCEX-II)*, IEEE, Anaheim, CA, June 12–14, 2001.

[77] Valdes, A., and K. Skinner, "Probabilistic Alert Correlation," *Proc. 4th Int. Symposium on Recent Advances in Intrusion Detection (RAID 2001)*, Vol. 2212 of LNCS, Davis, CA: Springer-Verlag, Oct. 2001, pp. 54–68.

[78] Vigna, G., R. A. Kemmerer, and P. Blix, "Designing a Web of Highly-configurable Intrusion Detection Sensors," *Proc. 4th Int. Symposium on Recent Advances in Intrusion Detection (RAID 2001)*, Vol. 2212 of LNCS, Davis, CA: Springer-Verlag, Oct. 2001, pp. 69–84.

[79] Carey, N., A. Clark, and G. Mohay, "IDS Interoperability and Correlation Using IDMEF and Commodity Systems," *4th Int. Conference on Information and Communications Security (ICICS 2002)*, Singapore, Dec. 9–12, 2002.

[80] Gabrilovich, E., and A. Gontmakher, "The Homograph Attack," *CACM*, Vol. 45, No. 2, Feb. 2002, p. 128.

[81] SilentRunner, SilentRunner Inc., http://www.silentrunner.com, visited Sept. 2002.

[82] King, N., and E. Weiss, "Analyze This!," http://www.infosecuritymag.com/2002/feb/cover.shtml, visited Sept. 2002.

[83] Schneier, B., "Securing Network Audit Logs on Untrusted Machines," RAID'98, University of Louvain-la-Neuve, Belgium, Sept. 1998.

[84] Schnackenberg, D., K. Djahandari, and D. Sterne, "Infrastructure for Intrusion Detection and Response," *Proc. DARPA Information Survivability Conference and Exposition (DISCEX) 2000*, Hilton Head, SC, Jan. 25–27, 2000, http://download.nai.com/products/media/nai/pdf/DISCEX-IDR-Infrastructure.pdf, visited Sept. 2002.

CHAPTER

7

Contents

7.1 Introduction

7.2 Forensic data mining— finding useful patterns in evidence

7.3 Text categorization

7.4 Authorship attribution: identifying e-mail authors

7.5 Association rule mining— application to investigative profiling

7.6 Evidence extraction, link analysis, and link discovery

7.7 Stegoforensic analysis

7.8 Image mining

7.9 Cryptography and cryptanalysis

7.10 The future—society and technology

References

Research Directions and Future Developments

7.1 Introduction

The field of computer and intrusion forensics is rapidly changing both in form and content owing to the rapid evolution of hardware and software technology, the type of criminal activity (e.g., on-line fraud) and, more recently, due to the growing impact of asymmetric operations (e.g., events of September 11, 2001).

The convergence of information and communications technologies (ICT), together with the uptake of these technologies by larger and more diverse groups of increasingly technically savvy users, has significantly increased the complexity of a criminal investigation. For example, ubiquitous computing has introduced new challenges for investigating criminal activities due to the diversity of data types (e.g., logs, databases, and network packets) and data states (e.g., persistent, ephemeral, and volatile), wide range of technologies, and different jurisdictions.

Over the past decade, both computer and intrusion forensics have been evolving in form and broadening in application. The growing impact of asymmetric warfare on the information operations of critical infrastructure, such as power and telecommunications, and corporate fraud are recent examples of how the areas of computer and intrusion forensics

319

are changing. As discussed in Chapter 3, computer forensics has its roots in digital evidence recovery, which deals with data recovery methods from media such as hard disks. Intrusion forensics has evolved from the area of intrusion detection. More recently, computer and intrusion forensics have developed and branched out into several (overlapping) interest areas, such as digital forensics, network forensics, cyber forensics, forensic accounting, and near-real-time forensics. The application area has principally driven these developments. For example, in the traditional law enforcement area, the primary focus is on the post-mortem collection and preservation of the *chain of evidence* custody, data analysis, and interpretation subject to strict established evidentiary guidelines. In other application areas such as e-commerce where the continual availability of the on-line business service is of prime concern, the focus is on ensuring the continuity and survivability of computer networks. Therefore, the timeliness of the cycle of detection, forensic analysis, and reaction is of critical importance in these application areas.

A common requirement that underpins many forensic activities is that they should be able to handle evidence datasets that are very large (possibly Terabytes or more), heterogeneous (involving many different types of data objects, such as spreadsheets, e-mails and their attachments, network logs, database tables, call records, physical security access logs, and financial data), complex (data objects with different sets of attributes), interrelated (data objects that are linked by some type of relationship), embedded (objects spanning other objects, such as archive files), hidden (objects hidden within objects), and of varying granularity (e.g., from bit-level to organizational structure data). Furthermore, an order of magnitude additional meta-information or metadata could be generated during the course of an investigation (e.g., file timestamps, file author, and file permissions). A recent FBI report stated that up to 120 TB of computer crime data was processed by that organization in the 2001 FY [1]. A single fraud investigation involving a network of large companies can possibly expect to handle data volumes that exceed this amount. The forensic investigator will find the task of analyzing these data increasingly difficult and time-consuming to undertake. For example, a recent ISTS/RAND report of 151 U.S. law enforcement agencies and other federal organizations states that computer forensic investigators currently spend 23% of their time in a typical investigation undertaking a single activity, that of interpreting and analyzing computer logs (e.g., cross correlating logs from different computers) [2]. As mentioned in Chapter 1, a judicious choice of forensic tools will generally be able to quickly reduce these large complex datasets, in a semiautomated way, so as to filter out the irrelevant data and rapidly reduce these data to a manageable subset. This data subset will then be the focus of more advanced analytical and interpretation forensic tools. Tool components that deal with the analysis and interpretation of filtered data will need to be highly interactive (not only with the investigator but also be able to share data and metadata) and visual in nature. During the data reduction phase, it is important that the forensic tools (and the investigator) neither filter out any relevant data (false negatives) nor, if possible, generate too many irrelevant data (false positives). That is, it is important to minimize the misclassification of suspicious activities or events as nonsuspicious (i.e., a false negative) as well as to minimize the misclassification of nonsuspicious activities or events as suspicious (i.e., a false positive). In general, the investigator will trade-off the number of false negatives with the number of false positives since he/she is willing to tolerate a larger-than-normal number of false positives so as not to "let through the net" any evidence arising from suspicious activities.

Given the dynamics of form, size, and content, predicting how the field of computer and intrusion forensics will evolve is a difficult task as it is subject to a high degree of uncertainty. Nevertheless, we propose that there are some common core technological activities that will underpin most, if not all, future computer and intrusion forensics investigations. In this chapter, we outline how some of these core activities will evolve in the future, emphasizing the challenges faced and how new technologies are evolving to meet these challenges. The core activities include

- Time-lining, correlation and causal analysis;

- Evidence extraction;

- Link discovery;

- Text categorization and author attribution;

- Investigative profiling;

- Image mining;

- Stegoforensics (detection and extraction of hidden data);

- Cryptography and cryptanalysis.

Other technologies, such as the following, are increasingly becoming more important in computer forensics:

- Embedded systems forensics;

- Wireless forensics;

> Network forensics;

> Reverse engineering.

Embedded systems forensics deals with the analysis of computer-based systems embedded in devices. There exists a diverse range of such devices, including handheld or PDAs, monitoring systems, and household appliances.

Wireless forensics analyzes wireless network systems that include mobile phones, satellite phones, and pagers. Evidence may be sourced from the end user equipment (e.g., the hand-held phone) as well as intermediate equipment used in the communication system (e.g., base station). More recently, a trend towards the convergence of embedded and wireless devices has been observed, for example, connected PDAs (a PDA with mobile phone capability), and Bluetooth-enabled PDAs. Both embedded systems forensics and wireless forensics are covered in more detail in [3].

Network forensics generally refers to the collection, fusion, and analysis of information on networks. Several problems arise when dealing with network forensics. Firstly, the networks may span multiple time zones and multiple jurisdictions, necessitating the use of absolute trusted timestamps (to ensure the authentication and integrity of timestamps for each piece of network evidence) and ensuring that all jurisdictions collaborate. Secondly, the network data will be available in both off-line and in real-time modes, the latter requiring the ability to capture and analyze data on the fly. Thirdly, the data could involve many different protocols and the amount of data could potentially be very large due to the increasing size of network bandwidth. A protocol could also involve multiple layers of signal (e.g., Voice over IP (VoIP), HTTP tunneling). Fourthly, the current set of computer forensics tools will not be able to handle the real-time and data size/volume issues mentioned earlier. Finally, techniques are required for rapidly tracing a computer criminal's network activities (e.g., IP addresses) and for mapping a network's topology. There needs to be a paradigm shift for network forensic techniques to analyze the rate and size of captured data. Some tools have made some inroads recently, for example, NetIntercept from http://www.sandstorm.net, and SilentRunner (http://www.silentrunner.com, a subsidiary of Raytheon) which includes a three-dimensional visualization capability for viewing very large network diagrams.

Reverse engineering explores methods of extracting the structure, schemas and interrelationships that can be recovered from systems such as software, software engineering documents, and databases. It is a broad area that encompasses program and data migration between systems, program comprehension, and understanding. Database recovery using reverse engineering techniques can be useful in computer forensics. Such techniques can be used, for example, to facilitate the comprehension of the internal structure of a database that has been seized.

7.2 Forensic data mining—finding useful patterns in evidence

As mentioned in Section 7.1, computer and intrusion forensics can potentially generate very large and complex datasets. Data can be sourced from multiple computer workstations and servers, from local and networked peripheral devices, from network data and routers, from back-up media, and from e-mail servers. It is now not uncommon to deal with case evidence of the order of several hundred Gigabytes (10^9 bytes), or even Terabytes (10^{12} bytes) or more if we include network data capture (a thousand copies of the *Encyclopaedia Britannica* is approximately 1 TB). We are therefore confronted with the task of analyzing these large datasets and finding interesting and unsuspected patterns ("needles in a haystack") that are understandable and useful to the investigator. This task is known as *forensic data mining*, a topic we addressed previously in the context of forensic accounting in Chapter 4. We note that not all computer forensics analyses use such large datasets. For example, an e-mail authorship attribution analysis (Section 7.4) would typically only be interested in the content of e-mail files and associated metadata.

A *pattern* describes a structure and a set of relationships that characterizes a set of data records. The challenge in computer forensics is to find (discover) and describe forensically "interesting," "suspicious," or "useful" patterns from the large forensic datasets. Such interesting patterns could be those that describe a typical profile, or those that are atypical and deviate from the norm, or those that allows the investigator to make nontrivial predictions on new (unseen) data. For example, patterns could be

1. 55% of middle managers access the financial database of the company.

2. A subset of log data records reveals that certain types of users login after hours.

3. Specific log data records indicating user X is accessing system files to which the person is not authorized to access.

4. Wire transfers that exceed 10,000 euros to foreign accounts are suspicious.

We are particularly interested in discovering anomalous or unusual patterns. Such a pattern could be, for example, a sudden shift in login behavior by an employee. These patterns could be detected by specialized pattern-finding or discovery algorithms, or by identifying "outliers" or "deviants" from the normal patterns.

Finding interesting patterns usually involves two phases in sequence:

1. Generating the patterns;

2. Selecting the interesting ones.

In the first phase, one can simply generate the patterns by creating them and seeing if, and how often, they occur in the data and if they are significant. This is very time-consuming because if we had N attributes or variables (e.g., "user name," "day of week"), each one with, say, M possible values, then we would need to generate M^N patterns. So, for all but simple cases or in cases where the patterns are unrelated, this is not practical. Fortunately, and as expected, there generally exists some relationship or structure among the patterns that the algorithms can exploit. For example, some combinations of attributes frequently occur together (e.g., correlated attribute-value pairs: **User** = fredDagg and **TimeOfDay** = night as in "Fred logs in after working hours"), or one pattern is more general than another. The pattern structure can be exploited by using specialized pattern-finding algorithms. However, particular care in the choice of the algorithm needs to be made as the algorithm may miss potentially useful or unusual patterns that occur in the context of computer forensics. There are also other issues that need to be considered. Firstly, when dealing with very large datasets there is the problem of ensuring the algorithm's scalability. That is, can the algorithm cope with both small and large datasets? This can be resolved in part by using, for example, time-space reduction techniques. Time-space reduction techniques include limiting the number of passes through the dataset, and sample the dataset. The downside to data sampling is the possibility of bypassing rare patterns. Secondly, the algorithm should be robust to high dimensionality (i.e., a large number of attributes) and minimize the effect of over-fitting. Over-fitting occurs when the error on unseen data points actually increases with increasing number of attributes. That is, the results of an investigation will get worse as more attributes or features are used! Careful selection of the algorithm is thus required to circumvent this problem.

In the second phase of finding interesting patterns, which deals with selecting the interesting patterns, we can apply simple objective or subjective techniques like, for example, evaluate the frequency of occurrence (which, unfortunately, may miss out on rare or unusual patterns), apply statistical significance tests, and use clustering.

Describing a pattern means that the pattern itself should preferably have some structure that captures the decision or interpretation in an explicit manner. Typical examples of such structural patterns are rules and decision trees. These pattern representation schemes have been used extensively in the cognitive sciences and in knowledge engineering. A rule can be expressed as

If *<antecedent>* then *<consequent>*

or, in logic form,

<antecedent> $\Longrightarrow$ *<consequent>*

where *<antecedent>* is a set of conditions (e.g., "DayOfWeek is tuesday") and *<consequent>* is a set of actions, recommendations or results (e.g., "User is maxineSmith and TimeOfDay is afterHours"). An association rule is a simple form of rule pattern, a conjunction of propositional attribute-value pairs (called *items*) written in implicational logic form:

$$(\textbf{StaffType} = \text{academic}) \wedge (\textbf{DayOfWeek} = \text{tuesday}) \Longrightarrow (\textbf{User} = \text{freddDagg})$$
$$\wedge (\textbf{TimeOfDay} = \text{night})$$

These rules are generally correlational in form and do not necessarily imply any causality. That is, the *<consequent>* and *<antecedent>* are correlated by virtue of the rule but it does mean that *<antecedent>* is a cause of the *<consequent>*.

Decision trees are structured as sequences of decisions followed by a single recommendation. Counterexamples are neural networks and support vector machines (SVMs) which do not provide the investigator with an explicit description of the discovered pattern. This does not mean that they are not useful in finding patterns, as they can be used as effective data mining techniques in computer forensics (see Sections 7.3 and 7.4), but that they do not provide the investigator with a simple way of describing the patterns found in an investigation. For example, neural networks need additional knowledge extraction tools to generate some rules from the network contents.

A well-known algorithm for finding useful rules is association rule mining. Association rule mining originated in "market-basket data" applications (i.e., applications that record a basket of items where the item value indicates whether that item was purchased or not) and has proven popular as a technique for analyzing commercial databases. As we shall see in Section 7.5, association rule mining has applications in computer forensics for mining particular datasets. The basis for association rule mining is the Apriori algorithm which provides for a computationally efficient way of finding useful association rules from large datasets [4]. We define

1. A *k-itemset* pattern as a logical conjunction of k items that comprise the antecedent and consequent, that is

   $$antecedent \wedge consequent \equiv (A_1 = A_{V1}) \wedge (A_2 = A_{V2}) \wedge \cdots \wedge (A_k = A_{Vk})$$

2. The *support* of an itemset pattern as the frequency of occurrence of the itemset in the dataset,

   $$support = freq(antecedent \wedge consequent)$$

3. The *confidence* (or *accuracy*) of the association rule

   $$(A_1 = A_{V1}) \wedge \cdots \wedge (A_n = A_{Vn}) \implies (A_m = A_{Vm}) \wedge \cdots \wedge (A_k = A_{Vk})$$

   as the support of the rule divided by the support of its antecedent, or

   $$confidence = freq(antecedent \wedge consequent)/freq(antecedent)$$

The algorithm effectively finds all the association rules that have support and confidence values greater than some chosen thresholds (t_s, t_c). Careful choice of the thresholds is required to avoid an explosion in the number of rules generated. The algorithm comprises of a maximum of k passes: the first pass to calculate the support of all 1-itemsets and discard all those that have support less than t_s; the second pass to generate the 2-itemsets from pairs of 1-itemsets that survived the first pass; until all surviving itemsets have been exhausted. Note that the algorithm stipulates that a k-itemset can only be considered if *all* of its $(k-1)$-itemset subsets exceed the support threshold. This rapidly reduces the number of itemsets that are required for the next pass of the algorithm and greatly reduces the computational requirements of the algorithm. The resulting high support itemsets are then simply transformed into association rules—discarding those that do not exceed the confidence threshold t_c. Although it would appear that there would exist a very large number of possible itemsets (i.e., patterns), in reality the number of frequent itemsets will be manageable since the number of goods in the basket is generally small and those infrequent ones will be discarded by appropriately setting the threshold levels high enough.

A problem with the above association rule mining algorithm, in the context of computer forensics, is that low support–confidence rules are automatically discarded, that is, some of the needles in the haystack are unwittingly removed when the threshold levels are set too high. Techniques that partially overcome this problem include concept hierarchies and mining the itemsets. The concept hierarchy introduces domain-dependent background knowledge in the form of domain concept interrelationships to reduce the number of uninteresting rules and, at the same time, identify any suspicious rules. This is often referred to as generalized, or multiple-level, association rule mining [5, 6]. We review this technique in more detail in Section 7.5 for computer forensics investigative profiling.

7.3   Text categorization

One important low-level activity in computer forensics is to search a potentially large set of files stored on the computer system looking for suspect behavior. In many cases, evidence of such behavior might involve text documents (e.g., e-mails, inconfidence documents, chat sessions) stored as files or left behind (unknowingly by the suspect) in the disk swap area, unallocated space, or file slack space. Due to the existence of a potentially large number of text documents, some automated methods for identifying suspect documents would be particularly helpful in a computer forensic setting. For example, an investigator might need to extract all documents that deal with a particular topic, for example, finance and drugs.

Text documents may be searched using traditional document retrieval techniques based on matching a set of one or more queries with the document set, or by other more advanced techniques such as text categorization (also referred to text or document mining). Document retrieval methods assume the input of expert user queries based on the domain of search (see Section 7.8 for analogous techniques for image and video retrieval), whereas document categorization methods automatically categorize a set of documents. Text categorization supports a wide variety of activities in information mining and information management. It has found applications in document filtering and can be used to support document retrieval by generating the categories required in document retrieval.

The aim of text categorization is to assign natural language text documents to a set of categories, referred to as topics or themes, based on a metric such as text content, document structure, or text grammar. When text document categories are defined a priori , the text categorization scheme is referred to as *supervised* text categorization. Alternatively, when the text document categories are self-determined by the document set, the scheme is called *unsupervised* text categorization. Text documents can be classified into a single category, multiple categories, or none at all. Documents that cannot be classified into a unique category may indicate that a new category needs to be defined and/or the existing categories need to be refined.

Text categorization can be undertaken using various techniques. Firstly, the simplest method is to use domain experts to identify new documents and allocate them to well-defined categories. This can be time-consuming and expensive and, perhaps most limiting, is that the method provides no continuous measure of the degree of confidence with which the allocation was made. Secondly, the domain expert can establish a set of fixed rules that can be used to classify new documents. Unfortunately, in many cases, the rule-set can be large and unwieldy, typically difficult to update, and unable to adapt to changes in the text document content. Finally, categorization can be undertaken automatically by inductively learning the classifiers from training sample documents. This learning algorithm approach is useful in a computer forensics setting as it will, hopefully, automatically generalize well to new, unseen text documents and has the advantage that it should be able to adapt to a measure of drift in the characteristics of documents.

Most of the work undertaken in text document categorization has concentrated on classifying a large number of independent text documents with a relatively small number of words in each document, a large variation in the number of distinct document categories, and usually a large number of documents in each category. For example, text collections used by different workers include the 21,578 text documents with up to 267 document categories in five different document category sets in the Reuters—21,578 dataset [7], 50,216 documents with 23 MeSH disease categories from the Ohsumed corpus [8, 9], 8,282 Web page documents in seven categories in the WebKB dataset [10], and 2,815 e-mail documents with two classes [11]. Also, the number of words (i.e., the "dimensionality") used in the categorization learning algorithms is generally much smaller than the possible number of words found in the documents, which can easily result in several thousands of words. This potentially large number of words can impose large computational costs on most categorization algorithms as well require the estimation of many parameters which will induce a large variance in the results of the categorizer. Furthermore, it is assumed that most words do not contribute to the categorization of a set of documents (note, however, that it has been observed in one study that the removal of even a small number of features can lead to a degradation in performance [7]). A reduction in the number of words is therefore generally pursued to minimize any such effects and improve the generalization accuracy of the learning algorithm (i.e., reduce the error on unseen data), reduce the problem of over-fitting (where the generalization error increases with increasing number of words) and, at the same time, reduce the inherent sparseness of the document—word statistics. For example, frequent/ infrequent words are filtered out, the number of attributes is reduced, and N-grams are used (see Chapter 2).

Many text documents in the domain of computer security, particularly in computer forensics, have some typical characteristics. Firstly, most of the text documents in the different subdomains of computer security have a much smaller document set. The number of words, however, is just as large. Also, the number of documents per category can be small and there can be a large variation from category to category; some categories may consist of a few hundred documents, some only a few documents. Secondly, the correspondence between document word and category is often "fuzzy," that is, classes can be overlapping making the classification procedure more difficult. For example, documents that relate to the topic of "hacking" often have a subset of their word dictionary that intersects with the dictionary of those documents that relate to the topic of "warez." Thirdly, the dictionary of words and document categories are continually evolving, requiring efficient learning algorithms with reduced training times. Fourthly, the document categorizer must be computationally efficient to avoid long processing times. Finally, document categorization in the domain of computer security is subject to asymmetric loss. That is, the cost of misclassifying a suspect document as nonsuspect (i.e., a false negative) is higher than the cost of misclassifying a nonsuspect document as suspect (i.e., a false positive).

The principal document model used in flat-text document categorization involves representing the text document as a "term vector" or "bag-of-words." In this model, a text document is defined as an attribute vector where each vector component or attribute corresponds to a single word in the document. A given text document is thus spanned by the vector space model, to produce a unique instance vector. The word attribute can be boolean, indicating the presence or absence of the word in the document, or continuous, indicating the frequency or probability of occurrence of the word. To simplify the document model and also reduce the number of documents for effective modeling, the words in the term vector do not carry any positional information that indicates the location of the word in each document. When comparing documents, the term vectors are compared to each other and similar documents will tend to cluster in vector space.

Many different learning algorithms have been proposed for text categorization, including neural networks [12], the Naïve Bayes probabilistic classifier [13], and SVM [7, 8]. An important characteristic of these algorithms in the context of forensic text categorization is that they should handle a large term vector size and a small number of text documents. SVMs, which classify data by finding a hyperplane that separates the data that maximizes the distance between the nearest data vectors and the hyperplane, minimize the true error on test (unseen) data and are believed to not overfit in high-dimensions (several thousands of words). Neural networks encompass a large class of models that are effectively multistage classifiers with derived attributes that are a linear combination of the input data. Neural networks are generally robust to the presence of noise, but can take a long time to converge and can be overparameterized (i.e., they have a large number of derived attributes and associated weights) which can lead to overfitting (see Section 7.2). The Naïve Bayes probabilistic classifier is computationally efficient, uses a simple classification approach based on maximizing the probability of word occurrence, and gives good overall results—even for large term vector size.

One of the authors [14] has undertaken a study of the comparative performance of some learning algorithms in the context of computer security-related text documents. The author used 127 documents sourced from three Usenet document categories dealing with the topics: hacking, manufacture and use of explosives, and techniques in steganography. Results, based on comparing the microaveraged F_1 statistic (a compound measure of both the number of false positives and false negatives), show that the Naïve Bayes classifier performed the best, followed by the multiplayer Perceptron (a type of neural network model) and SVM.

There exist several commercial products for text categorization that may be useful for forensic analysis. The IBM Intelligent Miner for Text toolkit [15] consists of various components for text analysis and text search applications. Text analysis tools include an information and feature extraction tool (see Section 7.5), a clustering tool (arranges documents into subsets whose members are similar to each other), a summarization tool (condenses a document into a summary while preserving its information content) and a text categorization tool. The U.S. Department of Energy's Pacific Northwest Laboratory (PNL) has developed the Galaxies and Themescape tools [16] that graphically display images based on word similarities and themes in text. Galaxies computes the word similarities and patterns in documents and then displays the documents to look like a universe of "document stars." Closely related documents will cluster together in a tight group while unrelated documents will be separated by large spaces. It also incorporates a time-slicer to make the "document stars" appear as a function of time, thereby enabling an investigator to gain an understanding of what trends in document patterns have developed over time. In Themescape, themes in the documents are layered and are displayed as a relief map of natural terrain. The "mountains" in Themescape indicate where themes are concentrated in the underlying documents; and their shapes reflect how the thematic information is distributed and related across documents. Megaputer's TextAnalyst is a system for (1) navigating the textbase (a semantic network extracted from the text document), (2) creating summaries of documents (extracts sentences with high semantic weight), (3) clustering documents, and (4) semantic information retrieval (extracts and displays the concepts taken from an immediate neighborhood in the semantic network of the text of the words distilled from the query).

In summary, we have described some automated methods for searching a potentially large set of text-based documents or files stored on a computer system for suspect content. The basic tenet of these techniques is the ability to assign text documents to a set of topics, based on a metric such as text content, document structure, or text grammar. The ability to identify suspect documents is particularly helpful in computer forensics as it enables an investigator to extract all documents that deal with a particular topic such as finance and drugs.

7.4   Authorship attribution: identifying e-mail authors

In Section 7.3 we surveyed techniques used in categorizing a text document based on the topics or themes contained in the document. A closely related, but clearly separate, area of document analysis is determining the author of a document. Author analysis itself consists of different subareas namely, *authorship attribution* (identification or categorization of the document's author), *author characterization* (determination of the author profile or characteristics), and *plagiarism detection* (computing the degree of similarity between two or more documents without necessarily identifying the authors). The first two areas are the most relevant in the context of computer forensics though, in this section, we mainly cover authorship attribution.

The principal objectives of authorship attribution are to classify an ensemble of text documents as belonging to a particular author and obtain a set of characteristics that remain relatively constant for a large number of documents written by the author. The conjecture is that a given author's style is comprised of a number of distinctive features or attributes sufficient to uniquely identify the author. The most extensive and comprehensive application of authorship analysis is in literature and in published articles. Well-known authorship analysis studies include the disputed Federalist papers [17] and Shakespeare's works, the latter dating back over several centuries. In these early authorship attribution studies, stylometric features ("style markers") such as character or word-based metrics, vocabulary-richness metrics (e.g., Zipf's word frequency distribution and its variants), and word length were used. Unfortunately, it is possible that some of these stylometric features could be generated under the conscious control of the author and, consequently, may be content-dependent and/or are a function of the document topic, genre, and epoch. Other features such as prescriptive grammar errors and profanities are not generally considered to be idiosyncratic and discriminatory. However, it is thought that syntactic structure is generated dynamically and subconsciously when language is created, similar to the case of the generation of utterances during speech composition and production. That is, language patterns or syntactic features are generated beyond an author's conscious control. An example of such features is short, all-purpose words (referred to as *function words*) such as "also," "if," and "to" whose frequency or relative frequency of usage is unaffected by the subject matter. In fact, over 1,000 stylometric features have been proposed in the literature. However, no one set of significant style markers has been identified as uniquely discriminatory. There have also been many different classification algorithms used for author identification including, statistical approaches such as the cusum or QSUM [18], neural networks, genetic algorithms, and Markov chains. However, just as in the case of stylometric features, there does not seem to exist a consensus on a correct methodology, with many of these techniques suffering from problems, such as questionable analysis, inconsistencies for the same set of authors, and failed replication.

An example of document set that is topical in computer forensics is the e-mail as it has become the dominant form of inter- and intraorganizational written communication for many companies and government departments. E-mail is used in many different situations as, for example, in the exchange of messages, documents, and for conducting electronic commerce. Unfortunately, e-mail can also be misused as, for example, in the distribution of unsolicited ("spamming") and/or inappropriate (offensive or threatening) messages and documents, in the conveyancing of unauthorized sensitive information. E-mail evidence can be central in cases, such as sexual harassment or racial vilification, threats, and bullying. Therefore, the ability to identify the original author of e-mail misuse can be a contributing factor to the successful prosecution of an offending user. E-mail authorship attribution has some unique characteristics. Firstly, the identification of an author is usually attempted from a small set of known candidates, rather than from a large set of potentially unknown authors. This reduces the computational requirements of the attribution/classification task. Secondly, the text body structure and contents of the e-mail is not the only source of authorship attribution as other evidence in the form of e-mail headers, e-mail trace route, e-mail attachments, file timestamps can, and should, be used in conjunction with the analysis of the e-mail text body. Thirdly, e-mail is a genre. That is, e-mails are neither long formal written documents nor are they a transcription of a brief conversational speech dialogue, but a combination of both. They generally have a layout structure similar to formal texts but often incorporate some elements of a discourse structure such as replies and/or rebuttals.

The question then arises; can characteristics, such as language and layout of an e-mail be used, with a high degree of confidence, as a kind of author phrenology and thus link the e-mail document with its author? Also, can we expect the composition style of an author to evolve in time and change in different contexts? For example, work-related e-mails might differ from informal e-mails posted to friends or newsgroups, or chat-room session transcripts. Fortunately, in this case, humans are creatures of habit and have certain personal traits that tend to persist. All humans develop a multi-dimensional profile that includes unique (or near-unique) patterns of behavior, and biometric attributes. We therefore conjecture that certain characteristics pertaining to language, composition and writing, such as particular syntactic and structural layout traits, patterns of vocabulary usage, unusual language usage, stylistic, and substylistic features will remain relatively constant for a given author. The identification and learning of these characteristics with a sufficiently high accuracy are the principal challenges in authorship attribution.

Only a small number of studies in e-mail authorship analysis have been undertaken. de Vel [19] used a basic subset of structural and stylometric features and Anderson et al. [20] studied the effect of a number of parameters, such as the type of feature sets, text size, and the number of documents per author on e-mail author attribution performance. The latter study showed that text chunks larger than approximately 100 words had only a marginal improvement in attribution performance. Also, they observed that as few as 20 documents may be sufficient for satisfactory attribution performance. This is forensically significant as it suggests that satisfactory results can still be achieved with a small e-mail text size and a small number of available e-mails. de Vel et al. [21] have extended these studies and evaluated author attribution performance in the context of multiple e-mail topic categories. The attributes were selected so as to minimize any topic bias—that is, attributes such as N-graphs or the short word frequency distribution were not included in the attribute set. The study obtained some encouraging results. The micro-averaged F_1 statistic (see Section 7.4) for four native language (English) male authors, four e-mail topics with minimal overlap in topic content (three for training, one for testing) and a total of 170 e-mail attributes varied between 51.4% and 98.2%, with most performance values in the upper quartile. The same authors have also undertaken authorship characterization and, in particular, authorship gender (male or female) and language background (English as a first or second language) cohort attribution [22]. They used a set of gender-preferential language attributes, in addition to content-free structural and stylometric attributes described earlier, to perform gender cohort attribution. The gender-preferential language attributes were derived from a subset of previous gender-specific studies in written and electronic communications (for example, women's language makes more frequent use of emotionally intensive adverbs and adjectives). Initial results indicated that the selected subset of gender-preferential language attributes only marginally improved gender cohort attribution performance and, that author language background cohort categorization performance results were observed to be better than the author gender cohort results.

A version of the authorship attribution software is to be included in CFIT® (see Chapter 2). Further research is necessary with larger e-mails sizes, e-mails with attachments, minimizing the effect of topic bias, and attribution subset selection to identify the best set of attributes for each author.

Summarizing, we have outlined the techniques that classify a document based on its author and obtain a set of characteristics that remain relatively constant for a large number of documents written by the same author. We have highlighted the e-mail as the main document set for this purpose as it can be the source of many computer forensics investigations. E-mail can be misused for the distribution of unsolicited and inappropriate messages and documents, and for the conveyancing of unauthorized sensitive information. Therefore efficient authorship attribution techniques, such as those described earlier, will be of value to the forensics investigator.

7.5 Association rule mining—application to investigative profiling

Association rules mining provides a simple and useful technique for obtaining rules from large forensic databases such as logs, network packet data, and personal user file types as mentioned in Section 7.2. In this section, we describe an application of association rule mining to investigative profiling.

Investigative profiling is an important activity in computer and intrusion forensics that can significantly narrow the search for the perpetrator and reason about the perpetrator's behavior. This is analogous to criminal profiling which attempts to identify the type of person involved in the crime (e.g., serial killer) based on the personality characteristics of the offender. Chapter 4 describes a forensic profiling application area: fraudulent behavior. An offender profile consists of two components:

1. The *factual* profile (FP) which consists of factual background knowledge about the offender, such as his or her name, computer user name(s), and employee status;

2. The *behavioral* profile (BP) which incorporates knowledge about an offender's crime scene-related behavior such as, log file transactions, keyboard command sequence, header and body of e-mails, and telecommunication data patterns.

The BP can be modeled in different ways. For example, a BP can be represented simply as a vector of profile features F_j:

$$BP \leftarrow \{F_1, F_2, \ldots, F_N\}$$

or, as a union of multiple *profile hierarchies*, PH_i:

$$BP \leftarrow \bigcup_j^M PH_j$$

or, as set of association rules, R_i:

$$BP \leftarrow \{R_i \,|\, i = 1, \ 2, \ldots, N\}$$

A profile hierarchy is a knowledge representation scheme using a hierarchy of multislot frames, similar to a concept hierarchy described later, that characterizes a BP. Examples of profile hierarchies include, database access usage, e-mail authorship profile, file transfer profile, and log-in profile. A simple user-attribute type profile hierarchy with just two attributes (author, see Section 7.4, and employee) is shown in Figure 7.1.

When a BP is represented by a set of association rules, the rule attributes can be obtained from the raw data and/or selected from the profile hierarchy nodes. For example, the rule "If user Y is a system administrator and is currently employed, then the application $Z = SQL\ query$ executed" may be a valid instance of a BP rule in a system administrator profile (assuming that access to the Oracle database holding the company's financial records, as used in Y's current job context, has been authorized), but probably not in the profile of a previously employed middle manager (who has logged-in remotely since he has, hypothetically, been fired). The rule can be written as

$$R:\ (\textbf{StaffType} = \text{sysAdmin})\ \wedge\ (\textbf{DayOfWeek} = \text{weekDay})$$
$$\wedge\ (\textbf{ApplicationType} = \text{database}) \implies (\textbf{Access} = \text{valid})$$

Profiling, such as obtaining customer transaction profiles, is an important personalization activity in e-commerce. Personalization tailors the delivery of services based on customers' preferences and transactional behavior. Customer transaction behavioral information can be derived from the customer's transactional history on the company's Web portal (e.g., purchasing, browsing patterns, and clickstream), from credit card transactions etc. Many of the personalization applications published in the open literature are aimed at product recommendation systems or at improving the customer's access to the Web portal. Web access personalization uses various techniques, such as Web page predictive prefetching, and Web page clustering employed for modeling the Web access patterns [23–25].

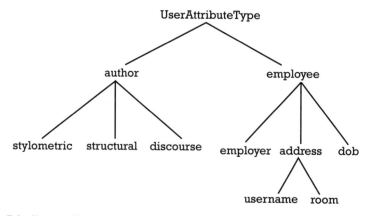

Figure 7.1 User-attribute type profile hierarchy.

However, many of these web personalization applications deal with aggregate user profiles rather than individual user profiles. Association rules have also been used for constructing personal profiles. Adomavicius et al. generate rules in two phases namely, apply the standard A priori algorithm to generate a large rule set, followed by a post-analysis phase to reduce the number of spurious and/or irrelevant rules from the potentially excessively large rule set. The post-analysis of rules involves the application of one or more rule validation operators such as similarity-based rule grouping and template matching to facilitate the validation of rules [26]. Aggarwal et al. introduces the concept of a customer "profile association rule" that defines an association between customer profile information (such as age and salary) and customer behavior information (such as buying a product). Rule clustering is used to present the profile association rules succinctly [27]. Customer profiling is aimed at obtaining normal profiles and generally not at anomalous profiles or identifying "outliers." Therefore, new profiling techniques need to be developed for computer forensics.

Abraham et al. have used association rules for investigative profiling in computer forensics [28]. They used event data from Linux *wtmp* log files as the data source (see Chapter 6). To guide the profile building and identify potential anomalous behavior, conceptual hierarchies were used. A conceptual hierarchy is a specialization–generalization hierarchy composed of background knowledge generated by the investigator (e.g., computer domain names and user organization hierarchy). The hierarchy allows for the production of high-level (i.e., more general) association profile rules and/or generalization of lower-level (i.e., more specific) profile rules permitting both hierarchy drill-down and drill-up. An example of conceptual hierarchy for a computer user organization hierarchy is shown in Figure 7.2.

Hierarchy drill-down allows interesting high-level rules to be further investigated by descending the concept hierarchies for some attributes, possibly lowering the support level requirement at the same time, producing more specialized rules. This would allow the investigator to quickly identify any suspicious behavior. Hierarchy drill-up allows higher level rules to be obtained that may have stronger support but lose some of the specific detail observable at lower concept levels. This would allow more higher-level rules to be presented to the user that would otherwise not be generated due to the lower support existing at the lower levels. The support threshold level can be set by the investigator to limit the over-generalization or over-specialization of a profile rule. An example of profile rule extracted from a *wtmp* log file is

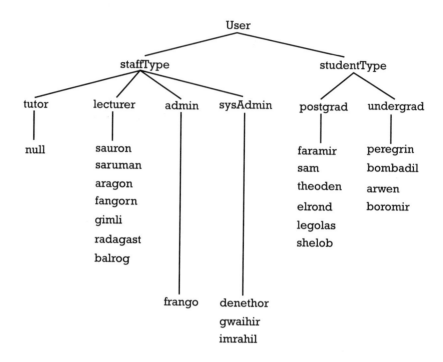

Figure 7.2 Computer user organization conceptual hierarchy.

$$(\textbf{StaffType} = \text{lecturer}) \wedge (\textbf{Source} = \text{salisbury.edu.au})$$
$$\wedge\, (\textbf{Status} = \text{employed}) \implies (\textbf{Duration} = \text{fewHours})$$

which indicates that a lecturer who is logging-in onto the server (located in Salisbury University, Australia, a fictitious university) via the ttyp console port is logged-on for a few hours, which is probably a typical behavior for local lecturing staff employed at that university. A rule that might highlight a shift from the above profile rule is

$$(\textbf{StaffType} = \text{lecturer}) \wedge (\textbf{Source} = \text{miami.edu}) \wedge (\textbf{Status} = \neg\,\text{employed})$$
$$\implies (\textbf{Duration} = \text{fewMins})$$

which may indicate suspicious behavior (the same staff member, who is now no longer employed by Salisbury University, is logging-in from a machine located in the miami.edu domain for a short time—a few minutes only, perhaps to install a backdoor program) since the lecturer concerned is not fitting the established profile. To find out whom the suspected lecturer (or masquerade) might be, the investigator would simply drill-down the user concept hierarchy (Figure 7.2) to identify the particular event in the log file and extract the user name.

We have described a basic investigative profiling technique that allows a BP to be represented by a set of association rules. These rules can then be extracted from forensic data and allow the investigator reason about the suspect's behavior and compare with other behaviors of computer systems. Better techniques need to be developed to detect deviations or anomalies from behavior profiles, and/or derive alternate profile representation schemes.

7.6 Evidence extraction, link analysis, and link discovery

In the preceding sections, we have concentrated on the problem of "finding one or more needles in a haystack." In this section, we focus on the problem of "extracting and reassembling fragments of seemingly unrelated needles located in many different haystacks." This is a much more challenging problem due to the existence of many complex patterns:

1. A large number of structured, semi-structured and unstructured objects (e.g., security logs, application logs, news text, people, call logs, bank accounts, and medical records).

2. A large number of object attributes (e.g., date, time, location, account numbers, telephone numbers, and photos).

3. A rich set of associations or relations (*links*) between objects (temporal, spatial, social, organizational, and transactional).

The volume of data (objects plus associations) could be enormous, but the volume of relevant data may be small. Also the available data may be uncertain and/or incomplete. There also exists a hierarchy of abstract ontological levels when reasoning with the data: the objects and links representing the lowest-level (e.g., security and call log activities, and timestamps), possible hypotheses at the intermediate level (e.g., installing rootkits on different computers—a rootkit is a collection of programs that a hacker uses to mask intrusion and obtain administrator-level access to a computer or computer network), and actions or functions at the highest level (e.g., undertaking a distributed denial-of-service attack). The investigator has to be able to reason within each ontological level as well as between levels. This can indeed be a complex process.

Applications of "needle fragments in multiple haystacks" include asymmetric operations, on-line fraud, money laundering, and illegal goods (e.g., weapons and drugs) trade. Details of some of these activities and the technology used to deal with them have been reported [29, 30].

We now outline the basic phases in this complex problem:

- *Evidence extraction*, where the objects and their interrelationships are extracted from the evidence;

- *Link analysis*, which explores and visualizes these relationships;

- *Link discovery*, which attempts to infer useful knowledge and derive new knowledge based on the existing relationships.

7.6.1 Evidence extraction and link analysis

The activity of evidence extraction involves extracting an ontology defining:

1. *Task-dependent objects:* for example, given Web pages on computer hacking, evidence extraction involves finding specific kinds of information such as the type of hacking attack, exploit names, and usernames;

2. *Syntactic relationships:* for example, object attributes (gender of person), object context (subject, verb, and object);

3. *Semantic relationships:* for example, organizational relationships between objects.

The object extraction process is analogous to information extraction (IE), that is, generating structured summaries of text documents. There is a large literature of works on IE, from both unstructured text (free text) and semi-structured text (ungrammatical, loosely structured text such as medical records or telephone call data). There exists a number of IE systems (e.g., RAPIER, AutoSlog, CRYSTAL, and TIPSTER [31]) and IE techniques are varied (e.g., simple multi-slot extraction algorithms that fill slots in a template with fragments of text from the document, and rule-based learning algorithms, such as inductive learning and Naïve Bayes learning, that learn to extract the correct information). Multi-slot extraction algorithms are often used for extracting domain-specific information from well-formed text, for example, finite state transducers map the free text to tagged text. Rule-base learning algorithms learn the set of rules for slot extraction, and are thus more domain-independent and flexible. The *Message Understanding Conference* (MUC) is where IE systems are evaluated with corpora in various topic areas. Examples of commercial software for text IE include ClearForest and IBM's Intelligent Miner. CFIT® (Chapter 2) has a facility for extracting forensically interesting metadata such as dates, times, addresses, and names of organizations from textual data streams.

When the extraction of relationships between objects is combined with their visualization, the process is sometimes referred to as *link analysis* (also known as *network analysis* or *entity relationship modeling*). Link analysis explores associations among the objects and generates the graphical or network model of the objects or entities (graph or network nodes) of interest in the domain, and associations or "links" (graph or network arcs) representing relationships or transactions. Figure 7.3 shows an example of graphical representation of the domain entities and their associations.

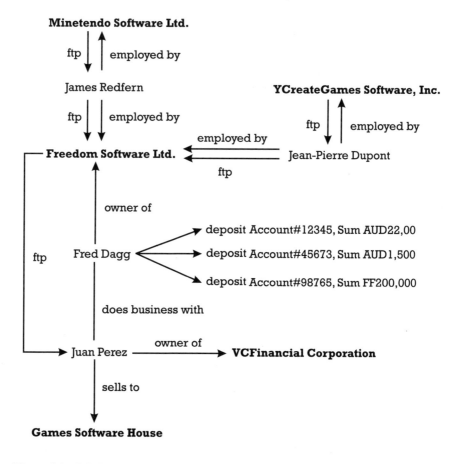

Figure 7.3   A link chart for a hypothetical software piracy scenario.

Network models inherently capture the set of entities in the domain of interest, their properties or attributes, and the relationships between them. Entities and links can represent bidirectional cash transactions between bank accounts (e.g., account numbers, holders, and date of transaction), file transfer operations between user accounts (e.g., user names and IP addresses), and chat-room sessions (channel, date, and time duration). The links can also have attributes indicating the strength of the relationship (e.g., the value of a cash transaction, and file name). The link analysis process involves matching attributes from different objects to establish the relationship. For example, it could establish that the same person has two user names on different computers. The generation of the graph can be computationally expensive as each object could have multiple attributes in common with other objects and the search would involve comparing all possible attribute pairs to ensure all relationships are extracted.

Link analysis can indicate where to focus an investigation and confirm suspicions [30]. However, it cannot reason or extract any meaning from the graph. For example, it cannot to discover associations based on the statistical characteristics of the nodes. However, link analysis can infer useful knowledge about the graph and can be used to examine questions, such as the following [29]:

- Which nodes are prominent in the graph (i.e., network centrality)?

- Which links can be severed (strengthened) to most effectively impede (enhance) the operation of the graph?

- Are there similarities in the structure of subgraphs of the graph which may indicate an underlying relationship (e.g., modus operandi)?

- What are the relevant subgraphs within a much larger graph?

A hypothetical and simplified "warez" (software hacking and piracy) scenario displaying the subnetwork of entities (e.g., people, companies, and bank accounts) and relationships is shown in Figure 7.3. "Warez" is software that has usually been reengineered so that it does not require a crack, patch, or key generation for installation. In this example, programmers employed by legal software companies (Minetendo and YcreateGames) for downloading software to an illegal software cracking company (Freedom Software), was financed by VCFinancial. Once the software is cracked, it is resold to a front company (Games Software) for redistribution. It is possible that the money earned from this venture (deposits made by Fred Dagg) could be linked to a parallel money laundering scheme. Link analysis allows the investigator to rapidly establish the prominent players involved in the "warez" case, as well as determine the dynamics between companies, employees, and bank transactions.

Link analysis has been used in the context of diverse applications, including detecting terrorist threats, retrieving and classifying Web pages, detecting nuclear proliferation, analyzing transportation routes, detecting money laundering, and finding previously undiscovered medical knowledge [31].

The Financial Crimes Enforcement Network (FinCEN) is one of the more well-known networked systems used for the detection, investigation, and prosecution of domestic and international money laundering and other financial crimes. It is a network of databases (e.g., law enforcement, commercial) and financial records maintained by the U.S. Federal Government. FinCEN handles more than 140 million computerized financial records compiled from 21,000 depository institutions and 200,000 nonbank financial institutions. Banks, casinos, brokerage firms and money transmitters all must file reports with FinCEN on cash transactions over US$10,000. As money laundering is based on actively exploiting relationships and on methods for hiding these, link analysis plays an important role in FinCEN. Link analysis is useful in identifying relationships hidden in the transactional data (e.g., disentangling multiple stages of money transfer or "layering"), exposing the structure and operation of organizations (e.g., relationships between businesses), and characterizing the roles of certain entities in the network (e.g., common identity of key actors) [32].

Examples of commercial software for link analysis include Alta Analytics' NETMAP [33], Xanalys' Watson [34], and I2's Analyst's Notebook [35]. The law enforcement community routinely uses some of these products.

7.6.2   Link discovery

Link discovery attempts to discover related objects, additional attributes, and other relevant relationships from a given set of objects, attributes and relationships. Given an ontology for the particular domain at hand, consisting of entities and relationships, and a set of knowledge base instances (the actual case evidence), we attempt to derive a knowledge base of understandable assertions. These assertions could simply be instantiations or more general rules (see Section 7.2).

For example, given an ontology for a "warez" cracking scenario (a simplified example is shown in Figure 7.3), we can obtain from the case the following (partial and simplistic) knowledge base of instances:

Person(fredDagg), **Person**(jamesRedfern), ...

Company(freedomSoftware), **Company**(minetendo), ...

EmployedBy(jamesRedfern,Minetendo), **EmployedBy**(jamesRedfern, freedomSoftware)

...

FileTransfer(jamesRedfern, minetendo.com.au, freedom.com.au), ...

and we can then derive the following knowledge base of logical assertions:

Rule 1: **Person**(A) $\wedge$ **EmployedBy**(A, B)$\Longrightarrow$**MembersOfCompany**(A, B)

Rule 2: **Company**(A) $\wedge$ **Company**(B) $\wedge$ **FileTransfer**(C, A, B)
$\wedge$ **EmployedBy**(C, A) $\wedge \neg$ **EmployedBy**(C, B) $\Longrightarrow$ **Warez**(C)

where the second rule states that, if an employee of one software company is transferring program code to another company (where it is known that this company is a front for crack software) in which he or she is not employed, then the software transferred is (most probably) "warez".

The derivation of these assertions is a nontrivial exercise (and potentially very time-consuming). In the simplest case, the investigator could create the network of the instances using link analysis, then try to derive the specific assertions from the network, and then finally attempt to generalize the assertions as rules as shown earlier. With this approach, the set of assertions can be large, difficult to derive from network, and difficult to update with new entity-relationship instances. A more flexible approach is to learn the assertions from the knowledge base instances. The learning approach can generalize well to new, unseen instances and has the advantage that it should able to adapt to a drift in the knowledge base of instances. We briefly mention two learning approaches that may be suitable for link discovery, namely inductive logic programming (ILP) and probabilistic relational modeling. ILP generates (induces) a set of assertions from the knowledge base of background entity-relationship instances. The induction of assertions can be undertaken by performing inversion resolution [36] or by using a top-down decision-tree based induction algorithm [37]. ILP is able to discover new entities and entity-relationships from the background instances. Unfortunately, ILP is not very robust to noise and has only been used in simple, constrained tasks. A possible extension to ILP is to combine relational learning with statistical learning [38]. Probabilistic relational modeling is an extension of Bayesian networks (a compact representation of complex joint probability distributions with conditional independence semantics between Bayesian network nodes [39]) by incorporating the relational structure with the Bayesian network [40]. A probabilistic relational model (PRM) does not actually learn and predict the relational structure (it is given as background knowledge) but rather learns the dependencies between the attributes of the related objects. A PRM represents the uncertainty over the properties of an entity, capturing its probabilistic dependence both on other properties of that entity and on properties of related entities. Thus, only the attributes participate in the probabilistic model and not the relational framework. A PRM is able to discover interesting dependencies between entities and their attributes and is more robust to noise than ILP.

7.7 Stegoforensic analysis

There are many reasons for hiding information. Some of these are for commercial reasons (e.g., proof of authorship and authenticity), others are for transmitting information while, at the same time, evading detection. There are two general directions within the field of information hiding:

1. Hiding the information such that no active adversary can remove it;

2. Hiding the information so that no passive adversary can detect its presence.

The former deals with issues such as watermarking, fingerprinting and tamperproofing, whereas the latter (called *steganography*) is more relevant to forensics investigations where the hidden mark is created so as not to raise any suspicion. In watermarking, the existence of the watermark in data, such as an image, is generally made known (i.e., its existence is public knowledge whether it is visible or not) and the watermark must be robust against its removal from the data. Fingerprinting is similar to watermarking, except that a different watermark (e.g., customer ID) is embedded in every distribution of the cover data to create a chain of distribution. In essence, watermarking and fingerprinting require robustness against attacks/ destruction whereas steganography requires robustness against detection.

Steganography (literally "covered writing") is the process of embedding information (called the *mark*, M) into a message (called a *carrier* or *cover*, C) to create a *stegocontent* (C′, with a mapping of the form $M \times C \rightarrow C'$) such that the presence of the mark cannot be detected. Cryptography, in contrast to steganography, generates messages from the information that are recognizable as encrypted messages, although the information content remains confidential. Steganography is becoming a significant issue in computer forensics as important evidence may be embedded in seemingly innocuous data, thereby making its detection and extraction potentially much more difficult [41]. This is in contrast with encrypted data where the investigator can easily detect the presence of the message (but perhaps have difficulty in extracting the relevant information). See Section 7.9 for more information on cryptography. Furthermore, there exist many possible carriers for hiding the information (e.g., text, audio, images, video, file slack, file systems, partitions, and network packets), as well as many different algorithms for doing so [42].

Steganographic algorithms generally exploit the redundancy of information in a cover C and fall into the following categories:

1. Substitute subsets of redundant parts in the cover with the mark M; alternatively, the transformed domain of C, such as the Fourier domain used in JPEG images, used as the embedding cover;

2. Statistical methods to modify the statistical properties of the cover;

3. Spread-spectrum techniques to spread the mark over the cover so that the mark is difficult to perceive but relatively easy to extract.

The steganographic process can also utilize secret keys (called *stegokeys,* SK) that can be used to control the embedding process (e.g., controlling the scattering of the mark in the cover) as well as cryptographic keys for encrypting the mark prior to embedding (the mapping $M \times SK \times C \rightarrow C'$). There is a large variety of available steganographic software available, including Jsteg, Steganos, S-Tools, EzStego, and Snow. Therefore, owing to the large variety of steganographic algorithms, diversity of media, and locations for hiding information (e.g., images on Web sites), steganography will become a significant challenge for forensics investigators.

We define *forensic steganalysis,* or *stegoforensic analysis,* as the science of discovering the presence of steganography in computer and intrusion forensic evidence. In stegoforensic analysis, we are generally interested in one or more of the following phases:

1. The *detection* of the presence of the mark in the stegocontent;

2. The *extraction* of the mark;

3. The *disabling* of the mark.

The detection phase attempts to determine if hidden information is present in the stegocontent. This phase is the most important forensically since detecting an embedded mark defeats the purpose of steganography, that is, concealing the existence of the mark to avoid raising suspicion. The detection phase can determine one of two possibilities—whether any type of embedded mark is hidden or if a specific type of mark is present. In the former case, no mark M is available to determine its presence in the stego-content (in this case we have the mapping $C' \rightarrow \{0,1\}$). In the later case, the mark M is available ($C' \times M \rightarrow \{0,1\}$) and can more easily be detected by, for example, cross correlating the stegocontent with M. Stegoforensic analysis deals mainly with the former case since the mark is rarely available (called *stegoonly forensics*). If the cover C is available then it is also possible to detect the presence of the mark M by cross correlating the stegocontent with the cover, i.e., the mapping $C' \times C \rightarrow \{0,1\}$, called *cover-stegoforensics*. It is possible to detect even minute changes in the stegocontent if the cover is available, so some effort should be made by the investigator to look for the presence of the cover in the evidence. For example, the investigator may need to search the file system or unallocated space for duplicates, search the Web browser cache and download the relevant media. If a stegokey is present then some sort of cryptographic key attack is also required (e.g., dictionary attack).

The extraction phase is generally much harder to implement than the first, since it requires knowledge of the embedding algorithm and possibly the stegokey. Once the presence of the mark has been identified (and possibly extracted), the third phase attempts to attack the mark and render it inoperative for the receiving party. The third phase can be exploited by crackers but is generally of little interest to the forensics investigator.

We briefly look at some of the techniques used for detecting the presence of a mark in stegocontent (a more thorough overview of steganographic techniques is given in [43, 44]). We note that a forensic investigator should not rely entirely on the use of steganalytic techniques to search for hidden evidence as there is potentially a large amount of additional evidence at hand, such as the presence of steganographic software (which may help in determining the best steganalystic technique to use) in the file system, metadata in the Microsoft Windows registry and deleted file space, the presence of multiple versions of the same image indicating some possible attempt at steganography (thereby allowing the investigator to try cover-stegoforensic analysis), and the presence of gray-scale images (which are good covers and indicative of the use of a steganographic algorithm that manipulates the least-significant bit of the color palette).

The detection of anomalous patterns in the stegocontent can point to the existence of embedded marks. These anomalies can arise from either

> • Weaknesses of steganographic software ("signatures");

> • The statistical properties of the stegocontent.

Such anomalies may be obvious and repetitive, with some perceptible to the human eye (e.g., color shifts) or human ear. Anomalies may arise from unusual ordering of the color palette, changes in the palette entries, presence of split or "twin" peaks in the color histogram, existence of duplicate entries, increase in file size, increase in noise, and presence of an echo in audio. For example, in the Snow white-space steganography tool, a mark is embedded in a text message cover by appending white-space to the end of lines. Such anomalies are not noticeable in text viewers but can easily be detected using a word processor or text editor.

Steganographic signatures arise from the unique characteristics that a steganographic tool produces when embedding a mark in a cover. The SysCop (System for Copyright Protection) [45] reduces the size of the GIF image palette to the actual number of colors used as well as using an unusually large number of black pixels (rather than shades of black). Hide and Seek [46] has image palette entries that are divisible by 4. Jpeg–Jsteg [47] has an erratic curve for the JPEG image coefficients distribution as well as the existence of duplicate coefficient values. These signatures can be automated for steganalysis. For example, Stegdetect is a tool for detecting steganographic content in JPEG images [48]. It is capable of detecting several different steganographic methods to embed hidden information in JPEG images. Currently, the claimed detectable schemes include Jpeg–Jsteg and Jphide.

Deviations in the statistical properties of the stegocontent compared with the cover may arise, even though these may not be imperceptible to the human eye. Statistical steganalytic techniques attempt to detect such changes in these statistical properties. Most of the techniques developed to date have focused on images and examining their first-order (mean and variance) and higher-order (skewness and kurtosis) statistical properties. One of the problems that arises is measuring the changes in statistical properties compared with some reference, otherwise the noise may be considered to be a part of the image and not be detected. In stegoonly or cover-stego forensics, no reference mark M or cover C is available and the changes must be examined from the stegocontent only. If no reference is available, some properties of the stegocontent may be exploited. For example, if the stegocontent of an image has similar adjacent DCT/JPEG coefficients, this may be due to the presence of an embedded mark M (e.g., [49] uses a χ^2-test to detect such distortions. Note, however, that the authors were not able to find any genuine hidden information in images; which may point to the technique's current ineffectiveness). Determining the properties of similar images may similarly be a difficult task as this may involve gathering large amounts of stegocontent statistics, such as luminance and hue. On the other hand, if the investigator has access to a reference image then it is much easier to evaluate the differences by comparing with the statistical properties of other similar cover images. For example, [50] uses both first-order and higher-order statistics together with a multiscale decomposition of the image and linear discriminant analysis (LDA) to determine the presence or absence of a mark M. The LDA classifier is able to differentiate whether or not an image contains a mark with a high detection performance rate and a low false positive rate. The results depend on the steganographic method used to embed the mark and on the image size (larger images have better detection rates).

In summary, few steganalytic techniques exist for detecting stego-content. The development of efficient techniques for computer forensics, particularly in the absence of a cover, is therefore an important area of research and development.

7.8 Image mining

Image and video production and use routinely occurs across a broad range of disciplines. For example, architectural, interior and engineering design; radiological diagnosis; pornography industry; art galleries and museums; photo-journalism; remote sensing; geographic information systems; interplanetary image analysis and discovery; scientific database management; weather forecasting; trademark and copyright database management; and image archiving. Much of the media can be downloaded from sites located on the Internet, obtained by means of e-mail attachments and scanned from print media.

Cases in computer and intrusion forensics can potentially generate very large image and video datasets, both from the point of view of the size of the media type as well as the quantity of images and video that may reside on a suspect's storage media. A disk can easily store several thousands of images and videos. This makes the following tasks potentially computationally intensive and time-consuming for the investigator:

> Searching the storage media for image and video data;

> Retrieving the forensically interesting image and video data from the set of extracted media.

Furthermore, the suspect may have deleted some of the images but have forgotten to scrub his/her unallocated space and/or Web browser cache. In the former case, the investigator has to reconstruct (if possible) the image file from block or cluster chains in unallocated space, or search the unallocated space for remnants of images or videos. In the latter case, the investigator has to search the cache for suspect files. Once the investigator has found a set of image and video media, he/she would wish to view the media in a convenient manner (perhaps as thumbnail groupings or hierarchy) to identify any suspicious images and/or videos, and also possibly investigate the media metadata (location of media, timestamps, application type, and owners) and their interrelationships. The search for, and retrieval of, forensically interesting image and video media is thus quite daunting. In some forensics investigations, such as child pornography cases, this can also be potentially a traumatizing exercise for the investigator. The use of some automated or semi-automated techniques for retrieving such media is therefore imperative.

Aside from text retrieval systems (see Section 7.3), almost all systems that search and retrieve media specialize in image media. Very few video retrieval systems currently exist. To deal with a variety of image contents and sizes, a number (currently at least 50) of content-based image retrieval (CBIR) systems have been developed. The main motivation for these systems is querying and searching for images and videos:

▸ That are semantically similar to another (e.g., perceptual similarity);

▸ That have objects in the image that are structurally similar (e.g., objects with the same shape);

▸ That have objects in the image with some spatial relationship (e.g., objects that are adjacent to each other in the media);

▸ That deal with some topic or description (e.g., find "images of political speeches");

▸ That involve some complex reasoning about the objects in the media (e.g., find "images illustrating drug dealing and money laundering");

▸ That search for, and track down, illegal copies of the media (e.g., watermarking and fingerprinting, see Section 7.7).

Some CBIR systems are available commercially, for example, Excalibur Visual RetrievalWare [51], ImageFinder [52], IMatch [53], QBIC [54], and VIR Image Engine [55], though many of these systems only handle relatively simple media queries and are not able to deal with any inference about the objects in a scene, nor with any complex reasoning about the importance of the objects in a scene. Some experimental systems have employed other techniques to improve on the search task. These include the use of relevance feedback and drawing on techniques from Gestalt psychology [56]. We consider these more advanced techniques shortly.

The goal of *image mining* is to combine effective image (and video) retrieval techniques with the ability to learn and generate a model of the objects embedded in the image and, consequently, to reason with these objects. Computer forensics has requirements that intersect with some of the motivations mentioned earlier. However, there exist some unique requirements that the above CBIR tools do not generally provide

1. Searching for specific types of objects ("things") as well as, though perhaps less frequently, similar types of objects ("stuff");

2. Searching for duplicates of media on multiple storage devices (see Section 7.7);

3. Searching for cropped media;

4. Searching for media blocks;

5. Performing link analysis of media metadata (see Section 7.6);

6. Learning and adapting to new objects.

Input from the forensic investigator is also an important source of user feedback in the search process, as this provides valuable, nonmalicious information for updating the parameters of the algorithm. User feedback should therefore be exploited by the forensic image mining system.

Image mining is an interdisciplinary endeavor that draws upon the areas of, for example, computer vision, image understanding, knowledge acquisition and modeling, and machine learning. As such it is a complex and multi-faceted issue. While it is generally feasible to state what an image consists of in terms of the objects it contains, one of the main difficulties arises from the subjective, individual interpretation of an image. While some degree of background knowledge and subject expertise is required on the investigator's part to identify the objects in the image, it is generally possible to state what the image contains. However, it is not necessarily clear what the image means, if anything at all. Often the semantics of an image is open to individual interpretation and could be subject to a number of external factors such as age, gender, and social grouping.

Most current CBIR techniques involve the retrieval of images based on matching some representation or characteristics ("features" or "attributes") of image appearance with a database of images. Examples of primitive image features include color, shape, edges, intensity, and texture. The color feature typically captures the global distribution of color in the image, the shape feature measures the relative orientation, curvature, and contrast of lines in the image, and the texture feature analyzes areas of the image for periodicity, coarseness, and anisotropy (directionality). The image query-retrieve process involves the following steps:

1. Computing a feature vector for the unique characteristics of the desired image;

2. Computing the similarity between the feature vectors of the image and the database of images;

3. Retrieving the image with the best similarity score.

The similarity score measures the visual distance between the two images represented by the feature vectors. An example of image query-retrieval operation is shown in Figure 7.4 (a, b). Here the investigator selects a subimage containing an object of interest—the wheel of the car in this instance. The system will then extract the relevant feature vector from the

(a)

Figure 7.4 The investigator (a) selects the object of interest (the wheel in this case) and (b) retrieves the set of matching images from the database. (*Source:* [57] ©1997, IEEE. Reprinted with permission.)

(b)

Figure 7.4 (Continued)

subimage, matches this vector with the image database vectors to obtain the image with the best similarity score, and displays the resulting set of similar images found [57]. The feature vector is generated by first filtering the image with Gaussian derivative filters at several subimage scales, followed by computing a set of differentials that are invariant to two-dimensional rigid transformations.

The main problems with the simple query-retrieval approach are

▸ It is difficult to decide which features should be included in the feature vector and how the matching should be undertaken. The primitive features (e.g., color and shape) can be useful in characterizing objects that look similar, but are insufficient for retrieving specific types of objects. Higher-level features such as regions and object parts may be more useful in understanding objects in an image and the image scene.

> There is little or no use of high-level knowledge; and there is a reduced ability to deal with incomplete or conflicting information.

> No ability to adapt to new situations, or learn new unseen objects.

Some of these problems can be overcome by using image retrieval techniques that are based on *semantic content* rather than low-level or primitive features. These methods include

1. *Model-based methods* that develop a model of each object (or class of objects) to be recognized. Objects are classified based on their constituent components or regions that, in turn, are characterized in terms of a combination of primitive features.

2. *Statistical modeling models* that use statistical techniques to assign semantic classes to different regions of an image.

3. *User relevance feedback methods* that require investigator feedback to drive and refine the retrieval process. The investigator is asked to indicate which retrieved images are, in the simplest case, relevant and which are irrelevant. The system is then able to derive rules from the feedback and generate better semantic classes of images.

Model-based methods extensively use knowledge about the object and are capable of reasoning about the nature of the object. However, the models created are often crafted and cannot easily improve their performance by learning. Statistical modeling techniques rely on statistical associations between image semantics and, as such, do not require the generation of any complex object model. Associations can also be learned using the statistical model. However, it is difficult for the investigator to interpret some of the results (e.g., "why are these objects in the image scene similar?") because statistical modeling techniques cannot easily reason with any high-level knowledge about the regions and image scene. User relevance feedback inherently captures continuous learning as the system is able to build up a knowledge base of past user feedback. Quite elaborate feedback mechanisms can be implemented, for example, ranking of images and input from collaborating investigators.

Computer forensics would probably benefit from a combination or hybridization of these methods. For example, the inherent interactive and continually changing nature of a forensic investigation would favor a user relevance feedback approach together with either a model-based or statistical approach that would capture knowledge about the objects in the image and better retrieve images with specific types of objects (rather than just similar images).

7.9 Cryptography and cryptanalysis

Cryptography provides a means of information hiding (see also Section 7.7). It provides a means by which a stream of symbols, such as standard English text or a sequence of digits, the plaintext, is kept confidential by being transformed into a different stream of symbols, the ciphertext, which is apparently unrelated to the plain text. This process is called *encryption* or *encipherment*. Cryptography also provides the necessary *decryption* or *decipherment* process which is the inverse of the above process and which recovers the plaintext from the ciphertext. The classic text in the field of cryptography is Bruce Schneier's "Applied Cryptography" [58] and readers are referred to that book for a wealth of information regarding a wide variety of cryptographic algorithms and their properties.

The power of cryptography stems from the fact that while the nature of the transformations employed by the encryption and decryption processes or algorithms are typically public, they make use of a *secret cryptographic key* which is not public and without which successful decryption is in general not possible. A key is typically simply a string of bits, variously 56, 128, or 1024 bits long or longer.

There are two forms of cryptography used in modern day computer systems:

1. *Symmetric* or *secret key cryptography*, which is a direct descendant of classical encryption techniques predating the computer era;

2. *Asymmetric* or *public key cryptography*, which dates from 1976 [59].

Encryption and decryption in *secret key cryptography* are characterized by the following relationships (p = plaintext, c = ciphertext, f_e = encryption function, f_d = decryption function, k_e = encryption key, k_d = decryption key):

- *Encryption:* ciphertext = f_e (plaintext, k_e);

- *Decryption:* plaintext = f_d (ciphertext, k_d).

where

f_e and f_d are public;

k_d is identical to k_e (although in some implementations it may be a trivial derivation from k_e);

k_e and hence k_d is secret.

The crux of secret key cryptography is that the same key is used for both encryption and decryption. Thus to ensure confidentiality of information or communication, this key must be kept secret but must be shared by both the sender and receiver of a message. This requires that the secret key be previously communicated between n communicating parties via some trusted channel. It follows also that systems that require direct point-to-point communication between communicating parties require a total of n^2 keys to be generated and then shared in this way.

Encryption and decryption in *public key cryptography* are characterized by the following relationships ($p =$ plaintext, $c =$ ciphertext, $f_e =$ encryption function, $f_d =$ decryption function, $k_e =$ encryption key, $k_d =$ decryption key):

- Encryption: ciphertext $= f_e$ (plaintext, k_e);

- Decryption: plaintext $= f_d$ (ciphertext, k_d);

where

f_e and f_d are public;

k_e is public;

k_d is secret (*private*);

it is "computationally infeasible" to derive k_d from k_e.

The crux of public key cryptography is that the key used for encryption is publicized so that anyone can encrypt messages to the known recipient who "owns" that public key. However, the encryption key will not allow successful decryption of a ciphertext. To do that requires the private decryption key k_d which cannot be discovered from the known encryption key k_e. Thus, every person, or *principal*, has their own pair of keys: their private decryption key k_d and their public encryption key k_e, and it is in the interests of each person or principal to make their encryption key k_e publicly available. This may be done via a Certification Authority, if the intention is to facilitate communication across the Internet at large, or it may involve some local protocol if the communication is to take place only across a private network. The benefit of public key cryptography is readily seen to be with regard to key management: only n private keys or key pairs are needed for n communicating parties and there is no need for a trusted channel for the previous communication of public keys. It is true that there needs to be some means of avoiding spoofing, that is, there needs to be some way of preventing a masquerader broadcasting a public key purporting to be the public key of someone else whose mail or other sensitive communications they wish to intercept. This is achieved by relying upon a Certification Authority (CA) and protocols which allow verification of the authenticity (i.e., true ownership) of a public key. Public key cryptography has some other benefits also: it provides a simple basis for digital signatures—a digital signature is a bit-sequence that can be used to verify the authenticity of a digital document. The downside of public key cryptography compared with secret key cryptography is that of computational performance: there is a margin here of several orders of magnitude, and as a result most cryptographically secured systems use a combination of the two, for example, public key encryption is used to share a secret (symmetric) key at the start of a session thus providing the trusted channel referred to earlier.

The advent of the Internet with its global communications and open computer systems led irresistibly to its use for any and all applications; before this, cryptography was a solution searching for a problem. General use of the Internet opened a Pandora's box of vulnerabilities. Cryptography, particularly, public key cryptography, provides a means by which those vulnerabilities can in general be managed. Cryptography is clearly the "silver bullet" needed to secure the Internet and the Web.

However, a bullet—silver or otherwise—depends upon the aim of the shooter for its effectiveness and, as with all technologies, cryptography is ineffective unless properly implemented and properly managed and used. Implementation problems may arise from simple errors in implementing and testing known cryptographic algorithms or they may arise from new, sophisticated attacks that exploit previously unknown and unexpected side effects. Such problems can lead an attacker to discover information either about the encrypted text itself or worse about the key required to decrypt that text. Management and individual errors may be as simple as allowing encryption keys to be "left lying around" or they may, on the other hand, relate to poor security policies regarding prescribed security and encryption protocols and their use. On top of this there is real a danger in the longer term that new technologies will arrive that make cryptanalysis of current cryptographic technologies computationally feasible, hence rendering the cryptographic algorithms themselves ineffective.

In the case of computer forensics—as long as the electronic evidence under investigation has been legitimately collected—the tables are turned and the attacker is the "good guy," that is, the attacker is the investigator and encrypted material is at best an irritation, at worst an obstacle to derive comprehensible and useful evidence. As a result, encryption has long been held up as one of the most serious problems facing computer forensic investigators in the future. It is the task of the investigator to recognize the existence of ciphertext when it occurs and—possibly without knowing the required decryption key—to attempt to derive the corresponding plaintext from that ciphertext or possibly the decryption key itself. This process is called *cryptanalysis*, a term which predates computer forensics and which extends to describe the methods used by cryptanalysts in attempting to derive plaintext or decryption keys in order to frustrate the use of encryption to hide information.

Schneier's book identifies a number of different types of attacks employed by cryptanalysts in order to develop methods for deriving plaintext or decryption keys from ciphertext. These include, inter alia,

> ‣ Plaintext only;

> ‣ Known ciphertext;

> ‣ Chosen ciphertext;

> ‣ Brute force.

The last relates to attack by key enumeration, that is, if the key is suspected to be 16 bits long then a brute force attack will test potentially all 2^{16} or 65,536 possible keys and eventually succeed as long as the key length has been correctly assumed. The time taken for a successful brute force attack will depend upon the hardware employed and the key length; in general the longer the key length the greater the brute force effort required to break the ciphertext so that for a key length of 17 bits (2^{17} possible keys) it will take twice as long as for a key length of 16 bits (2^{16} possible keys). Diffie and Hellman [60] speculated on the use of special-purpose machines built to crack ciphertext based on the U.S. Government's Data Encryption Standard (DES) and which would test 2^{64} keys in 214 days. This highlights the fact that secure encryption is a relative concept—whether an encryption is sufficiently secure depends upon the extent of the resources brought to bear by a potential brute force attacker. That in turn depends upon the expected value to the attacker of breaking the ciphertext. Conversely, the key length chosen by someone employing encryption will likewise reflect the value to that person of maintaining confidentiality of the data.

Besides, when faced with what may be encrypted text, a forensics investigation will first and foremost attempt to recover cryptographic key information "left lying around." Some packages actually store their encryption keys with the document being protected so that knowledge of the proprietary format of the document plus careful examination of the file image can yield the key. In addition, some software packages actually use quite primitive encryption algorithms and short key lengths—a legacy in some cases of U.S. export restrictions on encryption technology which have now been relaxed. In such situations, a limited brute force attack is likely to bring quick results. There are many utilities (see Chapter 3) available for recovering the plaintext of documents encrypted by some common document processing packages including versions of Microsoft Word. For their success in cracking the encryption involved, these utilities rely upon the weak proprietary encryption algorithms being used, or on the local storage of the passwords or keys within the application, or on the limited key length used. For instance, on account of the U.S. export regulations many packages use a 40 bit key length which means that a brute force attack will need to test only up to 2^{40} keys. (This is in contrast to 2^{56} keys that a brute force attack would need to test in the case of DES and to 2^{256} keys that would need to be tested in the case of the new AES standard. This has replaced DES as the NIST-approved symmetric encryption algorithm for use by U.S. Government organizations to protect sensitive information [61].)

Barak Jolish [62] reports two separate Al Qaeda related cases in which encrypted information was relatively easily recovered in such circumstances. Ramsi Yousef had apparently left his file protection password lying around on disk, while two other Al Qaeda files captured in Afghanistan had been encrypted with a weak version of the new AES DES.

Finally, keys are often stored either on a separate file on disk, or on floppy disks, all of which bear examination in such circumstances for suspicious files that may contain such information.

Timing attacks have been shown to be successful against both public key and symmetric key algorithms [63, 64]. This illustrates the fact that the exploitation of previously unknown side effects can unexpectedly undermine the strength of accepted encryption processes or algorithms. These attacks while of importance in certain situations have limited applicability in cryptanalysis and forensics in general but are nonetheless interesting in as far as they serve to flag the fact that there are no guarantees about the long-term security of any encryption process. They provide an important insight into the nature of such algorithms for further cryptanalysis research.

Whitfield Diffie, who together with Martin Hellman brought public key cryptography to the world, was a contributor to the March 2001 special issue of Communications of the ACM entitled "The Next 1000 Years." In his article "Ultimate Cryptography" [65], he wrote of both long- and short-term possibilities and that advances in cryptography in the more immediate future were likely to be in one of the three areas:

- Mathematics;

- Computational complexity;

- Computing technology.

It is in the latter area that we already have promise of very significant developments albeit not in this decade, nor perhaps the next one or the one after that. Quantum computing which was first proposed in the early 1970s, differs almost totally from conventional computing [66, 67]. Quantum computers consist not of binary state devices which have orthogonal states conventionally labeled 0 or 1 but rather they consist of *qubits* (quantum bits)—devices which may be partially in the one state (0) and partially in the other (1) simultaneously. This aspect of quantum computing is inextricably coupled with the ability of quantum computers to make parallel computations producing multiple simultaneous results. Once successful full-scale implementations are developed, this will enable the fast factoring of prime numbers which has formed the basis of recent advances in cryptography, thus presenting a threat to some public key encryption systems. Grover reports [67] that at the time of his article progress had extended to developing a quantum computer with seven qubits which operated only for a few microseconds. So much remains to be done.

In summary, secure encryption is a relative concept—whether an encryption is sufficiently secure depends upon the extent of the resources brought to bear by a potential attacker and the value of the protected information to both parties. Modern encryption is the cornerstone of computer security and as such research will continue to focus on the development and deployment of ever stronger encryption algorithms. There is an acute irony here not found to this extent elsewhere in security. There is a very strong reactive coupling between advances in strengthening security (stronger encryption) and advances in penetrating that security (cryptanalysis). Computer forensics and intrusion forensics is where this spiral of action and reaction plays out.

7.10   The future—society and technology

Computer and intrusion forensics is rapidly becoming a mainstream activity in an increasingly on-line society due to the ubiquity of computers and computer networks. We use computers daily either for communication or for personal or work transactions. From our desktops and laptops we access

Web servers, e-mail servers, network servers whether we know it or not, business and government services, and then—unknowingly—we access a whole range of computers that are hidden at the heart of the embedded systems we use at home, at work and at play. While many new forms of illegal or antisocial behavior have opened up as a consequence of this ubiquity, it has simultaneously also served to provide vastly increased opportunities for locating electronic evidence of that behavior.

In our wired society, the infrastructure and wealth of nations and industries relies upon and is managed by a complex fabric of computer systems, computer systems that are accessible by the ubiquitous user but which are of uncertain quality when it comes to protecting the confidentiality, integrity and availability of the information they store, process, and communicate. Government and industry have in turn focused attention on protecting our computer systems against illegal use and against intrusive activity in order to safeguard this new fabric of our society.

Computer and intrusion forensics is concerned with investigating crimes for which there is electronic evidence, and with investigating computer crime in both its manifestations—computer-assisted crime and crimes against computers. As computer intrusions using current technology become harder to achieve (due to a variety of factors including increased awareness of computer and network security, improved software and hardware computer security), we may expect unauthorized activities by authorized users ("insider" activity) to become even more prevalent and important. This too is intrusive activity and subject to intrusion forensics.

There are a number of possibly nonorthogonal influences that will direct future developments in computer forensics and intrusion forensics:

- The diversity of the "application" areas, such as law enforcement, cyberterrorism, international/domestic terrorism, critical infrastructure protection, white collar crime (e.g., fraud, ID and IP theft), organized crime (e.g., money laundering and drug trafficking), auditing, each with different requirements;

- The increasingly large and complex datasets sourced from multiple platforms, datasets that are heterogeneous, relational, of varying granularity;

- The varying time and other requirements regarding the cycle of detection, forensic analysis and—in some cases—reaction, a variation which covers the spectrum of forensic activities from near real-time forensics to postmortem forensics;

- The ongoing evolution of hardware and software technology, combined with ICT convergence: the size (e.g., disk sizes are growing faster than Moore's CPU speed law, and network connectivity is rising) and diversity (e.g., wireless, embedded, and ubiquitous/ pervasive systems) of data/evidence will grow significantly, necessitating novel and faster techniques and algorithms.

August 2001 saw the coming together of a select group of academic researchers and digital forensic investigators and practitioners at the First Digital Forensic Research Workshop (DFRWS) in New York. The Workshop was sponsored by the AFRL's Rome Research Site in New York [68] and included a number of addresses by key figures followed by four individual workshops. The first workshop was concerned with "Defining a Framework for Digital Forensic Science," the other three with

- The Trustworthiness of Digital Evidence;
- The Detection and Recovery of Hidden Data;
- Digital Forensic Science in Networked Environments (Network Forensics).

The last topic relates to the forensics of live networks, a specialization of intrusion forensics. One of the keynote speakers was Eugene Spafford who has kindly contributed the Foreword to this book and whose contribution to computer forensics—in addition to other contributions referred to elsewhere in this book—includes seminal work on identifying program code authorship ("software forensics") [69, 70], an area which shares some common ground with the determination of email authorship, a topic addressed in Section 7.4. In his address, Spafford noted the need for "technology that isn't so easily compromised." More recently, a second DFRWS meeting was held in August 2002 with an expanded set of special workshops, including

1. Complex recovery and data reduction;
2. Digital crime scene (e.g., incident response and data recovery);
3. Steganography;
4. Encryption;
5. Network forensics (tools, processes and legal issues);
6. Operation forensics analysis;

7. Standard testing;

8. Time and computer forensics;

9. Training and certification;

10. Unique identification, profiling and attribution methods.

This second DFRWS highlighted the increasing breadth and depth of issues, and challenges confronting computer forensics.

The need for better system security has been recognized for a long time but typically the cost of implementing secure systems, and doing so properly rather than attempting to secure systems by retro-fit, has been regarded as unacceptable so that security has been sacrificed for economy and convenience. Recent government and industry initiatives indicate that this may be about to change. Both Microsoft Corporation (e.g., [71]) and the USA Government (e.g., [72]) have foreshadowed the allocation of resources to the development of more secure systems. Only the future will tell if this translates into a continuing commitment that will both benefit preventive security and (hopefully) facilitate the task of computer forensics by providing computer systems environments that can be trusted. This will depend upon the extent to which society—government and industry—is persuaded that the cost of putting up with systems of uncertain security is too high and that better quality software is required in order that our systems can be secured and that the data they record can be trusted. A step in this direction is the recent announcement by Microsoft of a trusted computing and digital rights management (DRM) architecture initiative, code named Palladium (the Greek goddess of "wisdom and protector of civilized life"), that allows users to store encrypted information and only permits certain entities to see it—a sort of virtual vault residing within each PC. The Palladium architecture's advantage is that it is potentially attractive to enterprises or organizations that require the integrity and security of information and transactions. However, though this architecture purports to provide improved privacy, greater data security (e.g., reduce ID theft) and overall computer system integrity (e.g., only trusted code can be executed), it might (will?) make the task of computer forensics very much harder due to increased difficulty in accessing user information.

The second societal influence that will impact heavily on the future of computer forensics as practiced by law enforcement and national security relates to the regulated surveillance of Internet use and communications in general, and how pressures for data retention (in the United Kingdom and the EU) and data preservation (in the United States) will develop. Chapter 2 describes some recent legislation in this regard in the United Kingdom and the EU. One part of related U.K. legislation known as Regulation of Investigatory Powers Act (RIPA) 2000 focuses on access to communications data. Recent developments in that regard reflect just how easily tensions between the perceived public good on the one hand and the privacy rights of the individual on the other can lead to unexpected vacillations by government [73]. The difficulty of balancing these two important but competing goals and the resulting uncertainty of regulatory outcomes mean that there is some corresponding uncertainty as to how computer and intrusion forensics will develop into the future.

References

[1] Vanzant, D., FBI CART, *Law Enforcement Information Management Conference*, http://www.nlectc.org/nlectcse/download/vanzant_may2002_leim.pdf, visited Aug. 2002.

[2] Vais, M., "Law Enforcement Tools and Technologies for Investigating Cyber Attacks: A National Needs Assessment Report," *Institute for Security Technology Studies*, Dartmouth College, NH, June 2002.

[3] Casey, E., *Handbook of Computer Crime Investigation*, London: Academic Press, 2002.

[4] Agrawal, R., T. Imielenski, and A. Swami, "Mining Sets of Items in Large Databases," *Proc. ACM SIGMOD Conference on Management of Data (SIGMOD'98)*, NY: ACM Press, 1993, pp. 207–216.

[5] Han, J., and Y. Fu, "Discovery of Multiple-Level Association Rules from Large Databases," in *Proc. 21st Int. Conf. on Very Large Data Bases (VLDB'95)*, U. Dayal, P. Gray and S. Nishio (eds.), 1995, pp. 420–431.

[6] Srikant, R., and R. Agrawal, "Mining Generalized Association Rules," In *Proc. 21st Int. Conference on Very Large Data Bases (VLDB'95)*, U. Dayal, P. Gray and S. Nishio (eds.), 1995, pp. 407–419.

[7] Joachims, T., "Text Categorization with Support Vector Machines: Learning with Many Relevant Features," *Proc. European Conf. Machine Learning (ECML'98)*, Chemnitz, Germany, April 21–24, 1998, pp. 137–142.

[8] Joachims, T., "Transductive Inference for Text Classification Using Support Vector Machines," *Proc. European Conf. Machine Learning (ECML'99)*, Bled, Slovenia, June 27–30, 1999.

[9] Yang, Y., and J. Pedersen, "A Comparative Study on Feature Selection in Text Categorization," *Proc. Int. Conf. Machine Learning (ICML'97)*, Madison, WI, July 26–30, 1998.

[10] "The Four Universities Data Set," http://www.cs.cmu.edu/afs/cs/project/theo-20/www/data/, Carnegie Mellon University.

[11] Sahami, M., et al., "A Bayesian Approach to Filtering Junk E-mail," *Learning for Text Categorization Workshop: 15th National Conf. on AI.*, Madison, WI, July 26–30, 1998, pp. 55–62.

[12] Ng, H., W. Goh, and K. Low., "Feature Selection, Perceptron Learning, and a Usability Case Study for Text Categorization," *Proc. 20th Int. ACM SIGIR Conf. on Research and Development in Information Retrieval (SIGIR97)*, 1997, pp. 67–73.

[13] Yang, Y., and X. Liu, "A Re-examination of Text Categorization Methods," *Proc. 22nd Int. ACM SIGIR Conf. on Research and Development in Information Retrieval (SIGIR99)*, 1999, pp. 67–73.

[14] de Vel, O., "Evaluation of Text Document Categorisation Techniques for Computer Forensics," *Journal of Computer Security* (submitted for publication), 2002.

[15] International Business Machines (IBM), "Intelligent Miner for Text Toolkit Components," http://www-3.ibm.com/software/data/iminer/fortext/index.html, visited Nov. 2002.

[16] U.S. DoD Pacific Northwest Laboratory (PNL), "Galaxies and Themescape Visualization Tools," http://www.pnl.gov/, visited Jan. 2002.

[17] Mosteller, F., and D. Wallace, *Inference and Disputed Authorship: The Federalist*, Reading, MA: Addison-Wesley, 1964.

[18] Farringdon, J., *Analysing for Authorship: A Guide to the Cusum Technique*, Cardiff: University of Wales Press, 1996.

[19] de Vel, O., "Mining E-mail Authorship," *Proc. Workshop on Text Mining, ACM Int. Conf. on Knowledge Discovery and Data Mining (KDD'2000)*, Boston, MA, Aug. 2000.

[20] Anderson, A., et al., "Identifying the Authors of Suspect E-mail," *Computers and Security* (submitted for publication), 2001.

[21] de Vel, O., et al., "Mining E-mail Content for Author Identification Forensics," *SIGMOD Record*, Vol. 30, No. 4, 2001.

[22] de Vel, O., et al., "Language and Gender Author Cohort Analysis of E-mail for Computer Forensics," *Digital Forensics Research Workshop (DFRWS2002)*, Syracuse, NY, Aug. 2002.

[23] Chan, P. K., "A Non-invasive Learning Approach to Building Web User Profiles," *Workshop on Web Usage Analysis and User Profiling (WEBKDD'99)*, B. Masand and M. Spiliopoulou (eds.), Lecture Notes in Computer Science Vol. 1836, Heidelberg, Germany: Springer-Verlag, 2000.

[24] Nanopoulos, A., D. Katsaros, and Y. Manolopoulos, "Effective Prediction of Web-User Accesses: A Data Mining Approach," *Workshop on Mining Logdata Across All Customer TouchPoints (WEBKDD'01)*, 2001.

[25] Mobasher, B. et al., "Discovery of Aggregate Usage Profiles for Web Personalization," *Workshop on Web Mining for E-Commerce (WEBKDD'00)*, 2000.

[26] Adomavicius, G., and A. Tuzhilin, "Using Data Mining Methods to Build Customer Profiles," *Computer*, 2001, pp. 74–82.

[27] Aggarwal, C., Z. Sun, and P. Yu, "Online Algorithms for Finding Profile Association Rules," *Proc. ACM Int. Conf. on Information and Knowledge Management (CIKM-98)*, 1998, pp. 86–95.

[28] Abraham, T., and O. de Vel, "Investigative Profiling with Computer Forensic Log Data and Association Rules," *IEEE Int. Conf. on Data Mining (ICDM2002)*, Japan, Dec. 2002, pp. 11–18.

[29] Goldszmidt, M., and D. Jensen (eds.), *Report on the DARPA Workshop on Knowledge Discovery, Data Mining, and Machine Learning (KDD-ML)*, Carnegie Mellon University, June 1998.

[30] U.S. Congress, Office of Technology Assessment, *Information Technologies for Control of Money Laundering*, OTA-ITC-630, Washington, DC: Government Printing Office, Sept. 1995.

[31] Jensen, D., and H. Goldberg, *AAAI Fall Symposium on Artificial Intelligence and Link Analysis*, Orlando, FL, 1998.

[32] Goldberg, H., and T. Senator, "Restructuring Databases for Knowledge Discovery by Consolidation and Link Formation," *Proc. 1st Int. Conf. on Knowledge Discovery and Data Mining*, Menlo Park, CA: AAAI Press, 1995.

[33] AltaAnalytics, "NetMap," http://www.altaanalytics.com, visited Nov. 2002.

[34] Xanalys Inc., "Watson," http://www.xanalys.com, visited Nov. 2002.

[35] I2 Inc., "Analyst's Notebook," http://www.i2.co.uk, visited Dec. 2002.

[36] Muggleton, S., and W. Buntine, "Machine Invention of First-Order Predicates by Inverting Resolution," *Proc. 5th Int. Workshop on Machine Learning*, San Mateo: Morgan Kaufmann, 1988, pp. 339–351.

[37] Quinlan, J., "Learning Logical Definitions from Relations," *Machine Learning*, Vol. 5, 1990, pp. 239–266.

[38] Slattery, S., and M. Craven, "Combining Statistical and Relational Methods for Learning in Hypertext Domains." In *Proc. 8th Int. Conf. on Inductive Logic Programming*, Madison, WI: Springer-Verlag, 1998, pp. 38–52.

[39] Pearl, J., *Probabilistic Inference in Intelligent Systems*, San Mateo, CA: Morgan Kauffmann, 1988.

[40] Koller, D., and A. Pfeffer, "Probabilistic Frame-Based Systems," *Proc. 15th National Conf. on Artificial Intelligence*, Madison, WI: AAAI Press, 1998, pp. 580–587.

[41] Johnson, N., J. Giordano, and S. Jajodia, "Steganography and Computer Forensics: The Investigation of Hidden Information," *Technical Report, CSIS-TR-99-10-NFJ*, George Mason University, Center for Secure Information Systems, 1999.

[42] Petitcolas, F., R. Anderson, and M. Kuhn, "Information Hiding—A Survey," *IEEE Proc.*, Vol. 87, No. 7, 1999, pp. 1062–1078.

[43] Katzenbeisser, S., and F. Petitcolas (eds.), *Information Hiding Techniques for Steganography and Digital Watermarking*, Norwood, MA: Artech House, 2000.

[44] Johnson, N., and S. Jajodia, "Steganalysis of Images Created Using Current Steganography Software," *Lecture Notes in Computer Science LNCS1525*, 1998, pp. 273–289.

[45] SysCop, "System for Copyright Protection," http://syscop.igd.fhg.de and also http://www.mediasec.com, visited July 2002.

[46] Hide and Seek, http://www.rugeley.demon.co.uk/security/hdsk50.zip, visited Jan. 2002.

[47] Jpeg–Jsteg, ftp://ftp.funet.fi/pub/crypt/steganography/jpeg-jsteg-v4.diff.gz, visited Jan. 2002.

[48] Stegdetect, "Steganography Detection with Stegdetect," http://www.outguess. org, visited July 2002.

[49] Provos, N., and P. Honeyman, "Detecting Steganographic Content on the Internet," *Ninth Annual Symposium on Network and Distributed System Security (NDSS'02)*, San Diego, CA, 2002.

[50] Farid, H., "Detecting Steganographic Messages in Digital Images," *Technical Report 2001-412*, Department of Computer Science, Dartmouth College, 2001.

[51] Convera (formerly Excalibur Technologies), http://www.convera.com, visited Nov. 2002.

[52] Attrasoft, http://www.attrasoft.com, visited Dec. 2002.

[53] Photools, http://www.photools.com, visited Dec. 2002.

[54] IBM, "QBIC, Query by Image Content," http://www.qbic.almaden.ibm.com, visited Jan. 2002.

[55] Virage Inc., "VIR Image Engine," http://www.virage.com, visited Nov. 2002.

[56] Melihac, C., and C. Nastar, "Relevance Feedback and Category Search in Image Databases," *IEEE Int. Conf. on Multimedia Computing and Systems (ICMCS'99)*, Florence, Italy 1999.

[57] Ravela, S., and R. Manmatha, "Retrieving Images by Similarity of Visual Appearance," *IEEE Workshop on Content Based Access of Image Databases*, Puerto Rico, 1997.

[58]   Schneier, B., *Applied Cryptography*, New York: Wiley, 1996.

[59]   Diffie, W., and M. E. Hellman, "New Directions in Cryptography," *IEEE Trans. on Information Theory*, Vol. IT-22, No. 6, Nov. 1976, p. 644.

[60]   Diffie, W., and M. E. Hellman, "Exhaustive Cryptanalysis of the NBS Data Encryption Standard," *Computer*, Vol. 10, No. 6, June 1977, p. 74.

[61]   Advanced Encryption Standard FIPS, http://csrc.nist.gov/encryption/aes/, visited Feb. 2002.

[62]   "The Encrypted Jihad," http://www.salon.com/tech/feature/2002/02/04/terror_encryption/index.html, visited Feb. 2002.

[63]   Kocher, P., "Timing attacks on implementations of Diffie–Hellman, RSA, DSS, and other systems." In *Proc. Conf. on Advances in Cryptology* (CRYPTO '96, Santa Barbara, CA), N. Koblitz (ed.), New York, NY: Springer-Verlag,1996, p. 104.

[64]   Hevia, A., and M. Kiwi, "Strength of Two Data Encryption Standard Implementations Under Timing Attacks," *ACM Transactions on Information and System Security (TISSEC)*, Vol. 2, Nov. 1999, p. 416.

[65]   Diffie, W., "Ultimate Cryptography," in 2001 Special Issue of *Communications of the ACM* entitled 'The Next 1000 Years', Vol. 44, No. 3, March 2001, p. 84.

[66]   Steane, A. M., and E. G. Rieffel, "Beyond Bits: The Future of Quantum Information processing," *Computer*, Vol. 33, No. 1, Jan. 2000, p. 38.

[67]   Grover, L. K., "Searching with Quantum Computers," *Dr. Dobbs Journal*, Vol. 26, No. 4, April 2001, p. 34.

[68]   "A Road Map for Digital Forensic Research," *First Digital Forensic Research Workshop (DFRWS2001)*, Nov. 2001, http://www.dfrws.org/dfrws2001/DFRWS_RM_Final.pdf, visited Aug. 2002.

[69]   Krsul, I., and E. H. Spafford, "Authorship Analysis: Identifying the Author of a Program," *Computers and Security*, Vol. 16, No. 3, 1997, pp. 233–248.

[70]   Spafford, E., and S. Weeber, "Software Forensics: Can We Track Code to Its Authors?" CSD-TR-92-010, 1992, Department of Computer Sciences, Purdue Unversity, West Lafayette, IN.

[71]   Tracy, P., "In the News," *The Bull Market Technology Monthly*, http://www.bullmarket.com/wire/020102.php3, Friday Feb. 1, visited Aug. 2002.

[72]   "Security Factoids," ip3.seminars.com, http://www.ip3seminars.com/security/factoids.php, visited July 2002.

[73]   BBC News, "'Snoop' Climbdown by Blunkett," http://news.bbc.co.uk/1/hi/uk_politics/2051117.stm, visited Nov. 2002.

Acronyms

ACE-CI	Allied Command Europe–Counter Intelligence
ACES	Automated Computer Examination System
ACID	Analysis Console for Intrusion Detection
ACL	Audit Control Language
ACM	Association of Computing Machinery
ACPO	Association of Chief Police Officers
ACPR	Australasian Center for Policing Research
ACTS	Automated Computer Time Service
AFIWC	Air Force Information Warfare Center
AFOSI	Air Force Office of Special Investigations
AFP	Australian Federal Police Force
AFRL	Air Force Research Laboratory
ATCSA	Antiterrorism, Crime, and Security Act 2001
ATM	Automatic Teller Machine
BEEP	Blocks Extensible Exchange Protocol
BeOS	Be Operating System
BIOS	Basic Input-Output System
BP	Behavioral Profile

CA	Certification Authority
CAATT	Computer-Assisted Audit Tools and Techniques
CART	Computer Analysis and Response Team
CBIR	Content-Based Image Retrieval
CCIPS	Computer Crime and Intellectual Property Section
CCS	Cusum Cumulative Sum
CDFS	CD-ROM File System
CD-ROM	Compact Disk–Read-Only Memory
CERT-CC	Computer Emergency Response Team Coordination Center
CESA	Cyberspace Electronic Security Act
CF	Computer Forensics
CFCE	Certified Forensic Computer Examiner
CFIT®	Computer Forensics Investigative Tool
CFSAP	Computer Forensic–Secure, Analyze, Present
CHS	Cylinder, Head, Sector
CIA	Central Intelligence Agency
CIA	Confidentiality, Integrity, and Availability
CIDF	Common Intrusion Detection Format
CIF	Common Industry Format
CIITAC	Computer Investigations and Infrastructure Threat Assessment Center
CIP	Critical Infrastructure Protection
CIPA	Classified Information Procedures Act
CLI	Calling Line Information
CMOS	Complementary Metal-Oxide Semiconductor
CNA	Computer Network Attack

CND	Computer Network Defense
CNE	Computer Network Exploitation
CNO	Computer Network Operations
CoE	Council of Europe
COPS	Computerized Oracle and Password System
COTS	Commercial Off-the-Shelf Product
CPU	Central Processing Unit
CRC	Cyclic Redundancy Check
CSP	Communication Service Provider
CSTC	Cyber Solutions Tools Center
Cusum	Cumulative Sum technique (also QSUM)
CVE	Common Vulnerabilities and Exposures
DARPA	Defense Advanced Research Projects Agency
DCCC	Defense Cyber Crime Center
DCFL	Defense Computer Forensics Laboratory
DCS1000	Digital Collection System
DCT	Discrete Cosine Transform
DE	Disk Editor
DERA	Defense Evaluation and Research Agency
DES	Data Encryption Standard
DFRWS	Digital Forensic Research Workshop
DIBS	Disk Image Backup System^{TM}
DISA	Defense Information Systems Agency
DMZ	DeMilitarized Zone
DNA	Distributed Network Attack
DNS	Domain Name Server
DOD	Department Of Defense

DOE	Department Of Energy
DOJ	Department Of Justice
DoS	Denial of Service
DOS	Disk Operating System
DRM	Digital Rights Management
DSTO	Defence Science and Technology Organization
DTD	Data Type Definition
DVD	Digital Versatile Disk
EFL	English as a First Language
ESL	English as a Second Language
EU	European Union
FAQ	Frequently Asked Questions
FAT	File Allocation Table
FBI	Federal Bureau of Investigation
FDI	Fixed Disk Image
FinCEN	Financial Crimes Enforcement Network
FIRE	Forensic and Incident Response Environment
FLETC	Federal Law Enforcement Training Center
FOIA	Freedom of Information Act
FP	Factual Profile
FTK	Forensic Toolkit
FTP	File Transfer Protocol
GB	Gigabyte
G8	Group of Eight
GPRS	General Packet Radio Service
GrIDS	Graph-based Intrusion Detection System
GSM	Global System for Communications

GUI	Graphical User Interface
GUID	Global Unique Identifier
HFS	Hierarchical File System
HIDS	Host Intrusion Detection System
HTCIA	High-Tech Crime Investigation Association
HTCN	High-Tech Crime Network
HTML	HyperText Markup Language
HTTP	HyperText Transfer Protocol
IACIS	International Association of Computer Investigative Specialists
IACP	International Association of Chief of Police
IAP	Intrusion Alert Protocol
IBM	International Business Machines
ICAT	Internet Categorization and Analysis of Threats
ICT	Information and Communications Technology
ID	Identification
IDE	Integrated Drive Electronics
IDEA	Interactive Data Extraction and Analysis
IDIP	Intruder Detection and Isolation Protocol
IDMEF	Intrusion Detection Message Exchange Format
IDWG	Intrusion Detection Working Group
IDXP	Intrusion Detection eXchange Protocol
IE	Information Extraction
IEEE	Institute of Electrical and Electronics Engineers
IF	Intrusion Forensics
IHCFC	International High-Tech Crime and Forensics Conference

ILP	Inductive Logic Programming
IMET	International Mobile Equipment Identity
IO	Information Operations
IOCE	International Organization for Computer Evidence
IP	Intellectual Property
IP	Internet Protocol
IRC	Internet Relay Chat
IRCR	Incident Response Collection Report
IRS-CID	Internal Revenue Service–Criminal Investigation Division
ISO	International Standards Organization
ISP	Internet Service Provider
ISTS	Institute for Security Technology Studies
JPEG	Joint Photographic Experts Group
KDD	Knowledge Discovery and Data mining
KLS	Key Logger System
LAN	Local-Area Network
LANSELM	LANguard Security Event Log Monitor
LDA	Linear Discriminant Analysis
LE	Law Enforcement
LKM	Loadable Kernel Module
MAC	Media Access Control (a MAC address is a physical network address)
MAC	Modification, Access, Creation
MACtime	Modification, Access, Creation time
MD5	Message Digest 5
MOB	Motive Opportunity Benefit

MOD	Magneto-optical Drive
MSC	Mobile Switching Center
MUC	Message Understanding Conference
NASA	National Aeronautics and Space Administration
NATO	North Atlantic Treaty Organization
NCCOSC	Naval Command, Control and Ocean Surveillance Center
NCIS	Naval Criminal Investigative Service
NCP	Network Control Protocol
NCTP	National Cybercrime Training Partnership
NDIC	National Drug Intelligence Center
NFAT	Network Forensic Analysis Tools
NIDS	Network Intrusion Detection System
NIJ	National Institute of Justice
NIPC	National Infrastructure Protection Center
NIST	National Institute of Standards and Technology
NNTP	Network News Transfer Protocol
NPRU	National Police Research Unit
NSRL	National Software Reference Library
NTFS	New Technology File System
NTP	Network Time Protocol
NWCCC	National White Collar Crime Center
NWFS	NetWare File System
OECD	Organization for Economic and Cooperative Development
OLES	NIST's Office of Law Enforcement Standards
OWHF	One-Way Hash Function

PACE	Police And Criminal Act
Patriot	Provide Appropriate Tools Required to Intercept and Obstruct Terrorism
PC	Personal Computer
PDA	Personal Digital Assistant
PGP	Pretty Good Privacy
PH	Profile Hierarchy
PIN	Personal Identity Number
PLAC	Portable Linux Auditing CD
PNL	Pacific Northwest Laboratory
POP	Post Office Protocol
PRM	Probabilistic Relational Model
PRTK	Password Recovery ToolKit
QSUM	Cumulative Sum technique (also Cusum)
Qubit	Quantum Bit
RAID	Recent Advances in Intrusion Detection
RAID	Redundant Array of Inexpensive Disks
RAM	Random Access Memory
RIPA	Regulation of Investigatory Powers Act
ROM	Read-Only Memory
SANS	SysAdmin, Audit, Network, Security (as in SANS Institute)
SATAN	Security Administrator Tool for Analyzing Networks
SCERS	Seized Computer Evidence Recovery Specialist
SCSI	Small Computer System Interface
SHA-1	Secure Hash Algorithm-1
SIM	Subscriber Identity Module

SMART	Storage Media Archival and Recovery Toolkit
SMS	Short Message Service
SMTP	Simple Mail Transfer Protocol
SOP	Standard Operating Procedure
SQL	Structured Query Language
SS	String Search
SVM	Support Vector Machine
SWATCH	Simple WATCHer
SWGDE	Scientific Working Group on Digital Evidence
TB	Terabyte
TCP	Transmission Control Protocol
TCT	The Coroner's Toolkit
UDP	User Datagram Protocol
USSS	US Secret Service
UTC	Universal Coordinated Time
VA	Vulnerability Assessment
VoIP	Voice over IP
VPN	Virtual Private Network
WAN	Wide-Area Network
XML	Extensible Markup Language

About the Authors

George Mohay is an adjunct professor in the Information Security Research Center and School of Software Engineering and Data Communication at the Queensland University of Technology (QUT) in Brisbane, Australia. Before, he had been head of the School of Computing Science and Software Engineering from 1991 to 2002. His teaching and research interests lie in the areas of concurrency, distributed systems, security, intrusion detection, and computer forensics. He has worked as a visiting researcher while on sabbatical leave at Stanford University in 1981, Loughborough University in 1986, Bristol University in 1990, and the Australian National University in 2000. He is a member of the IEEE Computer Society and has been a long-time member of the ACM, IEEE, and ACS. He received a B.Sc. (Honors) (UWA) in 1966 and Ph.D. (Monash) in 1970.

He supervises Ph.D. and master's students in the areas of security, intrusion detection, operating systems, and distributed systems and has recently been involved as chief investigator in the following research projects:

- *Commonwealth government-funded SPIRT research project (1999–2001):* "Component system architecture for an open distributed enterprise management system with configurable workflow support" in collaboration with Mincom (Pty) Ltd., one of Australia's largest developers and exporters of software.

- *Trusted and secure components:* One of several component technology-related projects hosted by the Programming Languages and Systems Research Center.

- *Ongoing projects in the area of computer forensics and authorship analysis in collaboration with the Defence Science and Technology Organization, Australia.*

Alison Anderson is a senior lecturer in information systems at QUT. Her research interests include information risk specification, critical information dependency assessment, and computer forensics, including perpetrator identification. Completed and current projects include risk evaluation in banking, computer forensics, and most recently, e-mail authorship identification by machine learning. She is a chief investigator in ongoing projects in the area of computer forensics and authorship analysis in collaboration with the DSTO.

Her qualifications are B.Sc. (University of Queensland), M.InfSys. (University of Queensland), and Ph.D. (QUT).

Byron Collie is currently a senior information security engineer in Minneapolis, Minnesota, with the Wells Fargo Bank. Prior to this, he was a federal agent for 15 years with the Australian Federal Police (AFP)–Technical Operations, specializing in national information infrastructure protection, computer intrusion investigation, and computer forensics. Between March 1998 and July 2000, he was seconded from the AFP to the Royal Australian Air Force (RAAF)–Directorate of Information Warfare at Headquarters Air Command.

He has represented the AFP and Australia at a number of international forums, including the Forum of Incident Response Teams Computer Security Incident Handling Conferences in 1996 and 1999 and the Interpol Asia-Pacific Region Computer Crime Workshop in 1997. He has presented on the subject of computer crime and security at a large number of industrial conferences in Australia.

With Kymie Tan, a Ph.D. student from the Computer Forensics and System Security Group at the University of Melbourne, he undertook research into the use of artificially intelligent neural networks to detect intrusive behavior in TCP/IP network traffic. The results of this research were subsequently published and presented at the Annual Computer Security Applications Conference in San Diego, California, in December 1997.

Byron has contributed to the Information Assurance and Computer Network Defense section of a U.S. military book entitled *Information Operations: The Hard Reality of Soft Power*, published through the National Defense University.

Olivier de Vel obtained an M.Sc. (Honors) from the University of Waikato (New Zealand) in 1974 and a Docteur de Troisième Cycle from the Institut National Polytechnique de Grenoble (France) in 1978. He is currently the head of the Operational Information Security Group, Information

Assurance Branch in the DSTO, Australia, where he is undertaking R&D in computer forensics. Prior to his current position, he was associate professor in the Department of Computer Science at James Cook University. He has also worked in industry and various governmental and educational institutes. His current interests include computer forensics and dynamic computer security.

Rodney McKemmish has worked as a computer forensic specialist in Australian law enforcement and the private sector for the past 10 years. As a member of the Victorian police force for 13.5 years, he specialized in computer crime investigations and computer forensics for the last 6.5 years of his service. In 1997, Rod took up an appointment as officer in charge of the forensic computer examination unit of the Queensland Police Service.

During his law enforcement service, he conducted computer forensic examinations on a wide variety of computer systems and electronic devices. In his work, he has provided expert assistance and testimony in a wide range of criminal investigations. These have included fraud, drug, homicide, computer misuse, and pedophilia investigations.

In 1998, he completed an international research project on forensic computing under the auspices of the 1998 Donald Mackay Churchill Fellowship. During this research project, he attended the leading law enforcement agencies actively involved in forensic computing in the United States, Canada, Europe, and the Middle East.

In August 1999, he joined the accountancy firm KPMG to help establish a dedicated computer forensic service within the Australian Forensic Accounting Group. During his time with KPMG, he has undertaken a wide range of computer forensics, including investigations into employee misbehavior, IT systems abuse, intellectual property disputes, as well as the delivery of technical training on computer investigation techniques. Additionally, he has given expert opinion and testimony in a number of civil disputes.

He holds a bachelor of business degree from the Royal Melbourne Institute of Technology and has actively developed a range of computer forensic software applications, which have been used by law enforcement agencies in Australia and overseas. Additionally, he has delivered to industry and law enforcement a number of papers on a wide range of topics centered on computer crime and computer forensics.

Index

A

Access rights, 83
Across-event analysis, 277
Active vulnerability scanner, 302
Activity graph, 271–72
Admissibility, evidence, 184, 308–10
Advanced Password Recovery Software Toolkit, 145
Aggregation, data, 275–76
Air Force Research Laboratory, 168, 265–66, 362
Alert correlation, 275–76
Alert storm, 276
Allied Command Europe Counter Intelligence, 169
Ambient data, 43, 53–55, 63, 142
Analysis
    attacks, 306–8
    computer intrusions, 276–85
    secured data, 140–42
    seven-stage model, 211–18
    techniques, 143–46
    tools, 61–64, 154–59
Analysis Console for Intrusion Detection, 276
Analysis methodology development stage, 217
Analysis model, 131–32
Analyst's Notebook, 62
Anomalies, stegoforensic analysis, 347–48
Anomalous behavior identification, 258, 260–61
Anomaly-based fraud detection, 205–6
Anomaly-based intrusion detection system, 260, 265–67, 269–70, 273
Antiterrorism legislation, 93, 98–103
Application attacks, 299, 300
Application proxies, 293
Applications, 33
A priori algorithm, 326–27, 337
Archiving media images, 140
ARPANET, 2
Artificial intelligence, 62
Association rule, 273–74, 325
Association rule mining, 325–27, 335–39
Assurance state, 35
Asymmetric attack, 20, 340
Asymmetric (public key) cryptography, 355–57, 359
Attachment Interface Specification, 53
Attack analysis report, 306–8
Attack computer, 299
Attack phase, 298, 299, 302–3, 308
Attack types, 295–303
Auction fraud, 11
Audit control language, 189
Auditing, 175–77
Audit trails, 36
Australasian Center for Policing Research, 164
Authentication, 27, 130, 140, 152–54
Authenticity, 35
Author characterization, 331
Authorship attribution, 321, 331–34

Automated Computer Examination System, 65, 156–57
Automated computer time service, 284
Automated intrusion response, 311–12
Availability, system usage, 35

B

Back-door access, 95, 299, 308
Backing store, 24
Ball of string, 208–9
Bartlett v. Weir Ors, 80–81
Basic Security Module, 83–84
Bayesian networks, 344–45
Behavioral profile, 335–39
Benefit of fraud, 178–79
Benford's Law, 188, 198–200
BeOS, 148
Beowulf clusters, 149
Bertillonage, 14
Best evidence rule, 124–26
Binary coding, 122, 141
BIOS setting, 138, 284
Bit-by-bit duplicate, 43, 50, 51
Bit-stream duplicate, 50, 51, 116
Blackmail, 7
Blind event logs, 287
Block cipher, 52
Blocks extensible exchange protocol, 275
Bogus products or services, 183
Bookmarks, 68
Booting system, 44, 52, 79
Bottom-up method, 62
Boundary router packet filter, 291–92
Browser history, 144–45
Brute force attack, 358, 359
Business analysis phase, 212
Business stakeholders, 19
ByteBack, 150

C

Cache, 135
Calling line information, 241
Card4Tools, 60
Cards4Labs, 59, 61
Carnivore, 94–96, 98
CD-ROM, 69, 135
Centralized intrusion detection system, 267

Central processing unit, 24
Certification, 164
Certification authority, 357
Chain header, 268
Chain of custody, 23, 139, 142
Chain of evidence, 27, 43, 44, 56
Chain option, 268
Challenge, 184
Chat room participation, 28
Checksum, 51
Chosen ciphertext attack, 358
Ciphertext, 355
Civil courts, 19
Civil liberties, 91
Civil rights, 30
Civil time, 283, 284
Classified Information Procedures Act, 226
Clementine, 63, 190
Client computer, 33
Clusters/clustering, 48, 55, 57–58, 184, 210–11
Coding faults, 286
Coefficient of correlation, 196
Coefficient of determination, 196, 221
Collection, evidence, 27
Command line tools, 135, 150, 152, 153–54
Commercial off-the-shelf products, 286
Commonality detection, 270
Common intrusion detection format, 275
Common vulnerability and exposures, 300
Communication protocols, 288–95
Communications content surveillance, 94–95
Communications data, 95, 96–97
Communications service provider, 94, 98, 101, 102
Communication tools, computers as, 9
Comparison analysis, 118
Competencies, 160–64
Computer Analysis and Response Team, 65, 114, 156, 168
Computer-assisted audit tools and techniques, 176, 211
Computer-assisted crime, 10, 12
Computer crime, 6–12
Computer Crime and Intellectual Property Section, 86–87, 99–100
Computer Emergency Response Team Coordination Center, 227, 257, 260

Computer Forensic Investigative Tool, 63, 64,
    65–66, 72–76
Computer forensics, 1, 3, 5–6, 11, 13, 17, 21
    future, 360–364
    origins and history, 113–17
    research directions, 258, 319–23
    versus intrusion detection, 257–63
    within tradition, 14–21
Computer forensic secure, analyze, present
    model, 128–33
Computer Forensics Investigative Tool, 62, 282,
    285, 334, 341
Computer Forensics Tool Testing, 51
Computerized Oracle and Password System, 302
Computer network attack, 165
Computer network exploitation, 165
Computer Network Operations, 164–65
Computer piracy, 11
Computer-related crime, 10, 12, 25
Computer security, 31–37
Computer storage, 17
Concepts, Ferret, 73–76
Conceptual hierarchies, 337–39
Confidentiality, 35
Confidentiality, integrity, and availability, 285
Connection laundering, 277
Connectivity, 3, 4, 144
Consensual monitoring, 227
Content analysis, 72, 94–95, 117–18
Content-based image retrieval, 350–52
Content identification, 141
Context, 132
Copying software, 142
Copying system files, 137
CopyQM, 116
Coroner's Toolkit, 45, 56, 305–6
Corporate crime, 19
Correlation analysis, 195–96, 275–76
Council of Europe treaty, 30, 90–91, 123
Countermeasures, 36–37
Court testing, 244–53
CreationTime, 82
Credit card fraud, 7, 183
Critical Infrastructure Protection, 167–68
Cryptanalysis, 145–46, 321, 345–46, 358
Cryptography, 321, 345–46, 355–60
Cuckoo's Egg, 165
Current previous variance, 203

Custom and excise, 18–19
Custom attack simulation language, 302
Customer profiling, 337
Cybercrime, 10, 30
Cyberspace Electronic Security Act, 226
Cyclic redundancy check, 71, 152
Cylinder, head, sector addressing, 48, 50
Cylinder, head, sector translation, 50–51

D

Darwin, 190
Data acquisition, 43–44, 139–40, 215–16
Data aggregation, 275–76
Data analysis stage, 217–18
Database software, 189
Data Encryption Standard, 358, 359
Data filtering, 64
Datalifter v2.0, 158–59
Data link layer, 34
Data mining, 184, 186–88, 189–90, 212, 264,
    272–74, 323–27
Data mining tools, 62–63, 64
Data preservation, 101
Data recovery, 43, 63, 71, 115–17
Data reduction phase, 321
Data retention legislation, 93, 101–103
Dataset size, 320–21, 324
DataTrail FacTracker, 56
Data type definition, 275
Date-based search, 71
Date documentation, 138–39, 140–41
Daubert standard, 15
dd file utility, 151
Deallocated clusters, 55, 57
Decision trees, 325
Deconstruction engine, 72
Decryption, 355
Defense Advanced Research Projects Agency, 2,
    265–66, 273, 312
Defense Computer Forensic Laboratory, 149,
    168
Defense in depth, 264, 285–88, 290, 291, 302
Defensive information warfare, 20
Defragmentation, 37, 47–48
Degaussing, 59
Deleted data files, 118
Demarc, 276

Demilitarized zone, 293
Denial of service attack, 7, 274, 299–300, 339
Deny (illegal) outcome, 265
Dependent variable, 191, 192
Destination Internet protocol address, 292
Detection phase, stegoforensic analysis,
    346–47
Digest, 52
Digital collection system. *See* Carnivore
Digital evidence. *See* Evidence
Digital Forensic Research Workshop, 362–63
Digital rights management, 363
Digital versatile disk recordable, 52
Digit analysis, 188, 198–200
Direct evidence, 146
Disabling phase, stegoforensic analysis, 346, 347
Disconnection, 275
Disk editor, 115
Diskettes, 28, 42–43
Disk file organization, 46–49
Disk geometries, 48
Disk Image Backup System, 56, 116, 151
Disk imaging and analysis, 49–51, 52, 63
Disk mirroring, 50–51
DiskSearch Pro, 153
Dispersion analysis, 196–98
Distributed anomaly-based intrusion detection
    system, 269–70
Distributed intrusion detection system, 267
Distributed Network Attack, 146
Diversion attacks, 271
Documentation, 136–37, 138–39, 140–41,
    142
Domain name server, 34, 290, 292
Dragon Sensor, 270
Dragon Server, 270, 271
Dragon Squire, 270
Drill-down analysis, 193, 337–39
Drive Lock, 139
Drugs, 183
Dr Watson error logs, 144
dtSearch, 153
Duplicate investigation, 201
Duplicate invoices, 183
Duplicate payments, 181
Duplication, evidence, 125–26, 130, 149–52
Dutch National Forensic Institute, 116
Dynamic storage, 24

E
Electronic evidence. *See* Evidence
Electronic Privacy Information Center, 96
"El Griton," 229–36
E-mail, 28, 64, 145, 158–59
E-mail author identification, 321, 331–34
Embedded devices, 24, 44, 116
Embedded system forensics, 321, 322
Embezzlement, 7
EMERALD, 269–70
Emergent faults, 286
EnCase, 45, 53, 56, 62, 65, 66, 67–69, 79, 141,
    147, 148, 150, 156, 282
Encryption, 243–44, 355
End of file overrun, 48
Enron Corporation, 59, 118
Enterprise Miner, 190
Entity relationship modeling. *See* Link analysis
Environmental factors, 184
Environmental scan stage, 214–15
Escript macro language, 68–69
Ethereal, 306
Ethical system, 5, 6
Europe, 164
European Commission, 93–94
European Union, 102, 363–64
Evaluation and validation stage, 212
Evaluation, intrusion detection systems,
    265–66
Event analyzer, 275
Event database, 275
Event generator, 275
Event lag, 284–85
Event log analysis, 81–84, 278–80, 287–88
Event response unit, 275
Event storm, 276
Event time frame, 282–85
Evidence, 3, 12–13, 15–16, 17, 20
    admissibility, 13, 184
    characteristics, 21–23
    contamination, 22–23
    continuity, 23, 27, 43, 44, 56, 139, 142
    discovery, 61–63
    documentation, 142
    forensic methodologies, 42–63, 123–28
    jurisdictional issues, 123
    legal requirements, 29–31, 121–22
    real and direct, 146

retrieval and analysis, 23–27
    sources, 27–29, 34
Evidence-destroying countermeasures, 37
Evidence extraction, 118, 132, 321, 339–43
Evidence file, 67–68
Evidence window, 70
Expense account misuse, 182
Expertise test, 128
Expert witness, 15, 155
Expert Witness software, 67
Exploits, 299
Extensible markup language, 275, 276
External fraud, 180, 182–83
Extortion, 11, 12
Extraction. *See* Evidence extraction
Extraction phase, stegoforensic analysis, 346,
    347

F

Factual profile, 335
False alert rate, 265
False financial reporting, 181–82
False identity, 183
FastBloc, 45, 150
Federal Bureau of Investigation, 94, 95, 114,
    166, 168, 320
    Handbook of Forensic Services, 117–18
Federal Law Enforcement Training Center, 115,
    116, 164
Ferret Discovery Engine, 73–76
Fictitious supplies, 181
File allocation table, 49, 67, 69
File-by-file analysis, 50, 56–57, 79
File conversion, 64
File deconstruction, 72
File deletion, 57–59
File extensions, 57
File Extractor, 154
File filtering, 68, 71
File imaging and authentication, 51–53, 63
File logical order, 47–48
Filemap, 47, 48, 49, 53, 55, 57
File shredder utilities, 59
File signature, 36, 141
File slack, 142
File slack space, 53–55, 142
File transfer protocol, 2, 34, 269
File viewers, 159–60

Filtering, 275, 321
Financial Crimes Enforcement Network, 343
Financial pressures, 184
Fingerprinting files, 52, 130–31, 345
FireChief, 139
Firewall, 291–92, 293
Firewall ruleset, 292
Fishing warrants, 80
Fixed Disk Image, 116
Flat-text document categorization, 329–30
Foremost, 154
Forensic accounting, 175–77
Forensic filtering process, 25–26
Forensic platforms, 45–46
Forensics, 6, 14–15
Forensic Toolkit, 56, 140–41, 142, 148, 155–56
Forensic Toolkit, AccessData Corporation, 65
ForensiX, 45, 157
Forged checks, 183
Format conversion, 118
Fragile data, 56
Frame header, 290
Fraud, 118
    definition, 177–80
    internal vs. external, 180–83
    understanding, 183–84
Fraud detection, 176–77
    techniques, 190–206
    technology, 184–90
Fraud rules definition stage, 216–17
Fraud triangle, 178–79
Freedom of Information Act, 96
Freezing the scene, 14, 22
Frequent episodes, 273–74
Functionality evaluation, 141
Function words, 332
Fuzzy logic, 64, 184, 204–5

G

G8 Meeting of Justice and Interior Ministers, 91
Galaxies tool, 330–31
Gambling, 183
Generic forensic capability, 66
GenX, 56
Global computer crime, 23
Global connectivity, 3
Global system for mobile communications, 61
Global unique identifiers, 159–60, 239–40

Graph-based intrusion detection system, 271–72
Graphical user interface, 135
Graphics files, 118
Greed, 183
Greenwich Mean Time, 283
grep, 154
GUIDClean, 159–60

H

Hacking, 7, 11, 117, 120–21
Harassment, 11
Hardware configuration documentation, 138
Hash analysis, 53, 68, 71–72, 152
Hashkeeper, 72, 153
Hashkeeper database, 141
Hash sets, 68, 71–72
Heuristic methods, 62
Hidden areas, 53–56
Hidden files, 144
Hiding information, 345
Hierarchical file system, 69
Hierarchy drill-down, 337–39
Hierarchy drill-up, 337–39
High low variance, 202–3
High next highest variance, 203
High-Tech Crime Investigation Association, 164
High-Tech Crime Network, 164
Historical trend analysis, 201–2
Homicide cases, 118–19
Host-based intrusion detection system, 262, 265, 266–67, 270–71, 293–94
Hostile system, 135, 138
Human behavior, 4–6
Hybrid network and host intrusion detection system, 270–71
Hypertext markup language, 2
Hypertext transfer protocol, 2, 34, 269
Hypothesis development tools, 27

I

Identification, crime, 22
Identification of data, 130
Identity, 132
iDetect, 235–36
IIT Research Institute, 95

Illegal outcome, 265
ILook Investigator, 65, 66, 69–72, 156
Image mining, 321, 349–55
Imaging, 49–57, 61
    categories, 63
    disks, 49–51
    files, 53
    hidden areas, 53–56
    logical analysis, 56–57
    origins, 115–16
    procedures, 79
    tools, 26
IMDUMP, 115
Incident response and investigation, 303–6
Incident Response Collection Report toolkit, 305–6
Incident response/incident handling, 259–60
Incident response team, 304
Incident/security incident, 260
Independent variable, 191, 192
Inductive learning, 340
Inductive logic programming, 344
Informal time, 282, 283
Information and communications technologies, 319–20
Information-capture tools, 26
Information extraction, 340
Information Operations, 164
Information revolution, 1–2
Information store type, 25
Information warfare, 257–63
Informing the court, 15
Insider activity, 260
Insiders, 31
Insider trading, 182
Intrusion forensics, 21
Insurance firms, 19
Integrity, 22–23, 35, 51, 56, 305, 311
Integrity checkers, 52–53, 63, 65
Intelligent Miner, 190
Intelligent Miner for Data/Text, 63
Intelligent Miner for Text, 330
Interactive data extraction and analysis, 189
Internal auditing, 176
Internal fraud, 180–82
Internally developed software, 189
Internal Revenue Service Criminal Investigations Division, 114

International Association of Computer Investigative Specialists, 116–17, 124–28, 143, 164
International Association of Police guidelines, 88
International crime, 19
International Criminal Police Organization, 123
International Hi-Tech Crime and Forensics Conference, 123–24
International mobile equipment identity, 61
International Organization for Computer Evidence, 89, 123, 160
International Organization on Computer Evidence, 11
Internet, 4, 32–34, 42, 92
   usage analysis, 143, 144–45
Internet categorization and analysis of threats, 300
Internet protocol, 34
Internet protocol address, 289–95
Internet service provider, 32, 93, 94, 97–98, 102–3
Interoperability, 274–76
Interpretation of data, 132
Interquartile range, 197
Intruder Detection and Isolation Protocol, 312
Intrusion, 12, 257, 259
Intrusion analysis, 276–85
Intrusion alert protocol, 275
Intrusion detection, 230–31, 257–63
Intrusion detection exchange protocol, 275
Intrusion detection expert system, 269
Intrusion detection message exchange format, 275, 276
Intrusion detection systems, 259–63
   evaluations, 265–66
   evolution, 264–67
   future directions, 310–12
   interoperability, 274–76
   in practice, 267–74
Intrusion Detection Working Group, 275
Intrusion forensics, 3, 5–6, 11, 13, 134
   attack analysis, 306–8
   case study, 308–10
   defined, 259
   future, 310–12, 360–64
   incident response, 303–6
   networks, 103–4
   research directions, 319–23

Intrusive operation, 44
Invasive operation, 44
Inventory theft, 181
Investigation, 3
Investigation management tools, 27
Investigative profiling, 321
Invisible Keylogger Stealth, 227
IRFanView 32, 159
iScript, 236
iWatch, 230–31, 236

J
Java Agents for Meta-learning, 272–74
Joint Photographic Experts Group, 57, 58
Jurisdictional issues, 123
Justice Department guidelines, 86–87, 96, 99–100

K
Kernal functionality, 303
Keyghost, 228
Keystroke logging system, 120, 224–29
Keyword folders, 68
Keyword search, 68, 73, 118, 141–42
Knowledge discovery and data mining, 62
Knowledge discovery stage, 212
Known ciphertext attack, 358
Known File Filter, 65

L
LANguard Security Event Log Monitor, 278
LANSELM, 278–79
Laptops, 24
LastAccessTime, 82
LastWriteTime, 82
Latent data, 55–56
Law enforcement, 30, 44, 58, 65, 69
   computer forensics model, 128–33
   computer forensics role, 117–21
Law Enforcement Computer Evidence Suite, 56
Lawyers, 19
Layered message handling, 34
Learning mode, 261
Legal considerations, 29–31
Legal outcome, 265
Legislation, 90–103

Lightweight intrusion detection system, 271, 274
Lightweight network intrusion detection system, 268–69
Limited examination, 143
Linear discriminant analysis, 349
Linear trend regression analysis, 192–94, 221
Link analysis, 340, 341–43
Link discovery, 321, 340, 343–45
Link (relationship) analysis, 62, 64, 145, 184, 207–9
Linux, 45, 46, 50, 148–49, 279–80, 303
Linux Forensic Toolkit, 45
ListDirectory command, 82
List index node, 49
"Little Nicky" Scarfo, 223–29
Live system processing, 133, 134–36, 137–38
Loadable Kernel Module, 303
Local-area network, 17, 32, 33, 268
Locard's Exchange Principle, 14
Logical analysis, 50, 56–57, 79, 140
Logical backup software, 115
Logical disk, 48
Logical file system, 135
Logical volume imaging, 137–38
Logs. See Event log analysis; System log analysis
Love Bug virus, 92

M

Mace Utilities, 114
Machine-learning techniques, 264
Macro viruses, 236–37
MACtimes, 82–84, 85, 138–39, 140–41, 145
MADAM intrusion detection system, 272–74
Magic number, 57, 71
Magnetic force microscopy, 56, 58
Magnetic media, 43–44, 139–40
Magneto-optical drive, 116
Malicious code protection, 140
Management fraud model, 179–80
Market-based data application, 325–26
Master copy, 131
Master file table, 49
md5sum, 152
Media access control address, 295
Media access control addresses, 138, 160
Media sanitization, 57–59
Melissa, 236–42
Memory, virtual and physical, 135, 142

Message digest 5, 52, 65, 67, 71, 152, 270
Messages log file, 74
Message Understanding Conference, 340–41
Metainformation, 76–77, 81–84
Metalearning agents, 274
Methodology test, 128
Microsoft Windows, 45, 49, 55, 56, 57, 67, 69, 82, 83, 84, 142, 144, 147–48, 278, 284
Misbehaving employees, 245–48
Misuse identification, 258, 273
Mitigating countermeasures, 36
Mitnick, Kevin, 120–21
Mobile switching center, 60–61
Mobile telephone, 43–44, 59–61
Model-based image mining, 354
Modeling stage, 212
Modification, access and creation date, 52
Monitoring networks, 288–95
Moonlight Maze, 166–67
Morris worm, 300
Motive for fraud, 178–79, 182, 183–84
MS-DOS, 147
Multimedia file identification, 118
Multiperspective analysis, 270
Multiple jurisdictions, 17
Multiple regression analysis, 195, 221
Multislot extraction algorithms, 340
Multi-step attack, 271

N

Naive Bayes probabilistic classifier, 330, 340
Narcotics cases, 119
NASA Computer Crime Division, 149, 166
National Center for Forensic Science, 89
National Consortium for Justice Information and Statistics, 164
National Crime Squad, 97
National Cybercrime Training Partnership, 88, 163
National Drug Intelligence Center, 65, 153
National Infrastructure Project, 18
National Institute of Justice guidelines, 88–89
National Institute of Standards and Technology, 51, 65, 153, 267–68, 300
National security, 18, 20, 30, 164–69
  organizations, 168–69
National Software Reference Library, 65, 141, 153

National White Collar Crime Center, 163
NATO Lathe Gambit Information Security
    program, 164, 169
Naval Criminal Investigative Service, 97
[N] Curses Hexedit, 156
NetAnalysis, 159
Netmap, 62
Net Threat Analyzer, 153–54
NetWare file system, 69
Network analysis. *See* Link analysis
Network applications, 145
Network attacks, 299
Network-based intrusion detection system,
    265, 266–67, 268–269, 270–72, 276,
    293–94
Network block device, 149
Network connection, 24
Network control protocol, 2
Network Flight Recorder, 236
Network forensics, 259, 322
Network forensics analysis tools, 306, 311
Network history, 118
Network host, 144
Network IDES, 269
Network interface card, 292
Network Intrusion Detector, 230–31,
    234–36
Network layer, 34
Network management system, 276
Network news transfer protocol, 34
Networks, 103–4, 285–303
Network time protocol, 284
Neural nets, 62, 330
New technology file system, 49, 67, 69
N-grams, 74–75
NMap, 301
Noise events, 278
Noninvasive operation, 51–52
Noninvasive system boot, 44
Nonlinear trend regression analysis, 194–95
Nonprintable characters, 53
Nonvolatiles, 122
Noop attack, 307
Normal activity, 261
Normal behavior model, 265
Norton Ghost, 151
Norton's Disk Editor, 115
Norton's Utilities, 114

O
Objective definition stage, 212–14
Offensive information warfare, 20
Office of Law Enforcement Standards, 88
Omnivore. *See* Carnivore
One-way hash function, 52–53, 65, 130–31,
    140, 141
Ontology, 73
Operating system attacks, 299
Operating system configuration information,
    140
Operating systems, 147–49
Operation Oplan Bojinka, 243–44
Opportunist fraud, 184, 248
Opportunity for fraud, 178–79, 180–81
Order of volatility, 135, 304–5
Organization for Economic and Cooperative
    Development, 123
Organized crime, 120
Outliers, 210–11, 337
Outsiders, 17, 31, 260
Overwriting file clusters, 58
Ownership, 132

P
Packet filtering, 291–93
Packets, 288–95
Packet switching, 288–95
Page-files, 55
Palladium, 363
Partition imaging, 50, 51
PartitionMagic, 53
Partitions, 48–49
Passive intrusion detection system, 267
Passive vulnerability scanners, 301–2
Password recovery, 145–46, 228–29
Password Recovery Toolkit, 65, 145,
    155–56
Passwords, 118
Patriot Act, 31, 98–101
Pattern and relationship analysis, 62, 184,
    200–3, 323–27
Pattern-matching network intrusion detection
    system, 271–72
Pattern Recognition Workbench, 190
Pay-off and exit phase, 298, 303
PDA Seizure, 59

Pedophilia, 119–20
Performance monitoring and recovery, 36
Peripheral memory, 135
Permanent storage, 25
Permit (legal) outcome, 265
Personal digital assistant, 43–44, 59–61, 116, 322
Personalization applications, 336–37
Personal level, 19–20
Philippines, 92
Physical analysis procedures, 79
Physical hard drive, floppies, and tapes, 135
Physical imaging, 49–51, 56
Physical layer, 34
Physical memory, 135
Picasso, 72–73
Plagiarism detection, 331
Plaintext, 355
Plaintext attack, 358
Platters, 47
Police, 19
Police and Criminal Act, 17
Port scanners, 301
Postevent analysis, 186, 190
Post office protocol, 34
Postseizure analysis tools, 27
Potential evidence, 22
Premeditated fraud, 184
Preparation of data, 131
Preprocessor functionality, 293
Prescanning evidence drive, 67
Presentation of evidence, 127, 132–33, 142, 146
Preservation of data, 130–31
Pretty good privacy, 120, 224
Preventative security, 264
Preventive countermeasures, 36
Previewing, 67
Prime directive, 133
Printed records, 28
Printouts, 135
Privacy, 30, 94–95
Probabilistic relational model, 345
Problem definition phase, 212
Procedures, standard operating, 133–43
Processing of data, 131–32
Process integrity principle, 305
Profile association rule, 337

Profile-based fraud detection, 205–6
Profiling, investigative, 335–39
Protected computer, 100
Protocols, 33
Protocol stack, 289–90
Proving a hypothesis, 22
Proxy server, 293
Psychological disorder, 184
Public key (asymmetric) cryptography, 355–57, 359
Pulling the plug, 303

Q
Qualified bit-stream duplicate, 51
Quantum computing, 360
Query-retrieval method, 352–54
Quick View Plus, 159

R
Range measure, 197
Ratio analysis, 202–3
Reactive intrusion detection system, 267
Read only memory, 25, 52, 55
Real evidence, 146
Real-time fraud detection, 205–6
Real (wall-clock) time, 282–83
Reconnaissance attacks, 299
Reconnaissance phase, 296, 298–300, 307
Recordable compact disk, 52
Recovered folders, 67
Recovery countermeasures, 36
Recovery processes, 27, 117–18
Red flags, 179–80, 185–86, 204
Redundant array of inexpensive disks, 50, 52, 67, 137
Redundant protection, 36
Reference data set, 65, 72
Registers, 135
Regression analysis, 191–95, 221
Regulation of Investigatory Powers Act, 30, 94–98, 364
Relationship analysis. *See* Link (relationship) analysis
Relevance, 132
Relevant public authority, 97
Remote connection, 32
Remote host, 32

Remote preview, 67
Repository for information, 9
Resource objects, 269
Resplendent Register, 159
Retail organization, 185
Retribution, 184
Revenue Canada, 114, 115
Reverse engineering, 322–23
Review of analysis results stage, 218
RIPPER, 273–74
Risk avoidance, 286
Risk management, 286
Role enforcement, 31
Rolling over logs, 36
Rootkits, 302–3, 339
Rule-based fraud detection, 205, 206, 292
Rule-based learning algorithms, 340–41
Rule induction, 62
Running processes, 135

S
SafeBack, 116, 147, 150
Safe deletion, 59
Safe file deletion software, 56
SANS Institute, 300
Scatter diagram, 192–93
Scientific Working Group on Digital Evidence,
        89, 123, 160
Screening router, 291–92
Script attacks, 299
Search and seizure procedures, 23, 28, 61,
        76–77
    standards, 86–89
    steps, 77–80
    time attributes, 81–84
    time-lining, 84–86
    training, 116–17
    warrants, 80–81
Search boot, 71
Search for relevant data, 131–32
Search tools, 64, 153–54
Secret commissions, 182
Secret cryptographic key, 355
Secret key (symmetric) cryptography, 355–56,
        359
Secret Service guidelines, 88, 136
Sector addressing, 46–49

Sector-by-sector analysis, 50, 79, 114–15
Secure, analyze, present model, 136, 304
Secure boot, 44, 52, 79
Secure hash algorithm 1, 52, 65, 71, 152
Securing evidence, 129–131
    computer forensics model, 129–31
    guidelines, 136–40
Security
    case study, 308–10
    computer, 31–37
    future needs, 363–64
    national, 164–69
    network, 285–303
    policies, 5, 20, 31, 32, 35–37
    software design, 264
Security Administrator Tool for Analyzing
        Networks, 302
Security event log, 84
Security flaw analysis, 285–88
Security incident, 260
Seized Computer Evidence Recovery Specialist,
        116
Seizure, 3, 77–81. See also Search and seizure
Semantic content, 354
Semantic relationships, 340
Semipermament storage, 25
Sequencing, 200–1
Server computer, 33
Session hijack, 34
Session logs, 36
Shadow data, 58–59
Short message service, 61
Shut-down guidelines, 77–78, 138
Signature analysis, 68, 205–6, 250–53, 258,
        260–61
Signature-based detection systems, 265–67,
        268, 273
Signatures, steganographic, 348
Simple mail transfer protocol, 34
Skill level, forensic examiner, 127–28
Small computer system interface, 52
Smart cards, 60
SMART Watch, 53
Snapback DatArrest, 150
Sniffers, 268, 270, 302, 303
Snort, 268–69, 276, 307
Software development, 66
Solar Sunrise, 165–66

Solo II, 151–52

Sound bit stream image duplication, 115

Source code, 118

Source Internet protocol address, 292

Special purpose devices, 24

Spoofing, 34

Spreadsheet software, 189

Stalking, 11

Standard deviation, 197–98

Standardization, computer forensics, 17–18

Standard operating procedures, 133–43

Standards, 76, 86–89, 123–28, 274–76

Statistical analysis, 191–200

Statistical modeling imaging mining, 354

Steganographic file system, 53

Steganography, 345

Stegdetect, 348

Stegoforensic analysis, 345–49

Stegoforensics, 321

Stegokeys, 346

Stegoonly forensics, 347

Stepping stone, 277, 299

Sterile conditions, 136

Storage flushing, 37

Storage media, 9, 118

Storage Media Archival and Recovery Toolkit, 157–58

String Search, 154

String term search, 71, 154

Subscriber information module, 60, 61

Successful outcomes, 84

Supervised text categorization, 328

Supporting props, 146–47

Support vector machine, 325, 330

Suspects, potential, 21

Suspicious activities, 276–77

Swap-file, 55

Symmetric (secret key) cryptography, 355–56, 359

Synchronization, computer system, 283–84

Syntactic relationships, 340

System clock, 284

System for Copyright Protection, 348

System information, 76–77

System log analysis, 29, 36, 42, 76–77, 81–84, 82–84, 137, 144

System swap-file, 55

System usage analysis, 143, 144

T

Target of crime, 8, 9, 12

Task-dependent objects, 340

Taxation agencies, 114

Technology test, 128

Teledisk, 116

Telnet, 2, 34

Temporal analysis. *See* Time-line analysis

Temporary storage, 25

Tender document, 248–53

Tender manipulation, 181

Term nodes, 74

Terms (keywords), 73

Terrorism, 93, 167, 242–44

TextAnalyst, 331

Text categorization, 321, 327–31

Text string search, 141–42

Themscape tool, 330–31

Time attributes, 81–84, 85, 281

Time-based search, 71

Time documentation, 138–39, 140–41

Time-line analysis, 62, 64, 68, 73, 77, 82, 84–86, 143, 145, 184, 209–10, 280–85, 321

Time-posted values, 86

Timestamps, 52, 68, 82, 84, 86, 281, 283, 284–85

Timing attacks, 359

Tipping off, 98

Toehold, 298, 306, 307

Toolkits

copying, 136

disk analysis, 56

file imaging, 52–53

forensic platforms, 26–27, 45–46, 61–76

fraud detection, 189

hashing, 71–72

intrusion forensics, 305–6

mobile phones/PDAs, 59–61

origins, 114–17

validation, 135–36

Traffic data surveillance, 94–95, 102

Traffic type, 292

Training and certification, 116–17, 163–64

Transaction analysis, 118

Transferring countermeasures, 36

Transmission control protocol/Internet protocol, 2, 34

Transnational crime. *See* International crime
Transport control protocol, 34, 289–90
Transport layer, 34
Trend analysis, 191–92, 201–2, 221
Trend regression analysis, 192–95
Triggers, 74–75
Tripwire, 52
Trojaned utilities, 298
Trusted command shell, 137
TULP, 59, 61
Tumbleweed Communications Corporation
    survey, 7
Type anomalies, 141

U

Unallocated space, 142
Unauthorized manipulation of data, 181
United Kingdom, 17, 30–31, 62, 92,
    363–64
    Association of Chief Police Officers,
        87–88
    Defense Evaluation and Research Agency,
        169
    guidelines, 123–24
    legislation, 31, 101–2
    Regulation of Investigatory Powers Act,
        96–97, 101
United Nations, 93–94
United States v. *Carey,* 81
United States v. *Upham,* 80
Universal Coordinated Time, 283, 284, 285
UNIX, 45–46, 49, 53, 56, 57, 67, 82, 83–84,
    85–86, 142, 278, 284, 286
Unmodify, 160
Unsuccessful outcome, 84
Unsupervised text categorization, 328
User datagram protocol, 289–90
User relevance feedback method, 354

V

Vagueness in fraud detection, 204–5
Verifiability, inferences, 27
Virtual folders, 70
Virtual memory, 135, 142
Virtual private network, 292
Virus cleaning, 37
Viruses, 17
Visualization, 184
    techniques, 206–11
    tools, 61–64
VMWare, 141
Vogon evidential hardware, 151
Volatile data, 122, 135–36, 137
    order of, 135, 304–5
Volatile memory imaging, 63
Volatile storage, 25
Vulnerabilities, 299, 300–3
Vulnerability analysis, 301
Vulnerability scanner, 301–3

W

Warrants, 80–81
Waste paper shredding, 37
Watermarking, 345
Web surfing, 28
White Glove/PLAC, 45–46
Wide-area network, 17, 32, 290
WinHex, 156
Wipeinfo, 59
Wireless forensics, 321, 322
World Trade Center bombing, 242–43
Worm attacks, 271, 300
Write blocker, 45, 52, 63, 115

Z

Zapping clusters, 58
Zert, 59

Recent Titles in the Artech House Computing Library

Advanced ANSI SQL Data Modeling and Structure Processing, Michael M. David

Advanced Database Technology and Design, Mario Piattini and Oscar Díaz, editors

Action Focused Assessment for Software Process Improvement, Tim Kasse

Building Reliable Component-Based Software Systems, Ivica Crnkovic and Magnus Larsson, editors

Business Process Implementation for IT Professionals and Managers, Robert B. Walford

Configuration Management: The Missing Link in Web Engineering, Susan Dart

Data Modeling and Design for Today's Architectures, Angelo Bobak

Future Codes: Essays in Advanced Computer Technology and the Law, Curtis E. A. Karnow

Global Distributed Applications with Windows® DNA, Enrique Madrona

A Guide to Software Configuration Management, Alexis Leon

Guide to Standards and Specifications for Designing Web Software, Stan Magee and Leonard L. Tripp

Internet Commerce Development, Craig Standing

Knowledge Management Strategy and Technology, Richard F. Bellaver and John M. Lusa, editors

Managing Computer Networks: A Case-Based Reasoning Approach, Lundy Lewis

Metadata Management for Information Control and Business Success, Guy Tozer

Multimedia Database Management Systems, Guojun Lu

Practical Guide to Software Quality Management, Second Edition, John W. Horch

Practical Process Simulation Using Object-Oriented Techniques and C++, José Garrido

Risk-Based E-Business Testing, Paul Gerrard and Neil Thompson

Software Fault Tolerance Techniques and Implementation, Laura L. Pullum

Software Verification and Validation for Practitioners and Managers, Second Edition, Steven R. Rakitin

Strategic Software Production with Domain-Oriented Reuse, Paolo Predonzani, Giancarlo Succi, and Tullio Vernazza

Successful Evolution of Software Systems, Hongji Yang and Martin Ward

Systematic Process Improvement Using ISO 9001:2000 and CMMI®, Boris Mutafelija and Harvey Stromberg

Systematic Software Testing, Rick D. Craig and Stefan P. Jaskiel

Systems Modeling for Business Process Improvement, David Bustard, Peter Kawalek, and Mark Norris, editors

User-Centered Information Design for Improved Software Usability, Pradeep Henry

Workflow Modeling: Tools for Process Improvement and Application Development, Alec Sharp and Patrick McDermott

For further information on these and other Artech House titles, including previously considered out-of-print books now available through our In-Print-Forever® (IPF®) program, contact:

Artech House
685 Canton Street
Norwood, MA 02062
Phone: 781-769-9750
Fax: 781-769-6334
e-mail: artech@artechhouse.com

Artech House
46 Gillingham Street
London SW1V 1AH UK
Phone: +44 (0)20 7596-8750
Fax: +44 (0)20 7630-0166
e-mail: artech-uk@artechhouse.com

Find us on the World Wide Web at:
www.artechhouse.com

influence, impact, succeed

A PRACTICAL GUIDE TO NLP FOR WORK

DIANNE LOWTHER

This edition published in the UK
in 2018 by Icon Books Ltd,
Omnibus Business Centre,
39–41 North Road,
London N7 9DP
email: info@iconbooks.com
www.iconbooks.com

Distributed in Australia and
New Zealand
by Allen & Unwin Pty Ltd,
PO Box 8500,
83 Alexander Street,
Crows Nest,
NSW 2065

First published in the UK
in 2012 by Icon Books

Distributed in Canada
by Publishers Group Canada,
76 Stafford Street, Unit 300
Toronto,
Ontario M6J 2S1

Sold in the UK, Europe and Asia
by Faber & Faber Ltd,
Bloomsbury House,
74–77 Great Russell Street,
London WC1B 3DA
or their agents

Distributed in the USA
by Publishers Group West,
1700 Fourth Street,
Berkeley, CA 94710

Distributed in South Africa
by Jonathan Ball,
Office B4, The District,
41 Sir Lowry Road,
Woodstock 7925

ISBN: 978-178578-326-5

Typeset in Avenir by Marie Doherty

Printed and bound in the UK by Clays Ltd, St Ives plc

About the author

Dianne Lowther first started learning NLP in 1992 after graduating in Psychology. Since then she has delivered or co-delivered almost 1,000 days of NLP training. Dianne qualified as a Master Trainer of NLP in 2009 and has presented the annual NLP Conference in London every year since 2004.

Applications of NLP in the workplace have always been Dianne's principal focus. Since forming her company Brilliant Minds in 1996 she has become a sought-after consultant and coach, working with organizations in a variety of sectors including technology, finance, engineering and leisure. She specializes in working with Leadership Teams to increase engagement and create organizational cultures that support business priorities.

Author's note

It's important to note that there are many frequently used stories, anecdotes and metaphors employed in NLP. Where I know the source I will be sure to reference it, but my apologies to the originators of any material if I have overlooked them here.

Contents

About the author   iii
Author's note   v

Introduction   1

1. The NLP language and communication model   11
2. The presuppositions of NLP   51
3. Outcome thinking   85
4. Rapport, pacing and leading   97
5. Rep systems   119
6. Perceptual positions   135
7. Submodalities   143
8. Strategies   151
9. Modelling   165
10. Language   173

A final word   193

Resources   195
Index   201

Introduction

What is NLP?

Have you noticed that people are different? Of course you have. Have you ever wondered how those differences are created? I can explain. If you begin with the fact that most of us have five senses and yet there is far more information available to those senses than we could possibly process – or would want to – then you have the key.

Given that our senses pick up more information than we can actually use, a **filtering** process must be going on in order to feed our conscious awareness with a manageable amount of data. Therefore each of us is aware of only a tiny proportion of what is going on at any one time. In effect, we all create our own highly subjective 'map of the world' inside our own minds. Small wonder then, that two people can emerge from an hour-long meeting with totally different views about what has been agreed or decided!

Neuro-linguistic programming (NLP) is built upon a systematic approach to understanding this filtering process, the nature of the filters – which may include beliefs, values, past experiences and languages – and the extent to which the resulting map of the world is effective.

An effective map of the world is one that supports the achievement of your goals. An ineffective part of the map is one that gets in the way of the achievement of the same goals.

NLP is also concerned with how an ineffective part of your map of the world can be adjusted to become effective. Hence NLP has generated many techniques for systematically changing behaviour, beliefs, habits, negative emotions and so on. It has also given us many equally useful techniques for eliciting information about maps of the world and the ways in which they operate, effectively or ineffectively. And ultimately, all maps of the world are effective at producing results – it's just a question of whether or not they were the results you wanted!

NLP is mostly concerned with **structure** rather than **content**. That means we're interested in the *way* that people think and act rather than *what* they're thinking or doing. This is important because there are observable patterns in the way that people think and behave. When we become familiar with the patterns we can plan our approach to take account of them.

The 'official' Society of NLP definition of NLP is that it is 'the study of the structure of subjective experience'.

The origins of NLP

NLP was developed in the 1970s by Richard Bandler and John Grinder at the University of Santa Cruz in California. Initially, Bandler and Grinder collaborated on a project to identify the structure of influential language used by some of the great psychotherapists of the day. That study yielded not only some groundbreaking insights into the way that language can change perceptions, but also a new method-ology – **modelling** – that could be used to distil the essence of the skills of any person. (Modelling is described in more detail in chapter 9.)

There are plenty of books available that describe the early stages of the development of NLP. My purpose here is to give you an experience of the practical uses of NLP in a 21st-century workplace, so rather than a history lesson, let me give you some food for thought.

THINK ABOUT IT If you were able to figure out the patterns in the way that the people around you behave or respond, and then accurately predict the best way to influence each person, what would that make possible for you?

And if you knew how to create the conditions so that you could work effectively and easily every day, what would that make possible for you?

Is it worth a few hours of your time to read this book and find out?

What's in a name?

With a name like neuro-linguistic programming, it's not surprising that most people prefer to use the abbreviation NLP. However, the full name gives us clues to what this subject is about:

Neuro: concerning the brain and the nervous system

Linguistic: concerning language

Programming: mmm, this is where a lot of people get a bit uncomfortable. When you see 'brain' and 'programming' in the same sentence, what do you think about? Brain-washing? Mind control? I think I'd better explain.

When we talk about programming in NLP, it refers to programmes of behaviour. Habits.

 What did you do between the time when your alarm clock went off this morning and leaving the house to go to work? The same as every other morning after the alarm clock goes off?

The sequence of activities that you go through every morning in getting ready to go to work can be thought of as a programme. Unless you have small children at home (in which case, anything might happen!) then it's likely that you do the same things in the same order every morning. It's a

4

programme that gets you from lying in bed to standing at the front door, bag in hand, ready to go to work. And like anything you can do well, you do it without thinking about it very much.

But think about it now. How good is that programme? Does it result in you standing at the front door, suitably dressed, having eaten a nutritious breakfast, well rested from a good night's sleep and feeling upbeat, positive and ready for your day at work? Or not?

If you're at the front door, still half asleep, not having had any breakfast, badly dressed and feeling stressed, then I suggest that this programme could be improved. Wouldn't you agree?

And even if you're facing the day in the best possible frame of mind but it has taken four hours to get ready, I might still suggest that the programme could be improved.

So a programme is a habit. A sequence of actions that we can do so well we don't have to think about it to do it. Consider some of the things you do well; is it true that you do them completely instinctively, without any conscious thought?

 Anything you can do really well, you can do without thinking about.

Without NLP, changing habits is a laborious business. Remembering to do something different, something that doesn't really come naturally, is hard. Finding the motivation to keep doing it until it becomes instinctive is even harder. So if I tell you that NLP gives us the possibility to 'reprogramme' our behaviour with a one-off intervention, would you be interested to know more?

Or would you be a bit sceptical? It's good to be sceptical. I hope that you won't take my word about the information and techniques in this book. I hope you'll experiment with them and find out for yourself. I want you to make your own mind up about how useful NLP can be for you at work.

NLP for work

NLP has applications in lots of fields. In the working environment, I would list four main areas where NLP is useful, although, of course, this is not exhaustive. Once you start to use NLP, you can discover even more ways to apply it to your own work.

Self-management: As organizations get 'flatter', that is, they have fewer levels of management in the hierarchical structure, more managers have more direct reports and hence less time for each one. As organizations grow, more people work from home or on a different site from colleagues, their manager or their team members and so the amount of input from managers reduces even more. To be effective, everyone must be self-managing to some extent.

NLP techniques and principles can support activities such as goal-setting, self-motivation, overcoming nervousness before an important event, conquering mental blocks, improving decision-making techniques and managing self-development.

Communication: This is probably the biggest challenge in any organization, and the size of the challenge increases the larger the organization is. In the face of increasingly 'flat' company structures, greater reliance on technology and the growing support for diversity in the workforce, effective communication is getting harder, and yet is also becoming more important than ever.

NLP can be applied to work out effective communication strategies for different individuals and groups. By using NLP models and questioning systems we can find out a lot about how others think and how best to structure communications to achieve the desired goal.

Learning and developing new skills: The pace of life and business is increasing. In the race for market share, most companies have to be highly flexible and adaptable in order to succeed. This means that the workforce must be highly flexible and adaptable too. People need to be able to learn new skills efficiently, manage change enthusiastically and constantly seek out new sources of competitive advantage.

NLP can be used effectively by organizations to 'model' the key skills demonstrated by the organization's top performers and then teach those skills to other workers. Modelling a skill means finding out not just *what* those top performers do, but also the *way* they do it.

Similarly, for individuals, NLP can help you to formulate a reliable strategy for your own development and learn how to model skills from colleagues, specialists or even public figures.

Dealing with 'emotional baggage': There are an increasing number of people in the workforce who have experienced redundancy, a takeover, relocation, a merger or some kind of violence or harassment at work. Even the best organizations do not always succeed in helping people come through these kinds of experiences with a positive attitude. At worst, a whole organization may end up demoralized and unproductive through a badly managed merger or change programme. Reluctant though we may be to admit it, many people carry these bad experiences with them and this can hinder them from achieving their full potential.

NLP offers some of the best and quickest techniques for acknowledging the personal significance of these experiences and shaking off the negative consequences such as mistrust of employers, anger and distress, low morale, lack of motivation and loss of self-worth. Having moved through the negative effects of the events, the individual is then free to consider the value of what they have learned and apply it constructively in the future.

Overall, NLP offers a structured, systematic approach and reliable, effective techniques for dealing with the most challenging aspects of managing people – and 'people' includes yourself.

1. The NLP language and communication model

I want to begin by looking at a diagram.

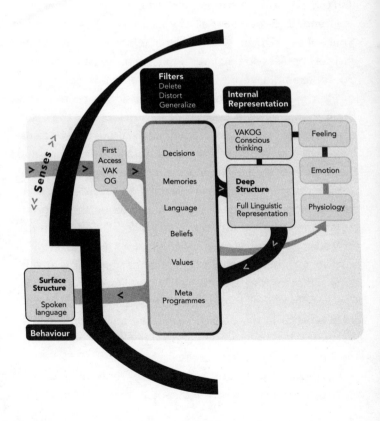

This diagram shows the 'NLP language and communication model'. This is a representation of how we process information via our senses and then respond to what we perceive.

Let's start in the top left of the diagram with information coming in from the outside world and into your senses. On the diagram it looks as if it's coming in at eye level but it is intended to represent all of your senses. The function of your senses is to translate external information into neurological impulses so that your brain can make sense of them. So, for example, your eyes transform different wavelengths of light into different kinds of neural signals so that your brain then creates a pattern of information on the visual cortex at the back of your brain. Each of our senses translates external information into neurological impulses and that's the first level of transformation that happens.

The result of this is the first level of representation, what John Grinder calls 'first access'. You'll see on the diagram that it's also labelled with the initials V A K O G. This stands for visual, auditory, kinaesthetic, olfactory and gustatory – our five senses. One thing it's important to realize at this stage is that the kinaesthetic here is purely sensory information, not emotional information. We don't detect our emotional feelings in the outside world, we only detect things like texture and temperature and weight, so it's purely sensory kinaesthetics at first access.

What we've got here is a first-level representation of reality that is purely sensory. Notice that there are no words at first access. This is the level of perception that a very small child would have before they learn language. But, of course, what makes us human, what makes us different from other species, is that we do have language, and our ability to 'make sense' of things using language is really what sets us apart.

Deletion, distortion and generalization

The second stage of transformation is when that sensory information is labelled with words and it becomes our internal reality. Referring back to the diagram of the NLP language and communication model, you will see the second phase where the sensory information goes through the linguistic filters. Below the 'Filters' heading is a short list of what the filters *do*: they **delete**, **distort** and **generalize**. Below that is the list of what the filters *are*. There are lots of different kinds of filters and the list on the diagram is a representative one: it's not completely exhaustive but it will cover most of the things that we need to talk about. So I'm going to come back and deal with that in a bit more detail, but first let's talk about the idea of deletion, distortion and generalization, because this is what language does.

KEY TERM

When we talk about **deletion**, it's not like you would delete a file from your PC and then it's gone. When we refer to deletion in terms of language it's more accurate to think about it as what we are not paying attention to.

13

THINK ABOUT IT

Deletion can also happen at sensory level. For example, if you live near a train station or an airport, you probably find that after a while, you no longer notice the sounds coming from the trains or the aeroplanes, but if someone comes to visit they really notice it. 'How do you live with the noise?' they cry and you say, 'What noise?' because it's just disappeared from your conscious awareness. In the same way, we delete things from language that we think we don't need to talk about. Sometimes that's useful, sometimes it can lead to misunderstanding.

KEY TERM

The same thing is true of **distortions**. Again, distortions can happen at a sensory level – things like optical illusions – but also distortions can happen in language because sometimes we have a tendency to want to make things fit with our reality. So we might slightly distort the way something is to make it fit in with our preconceived ideas. Then, when we talk about it, we talk about it as if it is the way we'd like it to be rather than acknowledging that in some ways it's slightly different. That's a distortion of the information we're getting from the outside world.

Generalization is the third thing that the filters do. As it sounds, that means making a principle or taking a general idea on the basis of a small amount of information.

When we generalize, we have one or two experiences, we make sense of those, and on that basis, we say, 'This must be true,' in a whole variety of different situations. We're taking information from one context and assuming that it would be true in a lot more. To do this is necessary because if you couldn't generalize, every time you changed your car you'd have to learn how to drive all over again. Every time you went to a different office, you'd have to learn how to open the door! However, generalization is also at the heart of all bias and prejudice, so it's not always a useful process.

So deletion, distortion and generalization have their uses and clearly that's why they have evolved as part of the way we interact with the world, but they also can cause problems if we're not aware of what we're doing.

When the information has come in from the outside world, it goes through two stages of transformation: the first one is at a sensory level and the second level is at a linguistic level where we label our experience, put words to it and literally make sense out of it. The end result of this is what we call our internal representation of reality.

Internal representation, conscious and unconscious

Your internal representation of the world has two elements to it: there's a conscious part and there's an unconscious part. Look again at the diagram on page 11 and see how this is represented. At the top, the conscious area, we've got V A K O G – visual, auditory, kinaesthetic, olfactory and gustatory – again to represent the sensory information that we're consciously aware of. We've also got an added element now, which is our conscious thought process. In other words, it's what we're saying to ourselves.

THINK ABOUT IT Do you talk to yourself? Some people think that they don't talk to themselves and some people know that they do and are a little embarrassed to admit it. So let me tell you that if you talk to yourself, inside your own head, that's completely normal. Some people even talk to themselves out loud and that's actually not as abnormal as you might think!

Obviously it's a hard thing to measure but informal research suggests that most people talk to themselves more than half the time that they're awake. In fact, the only time that we switch off that 'running commentary' inside our heads is when we're completely engrossed in doing something. That might be just watching a movie, it might be getting engrossed in a piece of work or it might be playing a sport. You've probably heard athletes talk about 'getting in the zone' which is when they stop thinking about what they're doing and just do it. So there are times when we switch off that commentary but often, when we're just going about doing our daily routines, we chat to ourselves inside.

 It can be an interesting experience to start paying attention to what you're saying to yourself because some of us are not quite as polite in talking to ourselves as we might be talking to other people!

What we say to ourselves, alongside all the sensory information, is a very important part of our conscious thinking. That conscious thinking interacts with the other part, which is the representation of reality that's outside of conscious awareness. The experts say that everything that happens to us, every thought that we have, every bit of information that we come across is recorded somewhere in the deep recesses of our brain. Some of that information we can get out again very easily and some of it is more difficult to access. But in terms of relative size, the unconscious part

17

of your mind is very much bigger than the conscious part. You can really only do one thing at a time consciously but always in the background, unconsciously, you're doing lots of other things. That's why sometimes you might go to bed wondering about how to tackle a particular situation in your job and the next morning, thinking that you haven't really moved on very much, you suddenly realize that you've got the answer to the problem because, while you were asleep, the unconscious part of your mind was still working on the ideas.

Before you go to sleep tonight, set the unconscious part of your mind a task to do. It might be a problem to solve, an idea to create or a decision to make. Explain the task to yourself in the same way as you would pass on a task to a colleague. Notice what has happened when you wake up tomorrow.

You have probably already realized that there's a lot going on outside of conscious awareness. The unconscious part of your mind is also the storage place for everything that you know.

If I was to ask you for your phone number, even though you weren't thinking about it until I asked, you would be able to answer me immediately because the information comes out of your unconscious storage and into conscious awareness instantly. You can tell me what the number is, and then it will leave your conscious thoughts again.

Thinking about the relative sizes of the conscious and unconscious part so your mind, the metaphor that I like the most is this:

Imagine you are standing in the middle of a large, round theatre, in complete darkness but with a very powerful pencil-beam torch in your hand. Think about how much of the theatre you could light up. It would only be a very small amount at a time. You could light up most of the theatre ... but not all at once.

That's what your conscious awareness is like in relation to your unconscious mind, because we can shine the light on a little tiny bit of information, and we can become consciously aware of a small piece in a given moment but, in order to be aware of something else, we have to let go of the first bit of information we were thinking of. There's a huge amount of information available at the unconscious level.

Thinking, feeling and physiology

We've looked at your internal representation of the world in terms of the sensory information and the conscious thought process. Let's have a look at how that interacts with the rest of your emotional experience and your body.

Once again, we need to refer back to the diagram of the NLP language and communication model on page 11. Look at the right-hand side of the diagram. You'll see that there is a connection between what you think and how you feel and then the physiological experience, lower down the diagram. You may also notice that there is a 'back door' route that goes from first access on the far side straight round to physiology. This represents reflex actions – the things that you can't do anything about, whereby as soon as you see or hear a certain thing, you react physically.

For example, imagine you're crossing the road and you suddenly notice there's a car racing towards you that you hadn't previously noticed: what are you going to do? You're going to jump out of the way. You're not going to go through some long thought process about how fast is the car going, the likelihood of it hitting you or whether you have time to move before it does. You're just going to jump out of the way. That's an immediate response: straight from the sensory information to a physical action. What will happen as a result of that? It's probably going to stir up a certain amount of adrenalin and that creates emotion. Emotion is just energy in motion around your body. Some people feel that, and some people don't. But suppose you've just jumped out of the way of a speeding car. Your heart's going to be pounding, and you might feel a bit afraid, or you might feel relieved because you got out of the way. You might also feel a bit cross with yourself for not having noticed sooner. So you're probably then going to say something to yourself. And, as I said earlier, you might say something kind and supportive to yourself, or you might say to yourself, 'Oh, you idiot!' I'm not going to ask you which!

There's a connection between what happens physically and how we feel, and what we then say to ourselves. But there's also a connection in the other direction: what we perceive in the outside world and we say to ourselves in response to that will also affect how we feel and will affect our physiology.

And that's how it is that by watching somebody's body language you can get some ideas about what's going on in their mind and about how they feel. Notice I say you can get *some ideas* because it's not an exact science. Yes, there are some generalizations we can make about what certain gestures and certain positions and types of body language might mean, but actually it's quite idiosyncratic. The people you know best – your nearest and dearest – you'll be able to tell what kind of thought is going through their mind or how they're feeling just by looking at them. The reason for that is that you have a lot of experience with those people; you've got memories of them telling you exactly how they feel that's associated with that particular expression on their face, or that tone of voice, or that gesture or that way of moving. Some people will say that they can walk into their house and look at the face of their loved ones, and know instantly what kind of a day has it been. Somebody even said to me once, 'I can tell by the sound of the footsteps approaching the front door whether or not my partner has had a good day.' We all have the ability to do that. We can all read body language but we're not always consciously aware exactly how we do that, or what it means.

THINK ABOUT IT

Let's go back to your own experience of this because you probably realize that there is a connection between what you think and how you feel and what goes on in your body, and intuitively you probably make use of it. There might be a time when you're working at your desk and your thinking grinds to a halt. You get a bit stuck and you don't know quite where to go next. What most of us do, quite instinctively without really thinking about it, is we get up and move around. Now we might rationalize that by saying, 'I'll go and get myself a glass of water', or a cup of coffee, or 'I'll go and smoke a cigarette', or 'I'll eat some chocolate', or maybe 'I'll just go and chat to somebody for five minutes'. The important thing is that we leave the desk and do something different. And then when we come back to the desk five minutes later, because the body is in a different state, the emotions are in a different state, and then the mind is in a different place, usually we can re-engage with the task and get on with it and make some progress in a way that we weren't able to before our thinking stopped. So we all instinctively know that there is a connection between what we think and how we feel and what goes on in our bodies. Perhaps we don't make use of it as much as we could?

There's also a connection here with physical health, which not everybody recognizes. You probably realize that if your body is not in a good state, it can affect your mood. So most people, if they have a bad cold, will feel a bit less enthusiastic about life than they would if they were feeling really fit and healthy. Have you thought about how that connection might work in the opposite direction? That maybe if you're feeling stressed or anxious, or a bit down about something, does that then make it more likely that you will physically succumb to some kind of problem, whether that's a slight headache, some muscle tension, or something more serious. That's a whole other branch of NLP – looking at how we can improve our health and well-being using NLP techniques. It's clear to me that the mind has something to contribute to the well-being of the body (more about this later).

Linguistic filters

So now that we've had a look at most of the areas of the NLP language and communication model, let's backtrack a little bit to the linguistic filters – that's the big chunk in the centre of the diagram (page 11). These are the filters that we use when we make sense of the world and they're basically to do with language because they're labels that we put on our experience.

Memories: Our memories are an important category of filters because they allow us to make sense of what's in front of us today, in the light of things that have gone before. This is how you can become really skilled in a particular type of work: you go into a new situation, you look at it and you remember dealing with things that are similar if not identical in the past, and that gives you confidence that you know what to do in the current situation. Memories also work on a visual level. For example, if you go somewhere where there are lots and lots of people that you've never seen before, as you look around the crowd, you will see people who remind you of people that you already know. You might think, 'Oh, that looks like Jo from next door,' or 'That person over there looks like Chris from work.' By comparing the faces in the crowd with familiar faces, we're able to make sense of the experience of the crowd more quickly.

If you got to know that person who looks like Jo from next door then they would stop looking like Jo. This is because although initially it's helpful to use memories to help us make sense of what's going on in front of us, at some point we would need to distinguish between Jo and this new person. At that point you start noticing all the ways that they're different from each other. But initially our memories help us to make sense of a current situation.

Memories are a very important category of filters and for a lot of people they're also a source of generalizations. This happens when there is a tendency to assume that what has gone before will happen again in the future. We use our memories to predict what we think will happen in the future and that's not always accurate.

Decisions: The next category of filters is decisions. Every time you make a decision it creates a new filter. You tend to ignore – you *filter out* – anything that might suggest that the decision you've made is not the best one, and you *filter in* anything that supports the decision that you've made. Now you might think that, on the face of it, that's kind of cheating, but consider the alternative. I have met people who do the opposite: they make a decision and then they notice all the reasons why it's not a good decision and they disregard anything that would support the decision they made. That is a really stressful way to live your life!

So, it's easy to see why, as human beings, we've evolved the ability to disregard things that would undermine decisions that we've made. I'm sure you can think of examples where having made a decision, afterwards you needed to justify it. This is known as post-decision rationalization and it's seen at its finest immediately after a bit of impulse shopping. That could be a gadget that just looked so shiny and interesting that you had to buy it, regardless of whether or not it was useful, or for certain sections of the population it applies to shoes. I'm sure you've got your own examples.

REMEMBER THIS!!! Our decisions act as filters because every time we make a decision we adjust reality to make sure that our decision was a good decision.

Language: The third category of filters is language itself. These are all linguistic filters but you may have noticed that people who can speak more than one language will experience more than just the words changing when they shift from one language to another. Sometimes you'll notice that when somebody shifts between one language and another that their whole personality seems to change; that maybe the gestures they use, the tone of voice they use, the level of energy can change along with the language.

Also you'll find that there are certain concepts that can be expressed in one language but can't be expressed in another. For example, we can translate the Spanish word *mañana* into English but it loses the nuance of the original and ends up with a different meaning. As a consequence, if you can't express something in words then you can't experience the idea. So that means that the cultures that define the languages very much stay separate until people start bringing new language into a culture.

In the English language we're not so aware of this because every time we come across something that we haven't got a word for, we just borrow it from somebody else's language, or we make one up. For example, in the last few years we've acquired a new verb in the English language which is 'to Google' – nobody ever heard of that twenty years ago because Google hadn't been invented but now people frequently say 'I'll Google that', or 'Have you Googled it?'

Our language is constantly evolving, but then there are other languages that are not so widely spoken that will have retained much more of their cultural integrity. And there are certain things that they can express and certain things that they just can't.

If you only speak one language, you might be thinking this has nothing to do with you but actually it's a demonstration of something that applies to all of us, which is that we choose the language we use depending on the circumstances.

THINK ABOUT IT

If someone asked you, 'How was your day?' you probably would answer differently depending on who was asking. So if your boss asked you, you might give a very different answer than if one of your friends asked you after you'd left work. And you might give a different answer again if a small child asked you the same question. So we adjust our language depending on the situation.

Beyond that, the language that you choose to describe your situation will affect how you feel about that situation. There's an old cliché: is the glass half full or is it half empty?

How you choose to describe it will affect how you feel about the situation.

 One of the things that I'd like you to notice is that there are always different ways of describing the same thing and yet we all tend to have habits in how we label our experiences. That's why we seem to keep repeating the same experiences over and over again – because we tell ourselves that it's exactly the same thing happening over and over again. But if we were to break out of that linguistically and call it something else, we might find that we break out of the experience and the emotions that go with it.

To experience this, try writing down in no more than three sentences a description of an incident where you had an argument with someone and you were sure that you were in the right. Then rephrase the description as much as possible without losing the accuracy of the facts. Notice how this affects your feelings about what happened.

Beliefs and values: So moving down the list brings us to beliefs and values. You could put a dotted line between language and beliefs: above the line would be the filters that are easy to bring into conscious awareness and below the line would be the things that are less easily brought into conscious awareness. Most of us are not that aware of our beliefs and values, but they do drive our behaviour so they are a very powerful category of filters. Technically, the difference between values and beliefs is that your values are what's **important** to you, and your beliefs are what's **true** for you. They tend to go together. If you have a value you also will have beliefs that relate to that value and the combination of the value and the beliefs will drive your behaviour. We are motivated to undertake activities that contribute to fulfilling our values.

I'll give you an example: if you value friendship then you'll have beliefs relating to friendship. And they can vary widely. I did a survey in a group of people I was working with and I got some very different answers: one person said, 'A friend is somebody who is always there for you,' and somebody else said, 'A friend is someone who, it doesn't matter if you haven't seen them for two years, you can pick up just exactly where you left off.' These are quite different beliefs, aren't they? Another person said, 'A friend is someone who makes you feel good about yourself,' and somebody else in the same group said, 'A friend is someone who will always tell you the truth.' I'm not sure those two things are compatible either! The most memorable was the man who said, 'A real friend is someone who will drive all night through the pouring rain with a pocket full of cash and not ask any questions.' I said to him, 'There's a story behind that, isn't there?' and he said, 'Yes, there is, but I'm not telling you what it is.'

THINK ABOUT IT — What do *you* believe makes a good friend?

We all have our own beliefs about friendship: what does it mean, what makes a good friend, how do friends treat each other and so on. It's the combination of those things that will drive behaviour, so if you believe that a friend is someone who always tells you the truth, then you will tell the truth to your friends. Ultimately that creates the kind of friendships that are important to you. Now, if you didn't have the value of friendship to start with, you wouldn't have the beliefs, you wouldn't behave that way and then you wouldn't have any friends. And that would be okay, because if you didn't value friendship, you wouldn't want any friends!

I'm sure you can see how the combination of the value, beliefs and behaviour gets you what you want in life. This makes our values and beliefs very powerful filters – especially considering most of us are not fully aware of what they are.

One way that you can start to get some ideas about what values you hold is to notice your emotions. Our emotional responses are a reflection of our values: when we fulfil our values, when we get what's important, we feel good. We get all the range of positive range of emotions: we feel happy, we feel satisfied, we feel excited, enthusiastic and so on.

When our values are not being fulfilled, that's when we get negative emotions. In fact I'd go further, it's when our values are being violated that we get negative emotions. So we get irritated, we get angry, we get upset, and whatever else comes to you when something's not going your way – whether you throw a tantrum or whether you just sulk or get depressed, those are all the result of your values being violated. And it could be you that's violating your values or it could be somebody else.

By monitoring your emotions you can start to get some ideas about what's really important to you; what matters sufficiently that it stirs up an emotional response. The situations where you don't feel very much – maybe you feel a bit bored, or apathetic, with no motivation – what that means is that there's nothing in that situation that is obviously connected to anything that you value.

 If there is a situation in which you need to get involved but you're currently not very interested, looking for a connection with something that matters to you is a good way to generate some motivation.

Back to beliefs and values as filters: the way that our beliefs work as filters is that what we believe will very definitely shape what we perceive. When people say 'I'll believe it when I see it,' that isn't really true because if somebody presents you with information that contradicts your own beliefs, you're more likely to reject the information than you are to reject your belief.

Let me give you an example: I was running a training programme a couple of years ago with a lot of people who were IT professionals. They were mostly project managers and there were a couple of more senior people – programme managers – within the group. As I was talking to them about subjective maps of the world and different realities, one of the more senior people said to me, 'Oh yes, I know exactly what you're talking about because we've got a poltergeist in our house and it affects people in really different ways.' He'd hardly finished speaking when I became aware of a ripple of reaction going round the room. It was almost as if people were recalibrating what they thought of this person. They were all very logical people; they worked in IT and were very good at process and very 'down to earth' for the most part. So now I could almost hear them thinking, 'Well, I thought this guy was okay really, but there must be something a bit strange about him!'

The information is coming in from the outside and if it hits a belief, in this case 'there's no such thing as a poltergeist', then rather than saying, 'Oh that belief must be wrong,' people judge that the information coming in from the outside world must be wrong.

That's quite an extreme example, but it demonstrates what happens with all of our beliefs that we're really committed to. When that senior person said, 'We've got a poltergeist in our house,' all the people around him were re-evaluating him because he believed something different from them.

THINK ABOUT IT Let me give you another example: think of a couple of people that you know; one who has a very strong personal faith and another who is a complete atheist. Imagine them having a discussion together about whether or not there is a God. Now that's probably fairly easy to imagine, isn't it? But can you imagine that discussion coming to a point where one of them said to the other, 'Actually, you know what, you're right, you've convinced me!' It's really hard to imagine, isn't it?

When it comes to matters of belief, people don't change their minds easily. Everything that you believe acts as a filter and that old saying of 'I'll believe it when I see it' really and truly works the other way round: generally, you see what you believe.

Values are even more deeply buried in our unconscious minds and are therefore even harder to bring into conscious awareness. Often we have values that have been with us since we were small children and that drive our behaviour, without us realizing that we have them.

The first time I went to work with my friend in Singapore he phoned me up beforehand and told me, 'Bring all your lightest clothing because its 36°C here and it's really humid.'

I was happy about that because I like warm weather and it was December in England. I couldn't wait to get out of the cold. So I did as he suggested, took all my lightweight clothing with me and when I arrived in Singapore, sure enough it was 36°C and really steamy.

The first day we were in his office and it was hot. I was grateful that I'd brought very lightweight clothing. The second day, we moved to a hotel conference suite to run our training programme and by about 10 o'clock in the morning I was really cold. I checked the air conditioning unit in the tea break and it was set to 18°C. My first reaction was, 'That's crazy – 36°C of sunshine out there and I'm freezing cold in here!'

Now I don't know about you but in the winter in England I have my central heating set to 21–22°C, and I don't wear really lightweight clothing! So it was three or four degrees colder than I'm used to and I was wearing much lighter clothing. It's not surprising that I was cold, is it?

The reaction 'that's crazy' is the sort of reaction you get when you come up against different values from your own. You get an emotional reaction to other people's behaviour when their behaviour is being driven by values that you don't share. So I thought to myself, 'Ah, this is a values clash, and I'm a trainer of NLP so I can figure this out.' Of course, we were running a training programme so I didn't have a lot of time to think about it immediately but I kept coming back to it over the days that I was there, and I kept thinking, 'What is it here that is so different? What's the

34

values clash? Why am I continually being subjected to these arctic conditions when actually I'd like to be warm because it's sunny out there?'

It took me about ten days and when I think back now, it surprises me that it took me such a long time to figure it out, because it seems so obvious now that I know. It's a good demonstration of how hard it is to get outside of your own values and beliefs and really fully understand somebody else's reality. When the penny finally dropped, I realized that since I grew up in the north-east of England, where it's cold, I have a value that being warm is good. If you grow up near the Equator, where it's hot and steamy, you have the opposite: you have a value that cool is good, which of course is where we get that Californian, 'Oh, cool!'

So now when I go to Singapore I understand intellectually that for the locals it's wonderful to go somewhere that's *so cold* that you can wear jeans and a fleece, in the same way that for British people it's lovely to go away on holiday somewhere where it's *so warm* that you can just wear shorts and a t-shirt. The difference is that, although I get it intellectually, I can't feel it emotionally because I don't share those values.

 Values are a very important category of filters. Emotions are always the clue to values. If you have an emotional response to another person's behaviour – such as feeling annoyed, amused, shocked

35

or embarrassed – it usually indicates a clash of values. Their behaviour is driven by a value that you don't share. Or sometimes you share the value but not the way that it's demonstrated.

Metaprogrammes: The last category of filters I want to look at is metaprogrammes, down at the bottom of the list. Metaprogrammes are different in that they're filters to do with the *way* that we think rather than *what* we think.

There are lots of them. Rodger Bailey and his team who did the original research in this area came up with over 70 different metaprogrammes. At Master Practitioner level we teach about 16 metaprogrammes, which account for a lot of the daily differences in the ways that people think and the way that they behave. I'm going to share with you just a couple that I think are particularly useful.

The first one has to do with motivation. A lot has been written about *what* motivates people. What's less well known is the *way* that people are motivated. It actually boils down to just one of two options: either I'll be motivated to do a task because I can see that there will be some benefit from doing it, or I'll be motivated to do the task because I can see that if I don't do it, it will cause a problem.

Towards motivation is when you're motivated to do something because you can see that if you do it there will be some benefit. The opposite is known as **away from** motivation, which is where you're motivated to do something because you can see that by doing it you'll avoid a problem.

These are two very different ways of being motivated but it's important to recognize that both get the job done. Both are effective. The difference has to do with the focus of attention. For someone who is 'towards' motivated, they'll be thinking, 'If I do this, what's the benefit?' And they'll go looking for potential benefits. They generally have quite an upbeat and positive frame of mind, although they may start projects without any consideration of what might go wrong. You really don't want someone who only uses 'towards' motivation in charge of a big project – unless they have a strong-willed risk manager as well!

Contrast that with someone who relies on 'away from' motivation: they're constantly evaluating what can go wrong. They think, 'If I don't do this, what problems will it cause?' This person will probably decide what to do from their 'to do' list by a process of elimination:

- Item one – if I don't do this today, will there be a problem? No, okay, I'll leave it till tomorrow.
- Item two – if I don't do this today will there be a problem? No, okay, we'll leave it.

And so on.

The net result of this is that someone with an 'away from' motivation pattern tends to do most of their work close to deadlines. Far out from a deadline there isn't much motivation to do anything, because nothing's going to happen if they don't do the work. But as they get closer to the deadline they can really see that it will cause a problem if they don't do the work and so they're very motivated.

THINK ABOUT IT You might think back to when you were in school and you were given homework to do. Did you do it as soon as you were given it because it would be great to get it out of the way? Or did you do it just before it had to be handed in because if you didn't then there would be a problem? That might give you a clue as to how you were motivated back then, and that may well be how you're motivated today, because this is a pattern that a lot of people carry through their lives.

Metaprogrammes can change from context to context so it's possible that the way you operate in your working life might be different from the way that you operate when you're at home. But in general most of us will have a kind of default, or a preference – the one that we do best – and then we'll use others where they are particularly useful to us, if we have a specific reason for doing so.

Therefore, in general, if you can understand how a colleague gets motivated at work, then they're probably going to use that way of getting motivated all of the time. The trick then is when you talk to them to use the same kind of process that they would use when they talk to themselves.

So, for example, if you want to get somebody to do a task for you and you tell them all the reasons why it would be really good for them to do the task, that's going to work if they're somebody who is 'towards' motivated. But if they're 'away from' motivated and you tell them all the great benefits of doing the task, it probably isn't going to make any difference. To that person you need to tell them what problems it solves.

The same message could be given in different ways – it all comes down to language. For example, I could say to you, 'There's a job I want you to do and really it needs to be done by Friday, because if it's done by Friday then first of all, you can probably leave early and relax for the weekend but also it will be really great because it will give you a flying start for Monday morning, and it's going to make you look good in the eyes of some of our customers who really aren't expecting this until next week. I think it's also going to take the pressure off some of the other people in the team so you get to be a bit of a hero in the team.' That's all the 'towards' motivation stuff.

Now if I'm talking to somebody who's 'away from' motivated I can tell them exactly the same thing but using different language. I might say to them, 'I've got this job I want you to do and it really needs to be done by Friday, because if it's not done by Friday then you might end up worrying about it all weekend, and then you're going to be under pressure on Monday morning because there's all the more work to do, and if you haven't got this one done then it's going to set you back a bit. There's also the danger that other people are going to feel let down if you don't complete the work as soon as you could and it might mean there's more pressure on the rest of the team if this particular job isn't done by Friday.'

You can see how I'm basically giving the same message but I'm using different language. And of course, if you're talking to a team where you've got both 'towards' motivated and 'away from' motivated people, then it's a good idea to do some of each. You might feel as if you're repeating yourself but for the people listening, they probably won't notice, they'll just pay attention to the bit that works for them.

 Next time you ask someone to do something for you, use 'towards' language first and notice if they seem motivated. If not, tell them the same thing using 'away from' language and see if it gets them more motivated.

(If neither motivates them, it is probably the case that you haven't touched their values.)

So, that's one example of a metaprogramme. The other one that I think is particularly useful and relevant for most of us is to look at the difference between people with a 'procedures' preference and an 'options' preference. This relates to how you go about doing a job. We've looked at the motivation to do it, now this is *how* you do it.

Someone with an **options** preference likes to experiment with different ways of doing a job. They won't always do it exactly the same way; they'll vary things to see if they can get different results, to see what works best and so on. Someone with this preference is likely to find it quite hard to follow an exact process that someone else has specified.

By contrast, someone with a **procedures** preference works on the principle that there is a best way or a right way to do something. They want to know what the best way is and then they will always do it that way. I'm sure you can imagine that people with those opposite preferences can drive each other crazy!

It's also sometimes a source of tension in a relationship between a manager and members of their team. When you give a task to someone with an 'options' preference they usually respond best if they're told, 'This is the end result: you work out how to make it happen.' That's great if you've got an 'options' preference because you can experiment a bit, but for somebody with a 'procedures' preference it might feel as if they're not being given the whole story. They might really want to know exactly the best way to do the job, so they can do it that way.

 I had an example of an 'options' boss who had a 'procedures' member of staff and the boss kept saying to me, 'It really irritates me because every other member of the team is quite happy when I tell them the end result I want and they just go away and work out and make it happen. But this one person keeps coming and asking me, "How do you want me to do it?"' and she said, 'I'm beginning to think that maybe this person isn't up to the job.' But then, when we looked at the metaprogramme differences it became apparent that it wasn't that the person wasn't up to the job, the real difference was just that the 'procedures' person had an assumption that the boss knew the best way to do the task and just wasn't sharing it.

Of course, which of these is the more suitable preference will vary from role to role. I'm sure you can imagine that in some jobs having a preference for 'procedures' is really helpful, especially where there are legal procedures that have to be undertaken and certain things have to be done in a particular way every time. Somebody with an 'options' preference might find that overly restrictive and might get rather bored. However, if you are looking for continuous improvement, somebody with an 'options' preference, who is willing to experiment a bit and look for ways to improve things, albeit in small ways, would be a really good person to have. So again, both of these preferences have their place, they're both useful in different circumstances, but they can also both have a downside.

The reason I think this is important is that it accounts for some of the differences in the way that people approach tasks. It also accounts for the differences in the way that people talk about what they're doing. If you asked two people, one with each of these preferences, 'How is your project going?' the 'options' person will probably just say, 'Oh, it's fine. We're on track, we're working on this at the moment and everything is going well.'

On the other hand, the person with the 'procedures' preference represents things in their mind in a way that they see the whole process. So they'll start at the beginning and they'll tell you everything that's happened up to today so that you can understand why they're saying that everything's okay today. For them, it would seem like incomplete information if they just told you the end result, so they have to give you the beginning and paint the whole picture.

Similarly, if they were on the receiving end of just, 'Yes, everything's fine,' the person with the 'procedures' preference could get quite frustrated because they might feel as if you're not telling them everything that they need to know. The next time somebody starts right back at the beginning of the story and tells you everything that they think you need to know, just bear with them because it isn't that they're doing it to bore you, it's just that for their process inside their mind, they need to do that. That's how they know that they've told you everything that matters.

REMEMBER THIS!!!

Metaprogrammes are a different type of filter. They filter the *way* that we think rather than *what* we think.

(For more on metaprogrammes I recommend *Words That Change Minds* by Shelle Rose Charvet.)

Linguistic responses, cause and effect

The only part of the diagram that we haven't looked at now is what happens when we respond to information coming in from the outside. This is when we go from inside of our minds back to the surface, for example when we answer a question or we comment on something that somebody else has done.

You'll see that on the diagram we've got the labels Deep Structure and Surface Structure. This comes from the work of Noam Chomsky. The deep structure refers to the full linguistic description of everything that goes on inside your head. All of your memories, as well as having pictures, sounds and so on associated with them, also have a linguistic description associated with them.

Notice that when you say something, the information on its way back out again is also passing through the linguistic filters. So that means it gets deleted, distorted and generalized all over again. That's why, sometimes when you try to explain an idea you had, you might realize that what you are saying isn't really doing justice to the idea you've got inside your mind. What we say is only ever a tiny fraction of the thought that we have inside our head. That's why we sometimes get miscommunications and misunderstandings, and sometimes it creates tension between people if they're not really grasping the ideas and understanding what somebody else wants to put across.

Overall, what we've seen is that between information coming in from the outside world and us reacting to it, there's a great deal of filtering and processing that happens. Sometimes when I look at the diagram I think it's a miracle that people ever have the illusion that they've communicated with each other at all!

THINK ABOUT IT

Before we leave this, I've got a question for you: suppose I'm in a situation that I'm not comfortable in, so in terms of our diagram, what's in the top right – what I'm thinking and feeling – would be, 'I don't think I like this, I feel uncomfortable, I want to get out of this situation.' Consider whether this experience, this discomfort in the situation, is being created to a greater extent by what's going on in the world around me, or by what I'm doing inside my own head? Clearly both are involved but which one is having the greatest effect?

Now if you think about it, I expect you'll come to the conclusion, as I did, that really the experience I have is created mostly by what I'm doing inside my head. And that's why two people can be in exactly the same experience and can react completely differently, and will take different learning from it, and will make different principles and generalizations as a result of that same experience.

I think the important thing here is to recognize that we always have a choice. If we're willing to take the responsibility for creating our own experience that means that we have the potential to create something different when we don't like what we're experiencing. Sometimes that just starts with being willing to describe it in a different way. Often we put labels on our experiences and other people's behaviour that are actually quite far removed from the reality of things that we can see and hear and directly sense.

For example, sometimes I hear people say, 'I've got a problem with a colleague. He or she just likes the sound of their own voice,' meaning that they talk too much. But even that is a judgement. Now if we go back to the sensory experience – if we go back to first access – what's really going on there? What is happening is that the other person, the one who is being labelled as liking the sound of

46

their own voice, talks when other people think that they shouldn't.

There could be lots of reasons for that. Some people talk when they're nervous; some people just don't like silence so if you're not speaking then they'll speak to cover the silence. Some people genuinely have a lot of thoughts they want to share but, actually, very few people talk because they like the sound of their own voice (in fact most people shy away from having their own voice played back to them on a recording). The words being used here are a poor description of either the experience or the speaker's opinion of the experience. The person who talks too much doesn't literally like the sound of their voice and the person who makes the comment doesn't believe that they literally like the sound of their own voice. It's a big distortion of reality, and one that promotes a negative view of what is happening.

So, by taking a different approach to the words that we use, by describing our experience in a different way, we can begin to get a different experience. So, if I say, 'Somebody's late and they've kept me waiting and that means they don't respect me,' that's one way of describing it. But I could also say, 'Oh, this person's late and they obviously trust me to wait for them,' so maybe that's a demonstration of our friendship, rather than the opposite.

There are always different ways of describing a situation depending on what you want to achieve. Sometimes people create the same difficult situation over and over again. Maybe you know somebody who does that: you see them one day and you say, 'How are you, how's it going, how's the job?' and they say, 'Oh well, I don't like my job too much, I don't get on with my boss very well, I don't really feel like the team values me so I don't think I'm going to get on. I think maybe I'll look for another job.' Maybe you then see them six months later and you say, 'Oh hi, how are you, how's the job going?' and they say, 'Well, I changed jobs.' So you say, 'Great, how's the new job?' 'Well, I don't really get on with the boss that well and I don't feel as if I fit in with the team and I don't know, I just don't think I'm going to get on.' And you think to yourself, 'I've heard that somewhere before.'

The reason why some people keep repeating the same experience is because they're not learning from their experience. The process of learning from experience is that you add new filters on the basis of experience, so if you do a job where you don't get on with the boss then you learn something from it. Maybe you learn something about what kind of job to go for, or what questions to ask at interview or maybe just what to do in the first few days in a new job so that you do get a good relationship with the new boss. But if you haven't learnt any of those things there's the danger that you recreate the same experience over and over again. And that definitely is going on inside your head, isn't it, because if you've changed everything in the outside environment – you've gone to a new boss, a new team, a new company, maybe even a new city or country – the outside world has changed completely but somehow the inside experience is exactly the same.

 We really do create our own reality and further-more we have choices about how we do it. If we're willing to take the responsibility for it then we can truly create the kind of experience that we want in life.

---

2. The presuppositions of NLP

The presuppositions of NLP are some basic principles that underpin NLP thinking and mostly have been modelled from successful communicators – people who are successful in influencing and persuading others. There isn't one single, definitive version of the presuppositions. I have listed the ones that are most often quoted and used.

Linguistically, a presupposition is something that must be true in order for what is said to make sense. So the presuppositions of NLP are principles that must be true in order for NLP to make sense. In other words, they are the assumptions upon which NLP is built. You do not have to believe that the presuppositions are true in order to use them.

You can think of the presuppositions as filters that you use to get a different view of a situation – a bit like putting on a pair of sunglasses. Like the sunglasses, you can discard them when they aren't useful. Using one of the presuppositions of NLP means acting as if it were true, and there can be some interesting results when you do that.

The map is not the territory

This is often regarded as the first presupposition of NLP. Indeed, if we did not hold this to be true then NLP would not make sense. It really is the fundamental assumption upon which most of NLP is built. But what does it actually mean?

 'The map is not the territory' is a direct quote from *Science and Sanity* by Alfred Korzybski, first published in 1933, pristine copies of which grace the bookshelves of NLP trainers all over the world. (I say 'pristine' because even if you managed to read it once, you'd be unlikely to wade through it a second time!)

The phrase can be taken to mean this: what we experience is a representation of the world, rather than the world itself. So when we say that we 'see' something, what that really means is that the sensory mechanisms of our eyes have picked up light reflected from that thing and created a representation – or map – of it inside our minds.

All of this would be trivial, were it not for the fact that the map is an incomplete map. And that every single person creates a map that is incomplete in ways unique to that person. That means that *you* don't see the world that *I* do; nor does the next person.

This is where it starts to get interesting.

The world that we live in is a very rich environment. As we have seen, there is far more information available to our five senses than our brains can process, so we have filters in place that enable us to be aware of what is most relevant and useful to us. We disregard the rest.

Of course, we can shift our awareness from moment to moment to take in a lot of information over time. But at any given instant, we are only aware of a tiny fraction of the data available to our senses.

On different occasions, and for different purposes, we might create different 'maps' of the same 'territory'. In the same way as it's possible to have a roadmap, a political map and a geological map of your home town, mentally you can also have multiple maps of the same activity or experience. This means that you can choose how to interpret events and what they mean.

That's what 'making sense' is all about. We take in the sensory data and we make meaning of it. The thing is, there isn't always a single 'correct' meaning. And the meaning I place on a particular experience is often affected by other experiences that are not related to it.

THINK ABOUT IT Can you remember a time when you met someone for the first time and, by coincidence, they had the same name as someone you know really well? Did you find yourself kindly disposed to that person if they shared a name with

53

one of your friends? Or were you just a little bit guarded if they shared a name with someone you didn't like so much? What if you meet someone who looks like someone you know? Does that have an effect on how you behave towards them? If you're human, it does!

These are examples of how we 'make sense' of situations we find ourselves in. Because of our limited ability to process data, our brains use short-cuts, filters and general principles to infer meaning from small amounts of sensory data.

The important thing to remember is that there is not usually just one, correct interpretation. Bear in mind that when you 'make sense' of a situation there are probably lots of alternative ways of making sense of it.

 The question to ask yourself is this:

Is the way I'm interpreting this situation useful or not? Is there another way of looking at this?

In my experience, there are always multiple ways of describing and interpreting a situation. We actually have a choice about how we perceive the world around us and about how we react to it. The trick is remembering to exercise that choice.

You cannot not communicate

This presupposition is usually paired up with 'the meaning of your communication is in the response it gets' (more on which to follow).

'You cannot not communicate' is an interesting notion. It's tempting to think that if you go to a meeting and say nothing, then you have communicated nothing. But it doesn't quite work that way.

Part of what makes us human and sets us apart from other species is the ability to 'make sense' of our experience. We constantly make meaning of our observations. We interpret and analyze, sometimes without realizing that's what we're doing.

THINK ABOUT IT

This is what people will do if I go to a meeting and say nothing.

Those who notice that I've said nothing in the meeting will make meaning of it. One person may say, 'Dianne didn't say anything today, I wonder if she's feeling okay.' (Well, it would be unusual for me to be silent for a whole meeting!)

Someone else may say, 'I guess she was out of her depth in that meeting and didn't have anything to contribute.'

A third person might think, 'She must know something we don't know.'

Whatever meaning they make of it, people are making some kind of sense of my behaviour even if I say nothing and that's why we say that you cannot not communicate.

Suppose I didn't go to the meeting at all. What meaning would people make of that? That I didn't think they were important? That I was sick? That I had forgotten?

In that sense, everything I do, or don't do, has the potential to communicate something to somebody. Everything I do can be interpreted and analyzed and people can make their own meaning of it.

That, of course, is the tricky part. People make their own meaning of what you and I do. They use the filters of their own experiences, beliefs, values and prejudices to make sense of the behaviour of others. That means that very often the meaning someone makes of another's behaviour is inaccurate or misguided. No wonder there are misunderstandings between people!

So, it must be best to talk, rather than risk others getting the wrong idea by 'making sense' of observed behaviour, mustn't it? I'd like to think so, although I'll temper that opinion with the observation that verbal communication is also open to misunderstanding. But at least you can discuss it when you notice that someone has 'got the wrong end of the stick'.

Also to bear in mind, people will make meaning out of not just what you say, but also the fact that you have chosen to say it *at that time* and *to that listener*. For example, if you heard me saying to a colleague, 'Don't forget the meeting tomorrow,' what meaning would you make of that? That I think there is a danger my colleague WILL forget the meeting?

KEY TERM

This is known in NLP as the **meta-message** in what I say. It's the message I communicate by the choice to say what I say in that context.

THINK ABOUT IT

Suppose your boss took you aside before a meeting and said to you, 'There's nothing for you to worry about.' What meaning would you make of that? The meta-message would probably be that there is something going on that might worry you. After all, if there was nothing going on why would the boss tell you there was nothing to worry about?

This is one of the aspects of communication that is particularly noticeable when there are reasons for people to be anxious or insecure. The tendency to make meaning of every small thing is usually exaggerated when people are under pressure. So if you're about to announce a restructure or you're recruiting for a key position, bear in mind that everything you do is a communication. The key is to give clear and consistent communication, both verbal and non-verbal. Consider 'how it looks' before you deliver any important messages. And here too, remember that 'the map is not the territory' and sometimes you have to give the same message several different ways to achieve complete understanding.

The meaning of your communication is in the response it gets

This is a very frequently quoted presupposition of NLP, and one that puts us firmly in a position of responsibility if we choose to accept it.

The underlying principle here is 'the map is not the territory'. You'll remember that first presupposition and how we looked at the way that experiences mean different things to different people. Two people can 'make sense' of the same events in very different ways.

Similarly, words can mean different things to different people. What you thought you meant might not be what someone else would understand by what you said. Or to put it another way, the words that you choose to express your thoughts might correspond to some rather different thoughts in someone else's head.

No word has an absolute meaning. Language develops and changes and the meaning of words can shift over time. What you understand by the words 'nice', 'disinterested' and 'gay' is probably very different from how those same words were understood 200 years ago.

In just the same way, words can have different meanings to people in contemporary times and alternative understandings can be equally valid.

This can be one of the reasons why misunderstandings arise in communication. It is especially true in written communications, because we are deprived of the voice tonality, facial expressions and gestures that would serve to clarify our meaning in a face-to-face encounter.

If you're one of the people who can spend a lot of time crafting the words in your emails and reports to convey the exact nuance of meaning that you intend, I have some bad news for you. The exact nuance of meaning that you intended to convey is probably lost on the readers of your carefully composed sentences.

Now, this where our presupposition – the meaning of your communication is in the response it gets – comes into play.

Have you ever found yourself saying to someone, 'No, that's not what I meant!' and feeling quite indignant that they have missed the point of your wonderfully articulate email? You have? Well, sorry, but the presupposition says that whatever meaning the listener or reader of your words ascribes to them *is* the meaning of your words.

And the meaning of your words can most accurately be identified by observing the reaction to them, because a response in words is subject to all the same vagaries as your original message.

So the meaning of your message is what others understand by your message, not what you intended them to understand. In other words, if you want people to get a particular message, it's your responsibility to create the understanding in their minds, rather than it being their responsibility to figure out what you intended.

 The managing director of a firm I was working with was very proud of his five-year plan and the clear objectives that he'd identified for each year. As I met different people in the organization, I asked them about the five-year plan and the current priorities. None were able to tell me what they were. When I relayed this finding back to the managing director he snapped, 'Well, they should know. I've told them.'

It was clear to me that whatever he intended and thought he had communicated, he actually had communicated nothing. Nobody knew the plan, therefore he had not communicated it.

So what's the answer? Well, for a start, ask yourself what is the response you want to your communication? To convey your message accurately, it's usually more effective to give the same message several times, in more than one way. Keep monitoring the response. When you get the response you were looking for, you may be justified in considering that you've communicated what you wanted to communicate. But then again …

There is no failure, only feedback

Of all the presuppositions of NLP, this is probably the most well known. But what does it actually mean and how is it useful?

At first glance, many people think, 'That's not true, there *is* failure: I failed my driving test' or 'I failed my maths exam' or 'I failed to win a contract' or 'I failed to get the job I applied for.' They might also think, 'One of my staff failed to meet a deadline' or failed to hit a target or failed to deliver a service to agreed standards. These are all experiences that can accurately be described as failure, so where does the presupposition become useful?

 Let's start with the example of 'failing' the driving test. I took the test when I was seventeen and I didn't pass first time. The examiner gave me a clear explanation of where my driving wasn't up to standard (I seem to remember it was reversing) and I went back to my instructor for some more lessons. Some months later when I took the test again I passed.

At seventeen, I was disappointed not to pass my driving test first time. I'd failed. But that wasn't the end of the story. I took the examiner's feedback and went for more lessons. Now I was better off because I knew what I needed to work on. I got quite good at reversing and then I passed the test. So the point is that the 'failure' actually left me better off because I knew where to focus my learning. And that's the sense in which it's feedback rather than failure.

When you have a goal in mind and you start working towards it, you might only do things that contribute to the achievement of the goal and you might achieve it very quickly. But if it's something you haven't done before the chances are that you might make some false starts, you might make assumptions that turn out not to be valid or you might get some unexpected results somewhere along the way. If any of these things happen, they're part of the learning process and they help you to decide how to proceed. That's feedback.

You could call it failure when something unexpected happens. If you did, you'd probably feel bad about it and that might get in the way of further progress. If you call it feedback when something unexpected happens, then you can feel good about that and progress is not hampered.

Now, if you've done some NLP training, or even if you haven't, you might think that this is obvious and you might be wondering why I'm writing about it. You already know that 'failure' carries negative feelings for the majority of people and that re-labelling the experience to something more constructive is helpful. Well, there's more.

KEY TERM

This isn't just about re-labelling the experience. It's also about **feedback**. In pure systems terms, feedback is defined as 'information generated in one part of the system that is relayed to an earlier part of the system to modify the behaviour of the system.' In other words, feedback is only feedback if it changes something. If nothing changes, it's just information.

So the point here is that this presupposition is not just about taking the sting out of 'failure'. It's about paying attention to everything you can that will help you to succeed next time. It's about recognizing that this experience of 'failure' can be valuable in equipping you to achieve what you set out to achieve.

At a really obvious level, you can say, 'Well, now I know that doing that doesn't work so I won't do that again.' That's okay, but it doesn't help you decide what to do instead. To really get the benefit of failure, you would examine in detail what went wrong and how, so that you could develop a more useful approach.

Going back to my driving test, if I'd simply resolved 'not to do that again' – i.e. not to repeat the whole experience in the same way – I might not have taken a test again!

Or I might not have gone to the same test centre again. Instead I listened to the feedback and adjusted the part of my behaviour that led to the 'failure'. So the information that the examiner gave me truly was feedback.

 There is no failure, only feedback – so long as you pay attention to the feedback and act on it!

You are not your behaviour

This is an interesting presupposition. Behaviour, that is, what somebody does, is often the thing that we're most aware of, and we make judgements about people on the basis of what they do and how they behave.

The thing to recognize is that there's always more to somebody than what they're doing right now.

You might dislike somebody's way of going about something but you might respect the knowledge and experience they have in a particular field. So, there may be a bigger picture where you can accept the person even if you don't like a particular aspect of their behaviour.

I think the ultimate example of this is Gandhi, who said he aimed to love his enemies while hating what they did – 'Hate the sin and not the sinner' is among his teachings. That really is making the distinction between the person and what they're doing!

You may also have noticed this principle at work in the way that people talk to small children. There might have been a time when it came naturally to say to a child, 'You're very naughty,' whereas now a lot of people have learned that it's more appropriate to say to a child, 'That's a naughty thing to do,' because phrasing it that way makes a distinction between the child and the behaviour.

The result of that is the child has some choice about whether or not they *do* the naughty thing, whereas if you told them they *are* naughty, they don't have any choice. If you tell a child often enough, 'You're naughty', it's likely that they will accept that as their role in life and keep looking for ways to be naughty.

In the work environment, lots of people learn to make this separation between the person and their behaviour as part of performance appraisal systems. Here you criticize the behaviour or the process or the end result, rather than the person: 'What you did was not helpful,' rather than, 'You're not helpful.'

Looking at the world in this way can be helpful to us because it draws attention to a bigger picture. It takes our focus away from the irritating or counterproductive behaviour of another person. By recognizing that what they are doing is not who they are, we may start to look for ways to engage the same person in different behaviour.

Remember as well that this principle applies to you. On the odd occasion when you make a mistake, do something inappropriate or say exactly the wrong thing to the wrong person, remember – you are not your behaviour!

People have all the resources they need to succeed

This is a great NLP presupposition if you work in an environment where you need to get the best out of people. It's particularly useful if, like me, you're involved in learning and development.

It's useful because the opposite is believing that some people just can't do certain things. And if you believe that somebody just can't do something, you are unlikely to be able to help them learn it.

The essence of this presupposition is that if somebody is having difficulty doing something in a given circumstance it isn't usually because they can't do it, it's usually because they haven't made the connection between skills they have in another context and the particular context in front of them.

One of the things a lot of people have struggled with at some point in their life is public speaking. The research says that it's the number one fear of mankind – most people would rather die than have to speak in public!

So does that mean that those people don't have the capability to speak in public? I don't think so! Because what do they need to be able to do? They need to be able to open their mouth and speak, string their words together into a sensible discussion or argument, or presentation, which most people probably can do in other circumstances. You get people who, sitting round the table in a restaurant with a bunch of friends, can hold forth for ten or fifteen minutes quite easily; can tell a story, can string an argument together, can make a really compelling case for their particular belief, and yet if you told them to stand up in front of a conference and do it, they'd say, 'Oh, I can't do that.'

They have all the constituent skills – they can stand up there behind the lectern, they can speak, they can string the argument together, but somehow when you ask them to do it all at once, they say, 'Oh no, I can't do that.'

They have the resources, they have those personal capabilities, the personal strength and so on to be able to do it, they just haven't brought it to bear in that particular context.

So, one of the things that we can do with NLP techniques is to make the connection between the context where you have a certain resource and the context where you need that same resource.

People say, 'Oh, I need more confidence,' and yet, even the least confident person in the world will have areas where they are totally and utterly confident. So somebody who really isn't confident in their capability at work may be totally confident in their ability to drive a car, or ride a bicycle, or play football, or whatever it is. They will have confidence in some arena of life, it's part of their repertoire. It's just about bringing it to bear in the place where you need it.

That's what this presupposition is all about: making the connections between the resources and the circumstances where you need them. There are no unresourceful people, just unresourceful states to be in.

The successful practice of NLP enables a person to manage their state so that they have all the resources they need in the situation they're in.

 The circle of excellence is a very simple and effective technique for getting yourself in the right frame of mind to do whatever you need to do. It is based on the presupposition that people have all the resources they need to succeed. If there's something in your calendar that you're not looking forward to, this technique can shift your feelings so that you are able to take it in your stride and even enjoy it.

Some examples of where it can be used:

- To deal with nervousness about making a presentation
- To boost confidence before an interview
- To relax at the end of the day
- To stay calm and alert when chairing an important meeting
- To overcome exam nerves
- To stop irritation at everyday distractions.

Here's how to do it:

- **Step 1:** Identify the situation or event coming up that you are nervous or anxious about or simply not looking forward to.

- **Step 2:** Remember an occasion in the past when you faced something similar to the situation you want to work on. Imagine a circle, about 1m in diameter, on the floor in front of you. Step forward into it as you relive the past situation. Make sure you're really feeling what it was like. Take your time and remember what you saw, what you heard and the way you felt in that past event. What was missing for you? What personal resources were you not able to access in that situation? Pick three.

- **Step 3:** For each of the personal resources required, anchor it into the circle as follows: Remember a specific time in the past (in any situation) when you had this resource. Go back and relive the memory in as much detail as possible. See what you saw, hear what you heard and when you can really feel the feelings associated with it, step back into the circle. As the feeling fades away, step back out of the circle. Do this for each of the three resources separately.

- **Step 4:** After anchoring each of the required resources in the circle, take a moment to clear your mind. Doing a mental exercise such as reciting your phone number backwards can be a good way to 'break state'. Then imagine yourself in the future event and step forward into the circle. Notice the difference.

If you identify any further resources that are needed, repeat step 3.

People are doing the best they can with the resources they have available (or, present behaviour is the best choice available)

When they first encounter this presupposition a lot of people say to me, 'I don't believe people always do the best they can.' It really depends on how you think about it. The important part of this presupposition is *with the resources they have available*.

Consider this: do you always do the best job you could possibly do? And if you think about a day when you know you didn't achieve as much as you could have, what was the reason? It's probably because there were other things going on. If you hadn't slept well the night before and you were feeling tired, then maybe you didn't achieve as much because you didn't have the energy that you would normally have. In terms of the resources you had available that day, there wasn't much in the energy bank. So you did the best you could, given the small amount of energy you had. That's what that presupposition means: given whatever you've got going on at that time, you will always do the best within those limitations.

For example, one day, you might be feeling really annoyed with somebody you have to work with and that's distracting you. That's what you've got available that day. And that will tend to go against you doing as much work as you might do on another day. So people are doing the best they can with what they've got *right now*.

It can be interesting to apply this presupposition to a situation where you think somebody is being obstructive or difficult. Sometimes I see this in organizations. I might go in to run some training programmes as part of a change initiative. Quite often, somebody who's trying to drive the change will say, 'This group of people (or this individual) is being obstructive, they're being difficult,' and you can take that at face value and go, 'Okay, people are being difficult.' But if you go back to this presupposition and say, 'Well people are doing the best they can with what they've got,' that will get you thinking differently. What has somebody got going on that means that this behaviour, that's being described as 'difficult', is the best they can do?

CASE STUDY

I had a good example of this in a company I did some work for – a small engineering firm that provided a service to quite a wide range of customers. The operations manager wanted to change some of their working practices and he said to me, 'Well, the guys in the workshop are being difficult, they're being obstructive, they won't do the new things.'

And when I got an opportunity to talk to them and find out a bit more about it, what they told me was that the new working practices were going to get in the way of them being able to do the best job for their customers. And they wanted to be able to do a good job for their customers. They cared about the work!

So, I went back to the ops manager and said, 'You need to sit down and talk with these guys because they're telling me that if they do what you want them to do, it will result in a drop in service. That's why they don't want to do it. Now I don't know whether they're right or wrong, but that's what they think and so that explains why they're not being very enthusiastic about your new working practices.'

So he went and talked to them and it turned out that in fact there was some information that he hadn't given them and so they'd filled in the blanks and made some assumptions that weren't actually valid (which is often what happens if people get incomplete information). And so, with a bit more conversation, they were able to come to an agreement where everybody was happy and the new working practices went in.

So, how about testing out this presupposition for yourself? If somebody is behaving in a way that you think is obstructive or difficult, ask yourself, 'What else have they got in this situation that means that their current behaviour is the best they can do?' According to the presupposition, their behaviour must seem the best choice available to them – if this seems inexplicable, you need to do some more talking to find out what's going on.

Give it a try – you may be pleasantly surprised how easy it is to deal with people who appear to be 'being difficult'.

The part of any system which has the most flexibility controls the system (Ashby's law of requisite variety)

This principle comes from systems thinking and, in terms of mechanical systems, is probably fairly straightforward.

THINK ABOUT IT Consider a central heating system: the one in my house has several radiators that just sit on walls, and yards and yards of copper piping that are fixed in place. It has a boiler that just sits there, although it goes on and off depending on what position the switch is in.

73

The thing that really makes the difference to what's going on is the thermostat, which has lots of different settings. It has flexibility.

So, the human equivalent of that is that the person with the most behavioural flexibility will have the most influence in a situation.

They won't necessarily always get their own way, because sometimes some of the most senior people are the least flexible and they just say, 'We'll do it my way because that's how I want it done,' and everybody else scuttles around and does it.

However, somebody who is flexible in the way they go about achieving what they want to achieve can often have a lot more influence than somebody who only has one way of doing things. If you've got more than one way of explaining your idea, then you stand a greater chance of people understanding it and coming on board than if you only have one way of describing what you want.

Note that I'm talking about flexibility in structure; flexibility in *the way you go about things* rather than in *what you do*. I remember somebody saying to me one time, 'Well, I'm very flexible, I'll fit in with anybody.' But that's being flexible in terms of content, rather than in terms of structure. So the more flexible you can be in how you go about things, then the easier it is to get your end result. The psychologist and philosopher William James (1842–1910) said that the definition of intelligence is having a fixed goal and multiple means of achieving it.

 Flexibility is not the same as compromise. When you see flexibility in content, that's compromise. However, if you're flexible in the way you go about something, you do not need to compromise your goals and indeed you will probably get your desired end result because you will be able to relate to different people, and encourage them to do things in different ways, and so on.

So true flexibility is having multiple ways of getting where you want to go, rather than compromising on where you want to go in the first place.

Every behaviour has a positive intention

At first glance you might be forgiven for thinking that this presupposition suggests that everything anyone does is intended to do good. But that isn't what it means. (And I'm sure that you can think of some behaviours that certainly don't do good to others.) Moreover, don't forget that the NLP presuppositions are filters. They're ways of looking at the world that can create more constructive results if we choose to use them. You don't have to believe that a presupposition is absolutely true in order to use it and get the benefit of it.

If you have a pet cat, you probably practise this presupposition regularly. Do you wander in to the kitchen in the morning to discover a half-eaten or half-alive mouse, bird or other small, helpless creature? And if you love your cat, do you shout at it and complain about the mess on the kitchen floor? If you're like most of the cat-lovers I know you probably don't do that. More than likely, you say something like, 'Ah look, he's brought me a present.'

If you do that, you're looking beyond the behaviour to its purpose. Excellent! That's what this presupposition is all about. Now, can you do the same when your colleague at work drops the equivalent of a dead mouse onto your desk and expects you to deal with it? Can you look beyond the irritating behaviour of people around you and focus on the purpose behind it?

I think that this presupposition is better if it's phrased 'Every behaviour is purposeful'. What people do is not random, it's usually not even aimed at annoying you, but it does have a purpose. Granted, the purpose may be of value only to the person who behaviour is in question, but it has a purpose nonetheless.

So next time someone around you does something that you consider to be inappropriate, foolish or annoying (or all three!), I challenge you to ask yourself, 'What is the purpose of this behaviour?'

By asking yourself this question you'll be achieving two things:

1. You may arrive at a better insight into the other person's motivation and thinking.
2. You'll distract yourself from being irritated and instead become curious, interested and maybe even more motivated to talk to the person about what they're doing. This can only be an improvement, can't it?

All behaviour is appropriate in some context

The underlying principle of this presupposition is that for anything a person may do, or be capable of doing, there will be some context in which it's absolutely the right thing, and others where it may be less appropriate.

Let's take an example where somebody is shouting in an office meeting, raising their voice and perhaps getting quite aggressive about it. You might think that's inappropriate behaviour.

First of all, separate the behaviour from the context. The behaviour is 'shouting' or 'raising the voice'. The context is 'during an office meeting'.

It's the context that makes the behaviour inappropriate.

There will be situations where that exact behaviour could be very appropriate. If somebody was in danger and you wanted to alert them, then shouting to get their attention could be the appropriate thing to do. So if you look purely at the behaviour – what the person is actually doing – and take it away from the context, then there will always be some context in which that's the right thing to do. And that's what that presupposition means. After all, why would a person develop a skill if there was no value in it?

I had a coaching client some years ago, who was regarded as being very inflexible and dogmatic and difficult to influence, and not very compromising or flexible with colleagues.

I discovered that he was the quality manager in his organization and that he was very definite about standards that needed to be met. In that context, it was a really useful skill to have, to say to people, 'No, we will not compromise on this.'

There were other circumstances in his working life, such as where to have a meeting, where it would have been perfectly all right for him to be a lot more flexible. But he'd got so much in the habit of being very fixed about his ideas that he was also doing it in other circumstances where it was inappropriate behaviour.

I tend to find that with a lot of organizations, whatever is the core competence of the organization tends to filter out into everything that they do, whether it's appropriate or not. It's as if there is a fundamental philosophy of the way that you do something, which is the thing that you're in business for, and everything else follows suit.

For example, I did some work for British Airways Engineering some years ago and some of the managers there said to me. 'It's very challenging managing aircraft engineers because they're very negative and they're kind of nitpicky, and we go to them with a new pay deal and they tell us all the reasons it won't work, and it just gets to be a bit frustrating.' And I remember saying to them, 'Well, what makes a good aircraft engineer?

Is a good aircraft engineer somebody who looks for problems and solves them? Do you want somebody who's going to really nitpick, to be maintaining that aircraft that's going to carry 300 people to their destination?'

And so, that's the core competence. It's that ability to spot the problems and solve them, so you can't really blame people for them doing it in relation to other things and not just their everyday normal work.

Around the same time I was also working at the BBC World Service and the managers there said, 'People act as if they don't trust us. You go to them and talk to them about changes in the working hours, and they all go, 'Yes, but what are you not telling us? What's going on behind the scenes?''

79

In the same way I said to them, 'Well, what makes a good journalist? It's somebody who doesn't take things at face value. So you should expect that you're going to get searching questions when you go and present information.'

After these two examples I started looking for other evidence of a 'core competence' in organizations that influenced the culture significantly. I remembered some training programmes I did for a fire service and, surprise, surprise, everything was done on a fire-fighting basis! Nothing was ever dealt with until it was really, really urgent.

I think that what this demonstrates is that we develop behaviour that is appropriate in some context, and if it works well we may continue to use it, even beyond the bounds of appropriate contexts. The challenge, therefore, is that even if our behaviour seems appropriate, in order to achieve our goals, sometimes we need to look for behaviour that's even more appropriate to the context.

The mind and the body affect each other

This presupposition of NLP has given rise to a whole area of application. There are lots of NLP practitioners who work primarily in health and well-being and whose skills stand on this most fundamental assumption. So what does it mean?

Quite simply, it means that the mind affects the body and the body affects the mind. Or if you prefer, it means that the mind and body are not separate, they are part of the same system.

For most people, it is quite obvious that the body affects the mind. If someone is feeling physically unwell, that is often accompanied by mental symptoms like reduced ability to concentrate, lack of enthusiasm for everyday activities or lower tolerance of others. Conversely, most of us feel mentally more alert, more creative and more interested when we're physically charged up, like after a long walk or run, a fast game or good work-out.

So, you've experienced how the body affects the mind and that's often a good motivation to look after your body. But what about the mind affecting the body? Not so many people take that for granted.

It's my belief and my experience that what we think affects our physical bodies. Many physical ailments have an emotional or mental aspect to them so that we may be more likely to suffer physical symptoms when we're in a particular frame of mind. The most common example of that is stress. When we're feeling overwhelmed by work pressure, that's when the headaches, backaches, upset stomachs and allergic reactions are most likely to appear.

It might be stating the obvious, but there is another assumption here that what's bad for the mind is also bad for the body. If you think gloom and doom, and see yourself as a victim of life's injustice, your body's well-being will be lessened. If see yourself in charge of your own destiny and think about what's important to you in life, then your body's well-being will be enhanced.

Neuroscience shows us that neurotransmitters – the chemicals that transmit messages from the brain through the nervous system – actually interact with every cell in the body. So it's possible that any thought you have can affect your whole body, albeit in a very small way. Deepak Chopra, one of the leading figures in mind-body medicine, says in one of his books that an average person has some 90,000 thoughts per day. He goes on to say that about 75,000 of those thoughts are the same thoughts as yesterday! So if every one of those thoughts has a miniscule effect on your body, it's easy to see how habitual thoughts might create chronic health problems, isn't it?

Now, here's the interesting bit: NLP gives us tools to get control of our thoughts and our habits. So if we change some of the 'bad' habits we have in thinking negatively, we can potentially change the way our bodies habitually respond. Or to put it another way, it's easier to recover from an illness or break out of a cycle of chronic symptoms if you change the way of thinking that contributed to the creation of the problem.

Now, I'm not saying that all illness is 'in the mind' or that 'positive thinking' is a cure for whatever ails you. What I am saying is that I believe the body is as intelligent as the mind in its own way and that physical symptoms are not random and do not spontaneously arise for no reason.

The mind and the body are one system. Physical symptoms are a way of drawing your attention to what is going on in your body so that you can do something about it.

Modern medicine has tended to foster the notion that what you do about symptoms is take a drug to relieve them. But a little bit of time spent on considering the reason for the symptoms can increase the effectiveness of the drug, reduce the likelihood of needing it again or even remove the symptoms without the use of drugs.

There is another presupposition of NLP that says 'all behaviour is purposeful'.

I believe that, in the same way, we can regard physical symptoms as 'behaviour' of the body and if we assume that that behaviour is purposeful, that can be the key to full recovery. Many people I know who are 'body workers' – massage therapists, chiropractors or reiki practitioners – confirm that the patients who make the quickest and most complete recovery are those who are also engaged in personal development – they're are working on the psychological as well as the physical.

I realize that not everyone will agree with my reasoning here. I hope I've stimulated you to think about your own attitude to your physical well-being, and the meaning of this presupposition, that the mind and the body affect each other.

All interventions should increase choice

This final presupposition becomes most relevant when you get into using some of the techniques for change. It's just recognizing that if you're going to reprogramme some of your behaviour, or you're going to help somebody else reprogramme theirs, the ideal thing is that we look for more choice in a situation.

Suppose somebody does something that they wish they didn't do, then you could say, 'Well, we'll reprogramme so that they just can't do that.' But if we believe the presupposition that *all behaviour is appropriate in some context* then there might be a situation where the person might want to behave that way. Currently they're saying, 'I want to stop doing it because it's not useful,' but there might be some circumstance in which it's absolutely the right thing to do.

So we prefer to keep that behaviour as a choice so that they could do it if the situation arose. And, of course, if we believe the presupposition that *people make the best choice on the basis of what's available* then that's going to be okay, because as soon as they have more choice they'll choose the most appropriate thing to do.

3. Outcome thinking

'Outcome' is the general term used in NLP to refer to some desired state, achievement, goal, target, wish, dream or event. 'Outcome thinking' means focusing your thoughts on what you want. It's important because it's hard to decide what to do if you're not certain about what you want to achieve.

THINK ABOUT IT At work, do you have a clear set of goals and targets? Can you put the tasks on your to-do list in the context of your overall goals? If there is a clear connection between what you're working on now and the results you want to deliver, it's usually easier to motivate yourself to work.

Some people are very good at focusing on their goals and are motivated by doing that. Other people find it more natural to focus on problems and how to solve or avoid them. Often a combination of both will be needed to move from your starting place to your desired end result.

Problem frame and outcome frame

This is a quick problem-solving technique that you can use any time. It works well on issues that are specific to you and that you can influence. It's not very good for stuff that's way beyond your control (world poverty, climate change etc.) and it might not yield anything new on a situation that you've been mulling over for weeks or months. But often it'll shift your thinking sufficiently to resolve the problem.

 Identify a problem that you'd like to resolve. Then ask yourself each of the following ten questions in turn. Answer each question fully before moving on the next.

Problem frame
1. What's the problem?
2. How long have you had it?
3. Why do you have this problem?
4. Who is to blame?
5. Why haven't you solved it yet?

Outcome frame
1. In this situation, what do you want?
2. How will you know when you've got what you want?
3. What resources do you already have that you can use to solve this problem?
4. When have you succeeded in something similar?
5. What is the next step?

Simple, but powerful!

This process is also a good one to use as part of a coaching conversation. Even if you don't fully understand the technicalities of the situation, you can use these questions to assist constructive thinking. It's a great alternative to giving advice or trying to solve the problem yourself.

Well-formed outcomes

Many people have bad habits in the way they think about what they want, and this can be a block to making any progress towards getting it. For example, some people have thought patterns that go like this:

> 'I'd really like a new car. Oh, but I can't afford a new car.'

> 'I'd really like to go out to dinner this weekend. Oh, but nobody will want to go with me.'

> 'I'd love to learn to play tennis. Oh, but I wouldn't be any good at it.'

Patterns like this prevent any kind of focused activity towards the desired state, by presenting the impossibility of having it.

KEY TERM

The meaning of a **well-formed** outcome is that it is represented in your mind in a way that makes it easy to achieve. It's the *representation* of the outcome that is 'well formed'.

An outcome is well formed when:

1. **It is expressed in the positive.** That means it's described in terms of getting what you want, rather than avoiding something you don't want.

2. **You can describe it in sensory-specific terms.** This will be a description of what you will be able to see, hear and feel when you achieve your outcome.

3. **It's under your control.** It is only well formed if you can make it happen without relying on other people.

4. **The context is clear.** You can define where and when you want it and have identified the exceptions.

5. **You can maintain the secondary gain of the current situation.** If you have been intending to make this happen for a while but not made any progress it could be because there is value for you in *not* succeeding in this. This is known as secondary gain. The key to progress is to maintain that benefit as well as moving towards your outcome.

6. **It is ecological.** There will be no unwanted consequences of you achieving the outcome.

The keys to making it work

The six criteria can be considered in the order that they
are listed, but they can also be looked at in whatever order
seems natural. Sometimes your thinking on one point will
naturally take you to another and it makes sense to follow
that flow instead of imposing a rigid sequence. It's also
quite usual to consider some of the criteria more than once.
If you find that your outcome is, for example, not under
your own control, then you may go back to the beginning
and define what you want in a different way.

Bear in mind that you are aiming for a crystal-clear
description of the end result you want, not creating an
action plan for how to achieve it. With a well-formed outcome in mind, you will naturally choose behaviour and
activities that move you towards the desired state.

There are considerations for each criterion that will
make a difference:

Outcome expressed in the positive

1. Sometimes the starting point with an outcome is what
   you *don't* want. To shift into the positive simply ask yourself, 'If I didn't have that, what would I have instead?'

2. Be aware that the outcome can involve specific activities, or it can simply be an end result. If the activity isn't

compelling – 'I want to go to the gym three times per week' – then it may be more useful to focus on what that will achieve. Ask yourself, 'What will that do/get for me?' The outcome may then become, 'I want to look slim and toned in my summer clothes.' For most of us, that is a more compelling idea.

3. Avoid comparisons and go for specifics. 'I want a better job' is not as well formed as, 'I want a job as a sales director that pays at least £100k.'

Sensory-specific evidence

1. Identify the moment in time when you will be able to say, 'Yes, I've done it!' What is going to be happening at that moment that will be your evidence of success?

2. The sensory description of that moment can be associated or dissociated. For example, it can be a description of what *you* will see or, alternatively, how you will look to another person.

3. Ensure that the evidence is specific. 'I'll feel better' is not as well formed as, 'I'll feel confident.'

4. If the evidence includes feelings, make sure that they are feelings that *indicate* success rather than the feelings of pleasure at having succeeded.

Under your control

Sometimes the outcome depends on the involvement of other people. This is only outside of your control if you are unable to influence or persuade someone to get involved or if it depends on one particular person. For example, if you are looking for a new job, it won't be you that decides who gets the job (unless you decide to become self-employed!). Your new employer is the person who makes that decision, but you can influence that person through the selection process and you can apply to another company if the first one doesn't take you on. So it is under your control to get a new job, but perhaps not a specific role with a specific company.

Context

As well as a clear context for the outcome, make sure you identify any exceptions or exclusions. Ask yourself, 'Where and when do I want this?' 'With whom do I want it?' and 'When do I *not* want it?'

Secondary Gain

1. The secondary gain is often key to understanding why a long-standing outcome has not been achieved. Look for benefits of not starting or not completing the activity. Also look at what is the result of current activity – what is being gained instead of the stated outcome?

2. Secondary gain isn't always logical. It can result from old beliefs and unconscious assumptions that, once you bring them into conscious awareness, cease to be a problem.

3. Remember that secondary gain means benefits of *not* achieving your outcome. Sometimes these are out of conscious awareness until you begin to ask the question.

4. If you find that the answer to the question, 'Why don't you have it already?' is something like, 'Because I'm lazy,' that indicates that there is internal conflict. One part of you wants something and another part isn't motivated by it.

5. If self-discipline or willpower seem to be a key part of achieving the outcome, dig a bit deeper. Self-discipline usually means that there is a lack of congruency. The parts that aren't committed to the outcome will eventually rebel.

Ecology

1. The key to this is to imagine the outcome achieved, associate into the future and check for any unwanted consequences of achieving your outcome. Notice any incongruence and use that to identify any areas of conflict.

2. Remember that time is an invention of the conscious mind. At the unconscious level there may be a sense of

conflict but if you introduce time, there may be a resolution. You can 'have it all' but maybe not all at once!

Think of something that you want to achieve. It could be a big goal, or it could be something as simple as the result you want from speaking to someone on the phone or sending an email.

Run through the six criteria with your outcome in mind and make it as well-formed as you can.

Does that make it clearer? Easier to achieve?

Like any skill, making your outcomes well formed will get easier with practice. It's well worth putting in some effort on this because when you have clear outcomes in mind, you will find that you work more productively and find it easier to focus on the priorities.

I recommend that you aim to develop ten to twenty well-formed outcomes per day. Not necessarily big, long-term goals, just whatever is part of your day. Before you attend a meeting, ask yourself, 'What's my outcome?' Before making a phone call ask yourself, 'What's my outcome?' Before deciding what to have for dinner ask yourself, 'What's my outcome?' and so on.

If you do this every day for two to three weeks you will install this pattern at the unconscious level and it will become your natural way of thinking about outcomes.

When you ask yourself the question, 'What's my outcome?' what comes to mind will already be well-formed.

Often the key to a well-formed outcome is in the secondary gain. Here is an example:

 Stephen had been working towards a professional qualification for some time and had twice put off taking the exam. He said that he didn't feel that he was up to the standard to pass. However, when we created a well-formed outcome about taking the exam we discovered that several of his colleagues had passed the exam first time and that Stephen wanted to do the same. The secondary gain of putting off the exam was, 'If I don't take the exam, I can't fail it.' With this new awareness, he was able to convince himself that he was ready to take the exam and passed first time!

You can use the well-formed outcome pattern to create team outcomes for a meeting, project or workshop. The elicitation of individual evidence criteria can be particularly useful in understanding different viewpoints and values.

The same principle can also be applied to company goals and targets. Goals, targets and objectives can all benefit from being well formed. Change programmes are often made easier by the consideration of the secondary gains. What do people perceive they will have to give up to make this change happen? Can it be maintained?

A medium-sized company had been through a period of significant growth, with the number of staff rising from 150 to 400 in two years. The new IT director inherited a team of around 40 IT specialists and was frustrated to discover that many of them were frequently interrupted during their working day by colleagues from other departments seeking help with PC problems.

The IT director decided to implement a formal helpdesk system and tracking process. They installed a suitable programme, designated two people to be the first line of contact and then notified the rest of the organization of the new procedure. You may not be surprised to hear that, as a result, nothing changed. Everyone continued to call on their friends for assistance with PC problems instead of using the new helpdesk.

The IT director considered his options. He could insist on the use of the helpdesk and discipline his staff if they stepped outside of the system. He could refuse assistance to requests that didn't come through the proper channels. Or he could look at it another way.

He asked himself what the benefit might be of not using the helpdesk. He realized that people wanted to ask for help from someone that they knew, that they trusted and that they believed could help. This was the secondary gain.

So he asked the two people who staffed the helpdesk to take two weeks and visit every department in the company and talk to people. Their objective was to find out which applications people used, whether they had any recurrent problems, what they might need in the future and generally understand how technology was used in each department. And also to make friends with everyone.

After doing this, the helpdesk began to be used. People stopped phoning random colleagues in IT, and rang the people who really could help – the helpdesk. This was because the helpdesk staff now fitted the criteria of being someone that they knew, that they trusted and that they believed could help.

The secondary gain was maintained and the helpdesk was operational.

4. Rapport, pacing and leading

Rapport

The term 'rapport' is one that is in everyday use. It's used to mean a state of friendliness and goodwill between people. In that sense, it's perfectly logical to refer to having rapport with someone you haven't seen for three weeks (or more). In NLP when we talk about rapport it usually refers to the state of play between people right now. It's important because it is the foundation upon which influence is built.

Rapport is a state of attention and responsiveness between two or more people.

So it's possible to be in rapport with the person you're talking with right now, even if that person is someone you have some reason to mistrust. It's also possible to be out of rapport with your closest friend if you meet up at a time when you've had very different experiences immediately beforehand.

THINK ABOUT IT

Imagine you meet your friend for dinner immediately after you both finish work. You've had a really satisfying day at work, finishing something that you'd been working on for a while and getting recognition for a job well done. Your friend, by contrast, has had a stressful day at work because two of his colleagues have had a major disagreement and they have both been trying to enlist his support on their side of the dispute.

In the first few minutes when you meet up, are you and your friend on the same wavelength? Are you in rapport (in the NLP sense)? Probably not. But if you're good friends it won't take long before you find the common ground again and settle into enjoying the evening together.

REMEMBER THIS!!!

Never assume that you have rapport with anyone, even your nearest and dearest.

It's also important to realize that there are degrees of rapport. In other words, there are degrees of attention and response that one person may give to another. The extent of rapport you have with another person will determine the extent of influence you will have over them.

If you approached a stranger on the street, how much rapport would you need to get them to tell you the time of day? Contrast that with how much rapport you would need to persuade them to give you £10.

Calibrating rapport

You may not realize it, but you can probably determine quite accurately the degree of rapport in an interaction between people. It's possible to gauge the level of responsiveness just by watching how people behave.

Imagine you went into a crowded restaurant. It's too noisy to hear what people are saying. Looking around, could you pick out a couple who had just had an argument? A couple on a date? Of course you could! The clues are all in the body language.

You can spot the couple on a date, because they're focused entirely on each other. They smile at each other, hold hands across the table, laugh at each other's jokes and pick up their drinks at the same time.

On the other side of the restaurant, the couple who just had a row are equally obvious. They are both sitting stiffly in their seats, not speaking. Their heads are turned from each other and they frown at their own thoughts. But they are still in rapport! They are a couple and they have been here before. Maybe unconsciously, they know what happens next. They know there is a 'programme' that gets them from the argument to kissing and making up. Soon one of them will take the next step in that programme and apologize, or storm out, or get the bill. They know what happens next. If that programme didn't exist, they wouldn't be a couple.

 Wherever you happen to be, put the book down for a moment and look around you. If there are other people around watch how they interact and gauge the extent of rapport between them. (If you're at home alone reading this, then do it next time you go out.)

 When people are in rapport with each other they tend to do the same things.

Creating rapport

In normal everyday encounters you are likely to create a degree of rapport quite naturally and without conscious effort most of the time. However, it may not always be sufficient for your purpose. Negotiating a pay rise may require a lot more rapport than requesting a few days off!

Barriers to establishing rapport naturally could be:

- You are starting from very different points of view
- You are preoccupied with your own internal world and not giving sufficient attention to others
- You are stressed or anxious
- There are cultural differences.

If the rapport doesn't happen naturally, it's very useful to be able to create rapport deliberately, or to deepen the rapport when it is insufficient to achieve your desired outcome.

 The simplest way to establish rapport is by matching and mirroring non-verbal behaviour.

Matching non-verbal behaviour is the most effective way to build rapport because it's usually outside of conscious awareness and because it's what people do naturally when they like and trust someone.

Having said that, it's not always the easiest strategy.

 Make yourself really comfortable in your seat. Notice how you are sitting: how far back on the seat are you? Resting against the back of the seat? Is your weight evenly balanced or leaning to one side? Are your feet out in front or tucked under? What is the position of your arms and hands?

Now have a look around and notice someone else who is sitting on a similar seat. Shift your position to match theirs. Adopt the same posture, position of hands, feet etc. and notice how you feel. Less comfortable? Awkward? Self-conscious?

So, if you are in the middle of a business meeting and you decide to improve the level of rapport by matching another person's behaviour, you run the risk of feeling uncomfortable, awkward or self-conscious. It's hard to see how that will help the meeting, isn't it?

The key here is that although in theory we can all sit and stand and move in an almost infinite variety of different ways, in reality we all have a limited repertoire of ways that we habitually sit and stand and move. When you move outside of your normal repertoire, you can be distracted by the unfamiliarity of behaviour even if it isn't physically uncomfortable. If it is uncomfortable as well, it will decrease rapport because you'll be paying more attention to yourself than to the other person.

The only remedy for this is practice. The first goal of your practice will be to expand your repertoire. That means becoming more flexible (sometimes literally!) in how you sit, stand and move.

Go for a walk through your nearest shopping area or park. Notice the people around you and experiment with walking the way they walk. Try sitting on a bench in the way that someone else does it. In the interest of not causing any offence, I recommend that you 'try on' the behaviour of people at some distance from you, not those in your immediate vicinity – so take your glasses if you need them to see at a distance!

If you practise in this way regularly, you will gradually increase your repertoire of ways to sit, stand and move. You should also notice a corresponding increase in the degree of rapport that 'happens naturally' in your interaction with other people.

Obviously there are some limitations on this. If you are matching the behaviour of someone who is physically much taller, shorter or bigger than you, there may be things that they can do that are impossible for you. Just do the nearest you can.

There are also some behaviours that it's not appropriate to match. For example, most women sit with their knees together. Matching the behaviour of a male colleague who sits with his knees wide apart would look odd and inappropriate. So the best plan is to match other aspects of the posture while keeping your knees together in a ladylike fashion. In fact, my experience has been that men rarely adopt such an aggressively male posture in dealing with female colleagues if they are focused on mutual benefit. It's only when there is a disagreement or power struggle that the extreme masculine behaviour comes into play.

 Practise matching behaviour in places where it doesn't matter whether or not you can carry on a conversation. To begin with you'll find that you can't concentrate on what someone is saying if you're also watching what they're doing. Obviously this is not a good strategy for important meetings. Practise at social events, at home with the family or when you're on the sidelines of a meeting.

One of my students told me that he practised by watching TV chat shows. He would sit on the sofa and match the actions of the host or the guests as they chatted. That's a very low-risk way to practise – as long as you're alone!

Matching behaviour

You can match someone's behaviour at the 'macro' level such as standing when they stand, sitting when they sit, leaning forward at the same time and so on. Overall posture is another example of matching the bigger picture. You can also match more specific aspects of behaviour such as position of feet and hands, facial expressions and details such as shoulder position, breathing rate, and gestures.

A word of advice about gestures. In normal conversation, most people use their hands to help convey the meaning of their words. Some people use big expansive gestures, others use much smaller movements. It's good to match the style of gesture, but only when you are speaking. In a normal conversation it's only the person who is speaking who you can see waving their hands around. If you match those gestures as the person is making them, you are likely to distract them and to reduce the level of rapport – because this is abnormal behaviour.

 Remember to match gestures only when *you* are speaking. If you match someone's gesture when *they* are speaking, you'll look strange and they'll probably think you're weird!

Matching voice

You can also match the way that someone speaks. This is especially useful on the telephone. The easiest things to match are speed of talking and loudness.

Have you ever noticed that when you speak on the phone to someone who has a very different accent from you, after a while you begin to adopt a little of their way of speaking? Some people are very self-conscious about this and consciously try to stop it. In fact, it is an indicator that there is good rapport and should be allowed to happen. Bear in mind that you will be very sensitive to any slight changes in your own voice, because it's totally familiar to you. What you may not have noticed is that the other person has also made a corresponding small shift in their own voice to bring it closer to your way of talking.

So, while I wouldn't recommend actively matching someone else's accent – because it might sound as if you're making fun of it – if it happens naturally regard it as a good sign.

 My husband can always tell if I'm talking on the phone to my parents. I grew up in the north of England and as I've lived and worked further south for many years my accent is less obvious than it was when I was growing up. Until, that is, I get on the phone to someone with a stronger northern accent. Then my accent flies north faster than I could hurtle up the motorway!

Matching words

In matching the way that someone is speaking, you can also match their choice of words. This can be helpful when building relationships by email as well as in spoken communication.

The reason why this matters is that words mean subtly different things to different people. If your colleague suggests that a piece of work is 'a priority' and you ask, 'Why is it urgent?' then you may confuse or irritate him. Asking, 'Why is it a priority?' will make complete sense to him. You're using the same linguistic label as your colleague and that creates a sense of being 'on the same wavelength'. It's a separate discussion as to whether 'priority' means exactly the same to each of you, but for now, you've deepened the rapport.

Measuring rapport

Rapport is a multi-sensory experience and so you can detect it in a variety of ways. With practice you'll be able to calibrate quite accurately the extent of rapport either between you and someone else or between two other people. With further practice you'll be able to tell what's going on in a group and who is most responsive to whom.

How rapport feels

When you are in rapport with another person you can feel it. It's very different from the feeling you get when you're *not* in rapport. Most people describe the feeling of rapport as a warm feeling. We have that in language – you might say, 'I met the new analyst today and I really warmed to her.' The warm feeling is usually in the upper body. Some people feel it in their chest, others in the stomach area.

How rapport looks

When people reach a certain level of rapport, it is significant for them. It creates a shift in the nervous system. It's as if the body is on full alert because something important is going on. You can see this happening, because the shift causes the tiny blood vessels under the skin to dilate slightly, which results in a subtle 'warming up' of the colour of your complexion. This is easier to spot in people who are fair-skinned, but it can be noticeable in those with darker skin as well. I'm not referring to the kind of bright blush that someone might get when they're embarrassed or nervous, nor the neckline flush that some people get when they are angry or stressed. This is much more subtle.

How rapport sounds

We've already established that people match voice tone, speed and volume when they are in rapport. What you might also hear are comments that indicate a significant level of rapport. Have you ever met someone for the first time and really 'clicked'? Did one of you say something like, 'I feel like I've known you for ages'? If you did, that's good indicator of rapport. It's less likely that someone would say, 'Oh, haven't we got good rapport?' so you have to listen carefully to spot this kind of clue.

If the conversation is with someone you work with and know quite well, they might say something like, 'I'm really glad you're working on this with me' or 'You and I are a good team aren't we?'

In a social situation your friend may simply say, 'We should do this more often!'

Pacing and leading

Once you are confident that you've achieved a suitable level of rapport, you can test to see if the other person is truly responsive to you by **leading**.

 Leading is when one person does something and the other person does exactly the same within a few moments, without realizing that they have done it. This is important because the person who follows is much easier to influence at that point in the interaction.

Generally you will achieve deeper levels of rapport through matching behaviour. That means that you are following the lead of the other person. Leading means that instead of doing what they do, you do something slightly different. If you really have deep rapport, they will follow your lead and do the same thing. If not, just come back to doing what they do and build up the rapport even further before you test again.

 Next time you're in a conversation with a friend (outside of work) have a go at leading them to do something new. For example, you could lean back in your chair or cross your legs. You could also try getting up or walking around to see if they will follow.

KEY TERMS

Pacing and leading is the process by which you can influence someone else to do something that you want them to do.

Pacing refers to any change you make in your state or behaviour, in order to establish or deepen your rapport with another person or group. It still counts as pacing whether you do it consciously and deliberately or whether you do it naturally and unconsciously.

Once you have established rapport you can **lead** the other person to different and new possibilities by subtly changing your own behaviour.

There are three main applications of this process. You can pace and lead someone to a new **state**, **behaviour** or **opinion**.

Pacing and leading states

This is a really useful thing to be able to do, especially if you work in a customer-facing environment and you have to deal with other people's problems.

State is the overall physical, mental and emotional experience of a person at a given moment. Technically, it's the sum of all the neurological activity at that time and it describes a person's current feelings. Everyday states include relaxed, irritated, concerned, confident, tired, anxious and excited. There are thousands of words for states.

Do you ever have to deal with someone who is in a state that isn't ideal for the situation? It could be someone who is overly nervous about making a presentation, or someone who seems to have no sense of urgency about a deadline, or someone who is angry because something has gone wrong. If you deal directly with customers, you'll almost certainly be familiar with the latter example.

In all of these situations, wouldn't it be great to be able to change the other person's state and make them easier to deal with?

The principle is very straightforward. First, you *pace* by matching their behaviour and matching the key aspects of the state. Then when you have sufficient rapport, you *lead* to a more appropriate state.

However, suppose you have an angry customer to deal with. She's shouting at you and is clearly very angry. Are you going to match her?

Common sense tells us that matching the anger is not going to help anyone and I agree entirely. What most people then do in that situation is the complete opposite. They speak very quietly and calmly and invite the customer to give them the facts.

This can work, but it can take a long time to calm down the customer and find out what you need to know to solve the problem. Sometimes it can make the situation even worse. Here's why:

REMEMBER THIS!!! When someone is angry, it's because one or more of their values are being violated. When you speak to the angry person in a very quiet and calm manner, there is a meta-message which says, 'This isn't worth getting excited about.' Clearly, to the person whose values are being trampled on, it *is* worth getting excited about and so your calm demeanour makes them even more annoyed.

This is the same thing that happens when you're having an argument with your partner or a family member and, just as you are working up to the thing that *really* makes you angry, the other person says, 'Calm down'. Usually that makes you even more angry than ever!

So, if you want to avoid your customer thinking that you don't really care about their problem and you don't think it's worth getting upset over, what do you do? You match.

Don't match the anger, but match the energy and emphasis in their voice and behaviour. Speak briskly and resist the urge to drop your voice. If you look the customer in the eye and say, 'Right then, let me see what I do about this' in a suitably urgent tone of voice, what will often happen is that the customer will immediately calm down. You have shown the customer that you care, and that is a huge relief to her and she will relax and let you help.

I realize that this is counterintuitive, so I recommend that you practise before trying it out on a real customer.

Practise saying something helpful in a brisk and loud voice. For example: 'Let me see what I can do to help', 'I'd like to resolve this for you' or 'Okay, tell me about the problem.'

I taught this principle to the customer services team at one of the call centres of a large IT company. Their main role was to deal with customer complaints and they had a very good track record in solving problems. The difficulty was that the average call time was around 65 minutes, meaning that sometimes there would be a queue of calls, with customers getting even more annoyed at being kept waiting. After the team started using a brisk and more energetic voice tone, the average call time dropped by half, without any change in the rate of resolving issues.

Pacing and leading behaviour

Pacing and leading behaviour is very easy to do. It may have fewer applications at work, but you can use it to test somebody's responsiveness when you want to influence them to help you, to co-operate on a project or to buy something.

The principle is again very simple. If someone is doing something and you want them to stop it, then you pace by doing the same thing, build the rapport and then lead by stopping what you were doing and doing something else.

For example, suppose you have a colleague who repeatedly clicks a pen open and closed during a meeting. You can match that behaviour and click your own pen (or tap it on the table at the same rate; that works just as well) until you have rapport and then stop. If you have enough rapport, your colleague will stop too.

To get someone to start doing something, use the same principle. Match their current activity or behaviour and gradually transition to what you want them to do.

 I was on site with a long-standing client, an IT director who had recently changed companies. I asked him to introduce me to the HR director of the company, so that I could make sure that what I was doing with the IT team wasn't at cross-purposes with anything else going on in the company. I was also aware of the potential for the HR director to offer me further business.

The IT director suggested that we all go out to dinner. I was sitting opposite the HR director and had lots of opportunity to match him and build a significant level of rapport. After we'd been at the table for a while the IT director asked me to move and allow him to go to the bar for some more drinks. I stood up to let him through and as I turned backed I found the HR director also standing up, with a slightly confused expression on his face. He'd unconsciously followed my lead (which wasn't intended to lead him) and couldn't figure out why!

Needless to say, he became a client.

Pacing and leading opinion

To pace and lead opinion takes more time and effort than pacing and leading behaviour or state. However, it's probably the kind of influencing that you'll most often want to do at work, so it's worth learning how to do it well.

Pacing opinion means expressing agreement. This might be tricky, because if you want to influence someone's opinion, that suggests that they currently don't agree with you. So the first stage is to thoroughly understand their point of view and identify points that you can agree upon. To do this might take time. You'll have to ask lots of questions, listen and clarify so that you really know where they are starting from. Look out for basic assumptions, outcomes or 'rules' that you can agree with. Tell the other person all the things you agree with and write them up on a flipchart if that's appropriate.

The next step is to ask yourself, 'What else would they need to know, so that the decision I want them to make, or opinion I want them to hold, is a logical conclusion from the information to hand?'

You see, when two people have different opinions about something, it usually means that they had different information to consult when they made up their minds. Each person had a different set of research, experiences, beliefs, values and outcomes. Their resulting point of view will make perfect sense when viewed from the standpoint of that information, but with a different set of information to consult, it may seem wrong.

If you remember the NLP presupposition that says 'people are doing the best they can with what they've got,' it's clear that a person's point of view is derived from the resources they have to hand. To influence their point of view, you have to manage the information available to them.

REMEMBER THIS!!! If you want to influence someone else's opinion, be prepared to have your own opinion influenced as well. When you start finding out about the other person's point of view, you may find out some useful information that shifts your own view of the situation. The more you share the factors that led each of you to your individual opinions, the easier it becomes to jointly decide on a course of action, plan or standpoint.

Pace... pace... pace... and then... *lead*.

5. Rep systems

We have six representational systems. They're often referred to as rep systems:

- Visual
- Auditory
- Kinaesthetic
- Olfactory
- Gustatory
- Digital

There is a rep system relating to each one of our senses, and then the digital system, which relates to non-sensory thinking such as facts, figures, concepts and language.

KEY TERM

The rep system is the internal mechanism for representing inside our minds the information relating to a particular sense. For example, the visual system is the way we see things in our mind's eye.

We all use all of our rep systems but few people use them equally. We each have habits and patterns in the ways that we use our senses and most of us tend to favour one or two rep systems over the others. Over time, the one that we use the most – our **primary rep system** – can also be seen in personality traits, behaviour and choice of language.

In working with rep systems in NLP, the olfactory and gustatory systems are often disregarded. This is because they are the least used for most people. And for most of us, they're not a significant part of the working day – my apologies to all chefs, tea tasters, sommeliers and perfumers!

I've always emphasized that sensory preference is just something people do and not who you are. I've tended to discourage comments like, 'I'm a visual'. However, I've recently read about research that suggests that there is a physiological disposition to use one rep system over the others and that this preference can be 'read' in the iris of the eye. *What the Eye Reveals* by Denny Ray Johnson or *Eye Yoga* by Jane Battenberg will give you more details.

We all use all of our rep systems. Our primary rep system is just the one most often used; it's not the *only* one used.

To find out your own preferences, you can use the following quiz.

Representational system preference quiz
For each of the following statements, place a number next to every phrase. Use the following system to indicate your preferences:

4 = Closest to describing you
3 = Next best description
2 = Next best
1 = Least descriptive of you

Statement 1
In making a decision I am likely to:
___ trust my gut feeling
___ talk it over with someone
___ look at the situation objectively
___ study the relevant information

Statement 2
During an argument, I am most likely to be influenced by:
___ the other person's tone of voice
___ whether or not I can see what other person means
___ the logic of the other person's argument
___ whether or not I can empathize with the other person's feelings

121

Statement 3

Learning is made easier for me if:

___ I'm shown a diagram or chart

___ I can have a go and experience something new for myself

___ theoretical models make sense of my own thoughts

___ ideas are explained to me and I can discuss with others

Statement 4

It is easiest for me to:

___ recall the sound of a person's voice

___ remember what someone said to me

___ remember someone's touch or handshake

___ see someone's face in my mind's eye

Statement 5

I get distracted:

___ by background noise

___ by inconsistent facts and data

___ if I'm physically uncomfortable

___ if I'm in a room that is untidy or painted an unattractive colour

Scoring

Copy your answers, in the same order you wrote them on the quiz, into the spaces below.

Statement 1

___ K

___ A

___ V

___ D

Statement 2

___ A

___ V

___ D

___ K

Statement 3

___ V

___ K

___ D

___ A

Statement 4

___ A

___ D

___ K

___ V

Statement 5

___ A

___ D

___ K

___ V

Each number now corresponds to a letter, V, A, K or D. Now look at the grid below and, for each statement, copy the numbers into the relevant columns to reflect the way they match up in the above listings.

	V	A	K	D
Statement 1				
Statement 2				
Statement 3				
Statement 4				
Statement 5				
Totals				

(Check that your totals added together = 50)

Interpreting the results

Comparing the total scores gives your *relative* preferences for each of the four major representational systems (remember V=visual, A=auditory, K=kinaesthetic and D=digital). The order of the four rep systems is more significant than the actual scores.

Unless you have a very marked preference for one particular rep system, it's likely that you will relate to some aspects of each. We will look at some of the typical behaviour relating to each preference below.

Just because you have a low score and low preference for a rep system, it doesn't necessarily follow that you are not skilled in using that rep system. It's a bit like left- or right-handedness – I'm left-handed, that's my preference, but it doesn't mean that I never use my right hand or that I can't achieve good results by using it.

Remember that it's a quiz and not a properly validated psychometric instrument. It's highly sensitive to context, mood and manipulation!

Indicators of preference

It can be very useful to be able to spot another person's sensory preferences. Often when communication is difficult or a relationship is strained, it's because of a clash of rep system preferences. Once you know somebody's preferences, you can adjust your method of communication, your style and speed of presentation and even the language you use to make it easier to get your message across.

Here are some of the obvious indicators for each preference.

Visual: If you have a significant preference for visual thinking, you'll probably relate to some of these indicators. Unless you scored 20 for visual when you did the quiz, it's unlikely that you'll relate to all of them. As you read the list, think about other people you know as well.

- Well-dressed and groomed
- Shiny shoes
- Neat hairstyle
- Upright posture

- Quick movements and speech
- Short attention span
- Finishes other people's sentences for them
- Notices how things look and comments on it
- Distracted by untidiness, decor and bright colours
- Keeps a tidy desk
- Likes diagrams, charts and pictures
- Will draw diagrams to help explain their thinking
- Prefers email to telephone, but probably likes Skype!
- Enjoys watching movies
- Uses the language of sight: 'vision', 'focus', 'perspective', 'viewpoint' etc.

Also bear in mind that a person's values will interact with rep system preferences. Someone who values tidiness may not be visually oriented, they may simply like to know where to find everything.

Auditory: There are fewer people in the general population who have auditory as their primary rep system. If this is you, it may explain why you sometimes sense that other people just aren't on your wavelength!

- Has a pleasant speaking voice
- Measured pace of speaking and activity
- Enjoys music
- Also enjoys silence
- Always knows whether they want to listen to music, what music it is and how loud

- Distracted by background noise
- Likes to wear earphones in an open-plan office
- Uses the telephone rather than email
- Likes to discuss ideas and plans
- Talks to self out loud
- Is a good mimic of accents and voices
- Enjoys listening to the radio
- Uses the language of sound: 'talk', 'tell', 'loud', 'clear' etc.

Someone with a high auditory preference is usually very aware of your emotions when you are speaking. If you don't want your auditory colleague to know that you're annoyed about something, wait until you've calmed down to call him. Even if you think you've masked the emotion in your voice, he'll be able to hear it.

Kinaesthetic: People with a high kinaesthetic preference aren't necessarily 'touchy-feely' types. Kinaesthetic includes physical movement and sensation, not always relating to emotions.

- Takes time and moves deliberately
- Dresses for comfort
- Touches people and things
- Speaks slowly
- Sits low down and leaning back
- Distracted by physical discomfort – temperature, hard chairs etc.

- Will set up an office chair or car seat to their exact needs (and be annoyed if you change it!)
- Can maintain concentration for a long time
- Gets their sentences finished by others!
- Walks around to think clearly
- Trusts their 'gut feeling'
- Prefers face-to-face interaction to email or telephone
- Likes demonstrations and interaction
- Likes live entertainment
- Uses the language of feeling: 'feel', 'move', 'warm', 'prickly', 'touch' etc.

Most people regard businesslike behaviour to be non-emotional, so some people who have a high kinaesthetic preference disguise it or detach from it at work in order to appear more professional. Therefore you might have to work harder to spot this preference at work. A lot of kinaesthetic types look more like digital at work.

Digital: This is the non-sensory preference. If it's not obvious how someone is thinking, it might be that they have a strong digital preference. This is also the conventional mode of communication in business.

- Tends to be a 'low-responder' – more interested in what you tell them than in you personally
- Low sensory awareness, finds it easy to concentrate
- Dresses for purpose and function rather than style or comfort

- Has a large vocabulary
- Chooses words carefully
- Often pauses in the middle of a sentence – to think about how to express their thoughts
- Likes facts and figures
- Gets distracted by incorrect or illogical information
- Organizes their environment to logical principles
- Rarely shows emotion
- Uses the language of logic and fact: 'think', 'know', 'calculate', 'deduce', 'infer' etc.

The most important thing for someone with a high digital preference is for things to make sense. If you have to present an idea or report to someone like this, make sure you've got all your facts straight and your numbers add up or they won't pay attention to anything else.

Language relating to sensory preferences

People tend to use language that relates to the rep system they are using as they speak. For example, someone looking at pictures in their mind's eye will use the language of seeing when they tell you their idea. A person who is thinking logically through a process is more likely to use digital language to describe their conclusions. The list below shows some common words relating to each preference, and of course, there are lots more.

VISUAL	AUDITORY	KINAESTHETIC	DIGITAL
see	hear	feel	sense
look	listen	touch	experience
view	tell	grasp	understand
appear	quiet	contact	think
show	sound	reach out	learn
focus	loud	running	process
reveal	discuss	ongoing	decide
perspective	clear	traction	perceive
picture	silence	instinct	consider

Communicating with different preferences

Once you've figured out a person's preferences, you can begin to adapt your communication so that it really works.

Why should I adapt my style to suit other people? Do I have to do all the work?

The purpose of adapting your style to suit the person you're talking to is so that you achieve your outcome. If you are conversing with no outcome in mind, it doesn't matter how you communicate. But if there's something you want to achieve, adapting to the other person's sensory preferences will make it easier to get what you want.

Below is a list of communication strategies for each preference. Some relate to your preparation before a meeting or conversation, some are tips to follow during the dialogue. I encourage you to experiment with these and notice what differences you can make to the success of your meetings.

Visual: How it *looks* is important
- Prepare notes before a meeting rather than scribbling on the board as you go along

- Maintain a clean and tidy environment

- Make sure that documents are well formatted and look professional

- Include pictures, diagrams, charts and models

- Cut to the chase, and keep the pace lively

- Give written as well as verbal feedback and recognition.

Auditory: How it *sounds* is important
- Choose meeting locations that are quiet and do not have distracting background noise

- Keep your tone friendly and pleasant

- Avoid meetings and phone calls when you are irritated or impatient – they will be able to tell

- Allow time for questions and discussion

- Give verbal as well as written feedback and recognition.

Kinaesthetic: How it *feels* is important

- Explain your plans a few days before a formal meeting and outline the involvement you want

- Take time to set the scene before turning to specific business

- Check that the environment is suitable for the purpose

- Allow time for decisions to be made – don't try to rush the process

- Do *not* finish their sentences

- Give feedback in person.

Digital: How it *makes sense* is important

- Check your facts and figures

- Check your facts and figures again

- Use an agenda that progresses logically through the business

- Keep your comments relevant

- Do *not* finish their sentences

- Give specific, factual feedback.

Speaking to a group

If you are preparing a presentation – or an email – to communicate with a group of people, it's unlikely that you'll know the sensory preferences of everyone concerned. The simplest strategy is to appeal to all four major rep systems at different times in your presentation.

This can mean preparing some specific sentences that express your key ideas in ways that appeal to all major preferences.

For example:

> "I know a lot of people have been a little anxious about the sales figures this month, but I want you to relax and let me tell you why you haven't been hearing good news so far. I want to show you the figures so you can see for yourself that we've had good reason to keep quiet until now. I'm sure you'll feel comfortable that it made sense not to shout about our success until we could demonstrate some concrete results."

Now let's look at the language used and how it might appeal to different primary rep systems (remember V=visual, A=auditory, K=kinaesthetic and D=digital):

> "I know a lot of people have been a little anxious (K) about the sales figures this month, but I want you to relax (K) and let me tell (A) you why (D) you haven't been hearing (A) good news so far. I want to show (V) you the figures (D) so you can see (V) for yourself that we've had good reason (D) to keep quiet (A) until now. I'm sure you'll feel comfortable (K) that it made sense (D) not to shout (A) about our success until we could demonstrate (V) some tangible (K) results."

I'm not suggesting that you script everything you want to say, just spend some time on key messages and make sure that you include everyone. The rest of the time, just use your normal style.

 If you regularly take time to craft sentences using language from all four major rep systems, it will become easier and easier until you can do it without thinking about it.

 To completely engage a diverse group will almost certainly mean also preparing some visual aids to *show* your ideas to the people with a visual preference, and it's going to be important to have a *logical structure* to your material if you want to keep the attention of those with a digital preference. Allow time before you speak to *connect* informally with people who have a kinaesthetic preference and include time for *questions and discussion* that will engage those with an auditory style.

6. Perceptual positions

The idea of perceptual positions will probably be a familiar one. It means taking different points of view on the same situation. In NLP we usually consider three different perceptual positions:

First position: Looking at the world from your own point of view, totally aware of your own feelings, needs, experience, agenda and thoughts, without taking account of anyone else's point of view – 'How is this for me?'

Second position: Imagining what it is like for another person in this situation; considering their needs, thoughts, experience, feelings and agenda – 'How is this for him or her?' (Notice that this is not the same as wondering 'What would I do in their situation?' It's more to do with wondering 'What would *they* do in their situation?' The stronger rapport you have with another person, the easier it will be for you to appreciate their reality and achieve second position.)

Third position: Seeing the world from an outside point of view, as an independent observer, someone with no personal involvement in the situation – 'How would this look to someone who is not involved?' From this objective viewpoint you can observe, evaluate and create new and useful choices.

All three positions are equally important; none is any better or worse than the others. Aim to be able to move between them freely and use whichever one suits your purpose at the time.

Many people find it quite natural to move between first and second position and are able to appreciate another person's point of view quite readily. The third position is a useful addition to that, because it is dissociated from the situation and provides a very different and more objective perspective.

Associated means totally immersed in an experience and fully aware of your feelings.

Dissociated means detached from the situation, seeing it as an outsider and removed from the feelings.

We can describe the first perceptual position as associated and the third position as dissociated. Second position is associated to another person.

Remember a time when you had an argument with someone and you *absolutely know* that you were in the right.

Take yourself back into the memory of the argument and fully associate into first position. Experience the feelings again and rehearse your arguments.

Next, take a mental step back from the memory and then imagine being the other person in the same argument. Fully associate into second position and become aware of the feelings and thoughts of the other person.

After that, take a mental step back again. Look at the argument from third position, as if you were a completely uninvolved observer. Observe these two people arguing. What do you notice that they do not?

Finally, come back to first position and fully associate into your own experience. Is it different?

This kind of mental exercise can be very helpful in understanding difficult interactions with others or long-standing tensions. It can also be very helpful in preparing for important meetings, presentations or interviews.

For an even more thorough approach to improving relationships using perceptual positions, try the technique known as the 'meta mirror'.

The meta mirror

The meta mirror is an excellent technique for addressing tension in relationships. You can use it after a disagreement, to explore a relationship that simply doesn't seem to be effective, or to prepare for an important conversation.

Unlike the mental exercise above, this one works best if you anchor the positions to the floor, walk around and literally see things from different positions.

Here are the step-by-step instructions:

1. Mark four positions on the floor in roughly the arrangement shown below.

2. Identify the person with whom you would like a better relationship.

3. Step into position 1 (first position); imagine the person standing at position 2 and fully associate into your feelings about him or her.
   Step out and break state

4. Step into position 2 (second position) and 'be' the other person. Fully associate into their experience of you.
   Step out and break state

5. Step into position 3 (third position) and become a dissociated observer. What do you notice? What insight do you have that they do not?
   Step out and break state

6. Step into position 4, from where you can see the 'you' in position 1 and the 'you' in position 3. Imagine the 'you' in position 3 moving to position 1 and replacing the 'you' in that spot.

7. Step back into position 1. How is it now?
   Step out and break state

8. Step into position 2. How is it now?
   Step out and break state

9. Finish by associating back into position 1.

Enjoy the results!

The principal key to making this work is to take your time and fully associate into each stage of the process. If you don't take the time to really experience the relationship from each of the perceptual positions, you won't get the full benefit of the process.

Along with that, it's important to break state completely as you move from one perceptual position to the next. Physically 'shaking it off' is a good 'break state', or you could use a challenging mental process such as reciting a phone number backwards or spelling a long word backwards. Experiment to find out what works for you. The 'break state' ensures that you get complete separation between the different perceptual positions and can be clear about which emotions and experiences belong in each position.

Let's look at each step individually:

Identify the person with whom you would like a better relationship.
You can put anyone into this technique – a difficult boss, troublesome co-worker, demanding customer or irritating family member. It really can make a difference to any kind of relationship issue.

Step into position 1 (first position); imagine the person standing at position 2 and fully associate into your feelings about him or her.
Visualize the other person standing about 2m away from you, facing you. Take time to imagine this fully, and to feel what it's like for you to be faced with this other person. Remember to break state.

Step into position 2 (second position) and 'be' the other person. Fully associate into their experience of you.
You may not think you know much about how the other person feels, but once you associate into this second position you might be surprised by how much you can experience as if you were the other person. Notice the view you get of yourself in position 1; how do you look from this perspective?

*Step into position 3 (third position) and become a
dissociated observer. What do you notice? What insight
do you have that they do not?*

Make sure that your third position is sufficiently far back
from the first two that you can easily 'see' both of the two
people without having to turn your head. Be open to whatever insights come in this position. It might be quite different from anything you expect. Now you have gathered
information from all three perceptual positions.

*Step into position 4, from where you can see the 'you' in
position 1 and the 'you' in position 3.*

Imagine the 'you' in position 3 moving to position 1 and
replacing the 'you' in that spot.

The only thing you do at position 4 is to visualize the
switch of places between the 'you' at position 1 and the
'you' at position 3.

Step back into position 1. How is it now?

Again, take your time and fully associate into the experience of facing the other person. My prediction is that it will
have shifted in some way. Notice how you feel and remember to break state when you're ready to move on.

Step into position 2. How is it now?

Go back to 'being' the other person and experience the
shift in the relationship from their perspective. Notice what
it's like now to be faced with a different attitude from 'you'.

Finish by associating back into position 1.
It's always a good idea to finish up by being yourself!

Once you've been through this process, I expect that you'll feel differently about the other person to some extent. What's interesting is how this affects your next actual encounter with that person. Because you've made a shift in your perception of that person, you'll behave slightly differently; and as soon as you do something differently, you open up the possibility that they can be different too.

 Think of a person who you don't get on with as well as you would like. Run the meta mirror with that person in position 2. Notice how it affects your perception of the other person and the relationship you have with them.

---

7. Submodalities

Representational (rep) systems are sometimes referred to as modalities. *Sub*modalities are the distinctions we make within each modality.

Submodalities are aspects of the structure of internal representation and are independent of content. They are significant because the meaning of a representation is encoded in its submodalities.

Working with submodalities can be a very quick way to get some control over what goes on in your mind. There are simple techniques we can use to increase motivation, change habits and attitudes or stop something from continually haunting our thoughts. The reason why submodalities techniques are quick to use is that they intervene directly with your internal representations.

Think of something that is on your 'to-do' list. As you think about that task, notice the picture in your mind's eye (it may not be a very clear picture – that's okay).

Notice if the picture is a still picture or a movie. Notice if it's in colour or black and white. Notice how bright it is and notice how big the picture is.

Now, just for fun, make the picture bigger. Put it in really vivid colour. Make it very bright. Make sure it's moving at a lifelike speed.

What effect does that have on how keen you are to do the task?

As you can see, when you change the way something is represented in your mind, you can change your attitude towards it. The more you practise this, the more you'll understand about how things work in your own head and the easier it will be for you to have choice about the way you think.

Submodalities checklist

In the example above, we used four different submodalities, all of them relating to the visual modality (or rep system):

- Size of picture
- Brightness
- Colour or black and white
- Movie or still

There are many more submodalities that can be used, including auditory and kinaesthetic submodalities. Here is a list of the main submodalities used in NLP techniques:

VISUAL	AUDITORY	KINAESTHETIC
Location	Location	Location
Distance	Distance	Moving/still
Size	Volume	Size
Framed/panoramic	Sustained/repeated/short/intermittent	Shape
Colour/black and white		Intensity
	Clear/muffled	Duration
Bright/dim	Fast/slow	Colour
Movie/still	Rhythmic or not	Pressure
Movie: Speed of movement	Constant/changing	Weight
Focus		Temperature
Associated/dissociated		Constant/changing

When we consider location in visual submodalities, it means the location in your visual field: is the picture straight ahead, for example, or above eye level, or off to one side etc. Obviously the picture is *actually* inside your head, but what we're considering is where it appears to be when we look at it with our mind's eye. Similarly, the distance is how far away we seem to be seeing the picture from, is it close up, at arm's length or further away?

Association and dissociation in this context can be detected from the viewpoint. An associated picture will be the view through your own eyes as if you are there, reliving the memory (in first position). In a dissociated picture you will be able to see your own face and/or body in the picture (as you would from third position). In general there will be stronger feelings with an associated representation.

The meaning of the representation is encoded in the submodalities. This means that the submodalities are your brain's way of tagging thoughts and filing them.

 Pick two memories of meetings you attended recently. Choose one that was a really interesting, stimulating meeting that achieved something tangible and that you feel good about having been part of. Choose another that was dull or difficult, that didn't achieve anything and that you feel was a waste of your time.

Now, one at a time, focus on the memory and notice how the internal representation is structured. Make a note of the submodalities as they relate to the two memories and then compare them. Pay particular attention to the location in your visual field, as this is often the key to meaning.

When you focus on representations of contrasting experiences, it's usual to find that there are differences in their structures. For example, the positive memory might be large, bright and/or colourful and straight ahead whereas the other one might be smaller, dimmer and/or in black and white and lower down.

What you have discovered in the exercise above are the 'filing codes' that your brain uses to tag 'memories of good meetings' and 'memories of bad meetings'. Everything you might think about is tagged in this way through its submodalities.

Since we've already seen that it's possible to change your internal representations at will, it won't surprise you to discover that we can use submodalities to change our attitudes to tasks, people, memories or anything else, simply by changing the 'filing codes'.

Mapping across

'Mapping across' means changing the submodalities of one representation into those of another representation that has a different meaning, for the purpose of changing your experience of, or attitude to, the first representation.

This technique is worth mastering, because once you've learned to do it, you can use it to make changes in a matter of moments. Here's how to do it:

1. Identify what you want to change (e.g. 'I want to stop eating chocolate' or 'I want to be motivated to make phone calls').

2. Specify the category in your mind that currently holds the thing you want to change (e.g. 'things I like to eat' or 'things I'm not motivated to do').

3. Note down the submodalities of the thing you want to change (e.g. when you think about chocolate, or making phone calls, do you have a picture in your mind's eye?).

4. Specify the category in your mind you want to move the thing to (e.g. 'things I eat occasionally' or 'things I don't like to eat' for the chocolate; and 'things I enjoy doing' or 'things I can do easily' for the phone calls). Be mindful of ecology in making your choice.

5. Think of something that is currently in the category identified in step 4 (e.g. ice-cream or Brussels sprouts – depending on the desired outcome – in the first scenario; and 'checking my email' or 'attending meetings' in the second scenario).

6. Elicit the submodalities of the thing identified in step 5.

7. Compare the two sets of submodalities; identify the differences.

8. Change the submodalities of the thing you want to change (elicited in step 3) so that they match the submodalities elicited in step 6.

9. Test.

You can use the worksheet opposite to help you through the 'mapping across' process.

Thing I want to change my attitude to: _____

Current category in my mind: _____

Category to move to: _____

Example from desired category: _____

Comparison of submodalities:

	Thing I want to change	Example from desired category
Location		
Distance		
Size of picture		
Colour or black and white		
Movie or still		
Associated or dissociated		
Focused or defocused		
Framed or panoramic		
Bright or dim		

149

8. Strategies

Back in the introduction to this book, we talked about 'pro-grammes' of behaviour that run our habits and skills. You'll remember the principle: anything you do really well, you can do without thinking about it. When we talk about strat-egies in NLP, we're referring to those programmes. Some strategies drive behaviour that can be observed on the out-side, some strategies relate to mental processes and can-not be observed directly.

A **strategy** is a specific syntax of external and internal experience that consistently produces a specific outcome. In other words it's a process of paying attention to a sequence of things – some in current experience, some in memory or imagi-nation. Following the process generates an end result.

Everything we do involves a strategy. Let's take an example. If I'm attending a meeting I'll have run strategies for:

- Remembering that I've agreed to the meeting
- Preparing for the meeting
- Turning up on time
- Choosing where to sit in the meeting room
- Building rapport with other people in the meeting

- Listening
- Asking questions.

… and that's before I get to the point of making any kind of suggestion or decision.

All our daily activity is generated and maintained by strategies. Whether or not we finish what we do is governed by a strategy. How successful we are at a specific task will depend on how good the strategy is. The things we're good at and can do without thinking about, they're run by very effective strategies. The things we mess up, they're being run by strategies that don't work very well.

Working with strategies can be very detailed and laborious, so I don't propose to include everything here. I just want to share with you some knowledge about one particular type of strategy, and in particular, one very useful component of it. This will be the key to getting things done with other people.

Decision-making

Your decision-making strategy is a strategy for choosing between options. It is used in a wide variety of situations, such as deciding who to employ in a job, which of several suggestions is the best, where to go on holiday or what to buy as a birthday present. This strategy is particularly easy to observe in buying situations and very useful for sales people to know about. Even if your job doesn't involve selling to customers, it probably involves 'selling' ideas and plans to other people, so you'll find this a very useful tool for influencing others.

A typical decision-making strategy has four stages:

1. Motivation
2. Choice
3. Convincer
4. Reassurance

I'll describe these stages as they relate to a buying situation; bear in mind that they will also apply whenever a choice is made.

1. Motivation: This is how a person knows it's time to buy. It could be an 'away from' strategy such as, 'I must buy a new briefcase before the old one falls apart' or it could be a 'towards' strategy such as, 'If I buy a new briefcase I'll look smarter and more professional.'

The result of this part of the strategy is that the person is ready to make a choice.

2. Choice: This is how the person chooses between the various options. This part of the strategy can vary a lot between people. Some use a set of criteria to evaluate each option, others simply go for the first one they like. Whatever the strategy, it usually has some benefits and some drawbacks. There is rarely a 'perfect' strategy. If the strategy has a visual component then the person will want to see something before they buy, even if it's just a picture of what they are getting. If the strategy has a kinaesthetic component, the person will probably want to touch or handle the product. If that's the case, they probably won't buy by mail order unless they have been able to handle the item before ordering. This can seem a little bizarre at times – why would you need to touch a DVD before buying it? The answer is just because it's part of the strategy. And the strategy remains the same no matter what the choice is.

The result of this part of the strategy is that the person knows which option they prefer.

3. Convincer: This is how the person convinces himself or herself to act on the decision. It takes them from, 'That's the briefcase for me' to actually purchasing it. Have you ever wondered what's going on when someone seems to have made a buying decision but then doesn't buy. It's their convincer pattern at work. This is such a key piece that I'm going to elaborate on convincers later in the chapter.

The result of the convincer strategy is that the person acts on the decision. They buy.

4. Reassurance: Having bought, the reassurance strategy is the process that reassures the buyer that they've made a good purchase. This might involve external validation by showing the purchase to another person, it might simply be a case of using the item as soon as is practical to experience the benefit of owning it or it could be something else entirely. Reassurance strategies vary a lot between people, but each person sticks to their own strategy.

The result of this part of the strategy is that the person is satisfied with their decision.

Can you see how useful it can be to a salesperson to know what to do between the choosing part and the actual sale? If you know how people get from the choice to the purchase, you can help them to do it their way instead if trying to do it your way. So let's look at convincers in more detail.

More on convincers

The convincer strategy is a way of processing information in order to become convinced. Its useful to know about because a person will use the same strategy regardless of the context and the importance of the thing they are to be convinced of. Have you ever considered what it takes to convince you? Some people appear to make up their minds quickly and others take a long time. If you want to convince someone to buy, to authorize, to agree or to change, wouldn't it be useful to know how that person can be convinced most easily?

One of the greatest benefits of this can be in reducing your stress. I spent some time working with the customer services team of a large organization. These were the people who took the telephone calls that no one else could resolve. The people they were speaking to were often angry, upset or rude but the team members were skilled in winning confidence and resolving problems. However, they had some questions. For example: 'I told this customer the same thing three times, but he just wasn't convinced. Then I transferred him to my boss and she told him the same thing again and he accepted it – why didn't he listen to me?' or 'I explained to the customer exactly what to do, but she kept asking if I could fax some instructions to her – does this mean I wasn't clear?'

Another company I worked with was a small exclusive bridal shop. They couldn't understand why they were able to close some sales easily and other customers came back to the shop repeatedly but didn't place their order. They were beginning to think they were missing out on something.

They were right. If you need to influence other people then it helps to understand how people become convinced. There are two elements to a convincer strategy: one relates to the way the information is presented; the other is the time factor.

The first is the **convincer representational system** or how the person has to have the information represented in order to be convinced: Do they have to see it, hear it or experience it for themselves? Think about this: suppose you have a new colleague. How would you know that the new person was good at the job? Would you have to see them do it, hear about it, or perhaps work alongside them and experience it? If you have to *see* it to be convinced that the person is good at their job, then chances are you also have to *see* it to be convinced of anything else. Doubtless, the person who coined the phrase 'I'll believe it when I see it' had a visual convincer. Someone with an auditory convincer has to *hear* it to be convinced. The person with a kinaesthetic convincer needs to experience it for themselves.

The convincer rep system accounts for the irritating phenomenon of people who receive memos or emails but don't act on them until someone calls them on the phone (auditory convincer). Or the people who agree to your proposal outlined in a meeting but do nothing until they see it confirmed in writing (visual convincer). Or the customer who has listened to your explanation but would much prefer it if you could fax something to her to look at (visual again).

However, the convincer rep system doesn't account for the difference between the snappy decision-maker and the person who prevaricates until you begin to think that there must be something wrong.

The second aspect of convincer strategies is the **convincer demonstration**. Going back to that new colleague, how often does the person have to demonstrate their competence for you to be convinced? Answers to this question will usually fit one of four categories:

Automatic: The person with an automatic convincer will assume that the new colleague is competent on the basis of a small amount of information unless the new person demonstrates otherwise. You can convince this person quite easily, but so can everyone else! They are easy to sell to and readily give their support for new ideas. They are probably the 'early adopters' of new technology. Being easily convinced, they can change their minds quickly and don't always stay committed to a decision once they have made it. Interestingly, lots of sales people have automatic convincers and find it difficult to understand why other people are not as easily convinced as they are themselves.

Number of times: This person has to have the demonstration repeated several times before they are convinced. In a sales situation they may have to run their 'choice' strategy a fixed number of times. If you're selling to someone like this, either show them the product this number of times, or show them this number of alternatives. For the customer services team I was working with, this knowledge made a big difference. Their comments changed from 'Why did I have to repeat myself FOUR times to that man – is he stupid?' to 'I've just had someone on the phone with a four-times convincer' – imagine how much this reduced their stress. As a general rule, the higher the number of times a person has to run their strategy in order to be convinced, the more committed they will be to the decision once it is made. Persuading them to change their mind could also involve the same number of repetitions to achieve the necessary conviction.

There is potentially a shortcut to this, although it will only work if you have good rapport with the person. Suppose you need to convince someone who has a five-times convincer but you don't have time to go through your proposal five times. After the second time, you say something like 'If I've explained this to you once I must have explained it five times – isn't it time you made a decision?' As I said, it only works if you have good rapport...

Period of time: This person needs a period of time to be convinced. It was probably someone with this style who invented the 'trial period'. It might be a few hours or it might be months, but this person always needs this period of time to be convinced. If you're seeking their support for your project and they say, 'I'll think about it', ask them how long they'd like before you meet again. Don't assume that this is a polite way of saying no.

Again, there is a possibility of a shortcut. Suppose you want to convince someone with a three-month period of time convincer. You present your proposal and they tell you they need to think about it. Do you wait three months? Unless you also have a three-month convincer, probably not. So, if you contact the person again after a few weeks, you can open the discussion with a comment such as, 'I've been so busy since we last discussed this proposal, it feels like at least three months since we spoke.' Again, it only works if you have good rapport but for those of us not endowed with the patience to wait three months for an answer, it can really speed up the process!

Consistent: This person is actually never convinced. They require repeated demonstration of your competence or the quality of your idea or your product. Nothing you can say or do will convince them. These people make tough managers, as you'll only ever be as good as your last result in their eyes. Paradoxically, they can sometimes they end up appearing to be inconsistent inasmuch as they may strongly support someone they believe to be doing well, but then if that person fails in any way to meet their standards, they will cease to support them and look for a new champion. Over a period of time it can appear that different individuals are 'flavour of the month' for a time, before falling out of favour and making way for the next one.

If you want to sell to someone with a consistent convincer, acknowledge that you can't convince them. One of the ladies in the bridalwear shop had a customer with a consistent convincer. She eventually told her, 'I know that I can't convince you that this is the perfect dress for you. You'll only know for sure when you're walking down the aisle in it on your wedding day.' She made the sale *and* the customer was happy.

To make full use of convincers, it helps to know your own convincer strategy. If you have, for example, a three-times convincer and you make a sale to a person with an automatic convincer, you might not be convinced that they have really bought unless they tell you three times! Save them the effort and tell yourself three times – it works just as well.

Of course, the main piece of information you'll need to be able to make use of all of the above is the convincer strategy of the person you want to influence. Most of them won't know themselves and even if they did, in the normal course of business it wouldn't be appropriate to ask. However, most people will tell you, if you know what to listen for.

For example, the customer services team quickly found that the way a person described their problem often gave some clues as to their convincer strategy. After all, they must be convinced that they have a problem! The kind of things customers said varied from, 'It hasn't been working properly for about a month, I knew it was time I got some advice' (period of time – one month) to 'I've tried it three times and it just won't work' (number of times – three) and 'I've never had any trouble with it and then today it completely packed up' (automatic).

By listening carefully to the way a customer describes the problem, it's possible to discover their convincer

161

strategy and then use it to convince the same person that you have a solution for them.

Managing change

From time to time, organizations engage in programmes of widespread change. For example, there might be a change in business systems, a restructure of departments or the introduction of new working practices, equipment or standards. Often the people involved in driving the change get frustrated and discouraged because of what they perceive as 'negative attitudes' from the people required to do things differently.

In general, the reason that people resist change is that they are not convinced that the change is justified, appropriate or useful. This doesn't mean they will never be convinced (unless they have a consistent pattern), it just means that they're not convinced yet.

Change programmes run more smoothly and stand a much greater chance of success when the changes are implemented through volunteers who are ready to learn new processes, use new systems or implement new working practices. I always recommend that when an organization needs to run a programme of workshops or training programmes, that people are invited to choose which workshop or training session they attend. That way, the automatic convincers and one-time convincers can be the first to attend. They are usually the 'early adopters' and will be convinced quite quickly of the merits of the change.

They will be followed by the people with two- or three-times convincers and those with a short period of time convincer.

The third wave will be the people with longer period of time convincers and convincers of more than three times. It takes longer for this group to be convinced that the change is beneficial and to be ready to act on that decision.

Finally, there will be people with a consistent convincer who might never be convinced. They might still be reluctant to attend, even when you get to the last workshop. But there's no point in forcing them to come sooner. They'll just cast an air of gloom over the event and sow doubts in the minds of others.

By allowing people to adopt the changes when they feel ready, you eliminate the power struggle that so often emerges from major change programmes. It may seem like a leap of faith when only five people volunteer for the first workshop and you need to average twelve at each one. It *is* a leap of faith, but one worth taking.

9. Modelling

Earlier in the book I highlighted modelling as one of the key aspects of NLP that are useful at work.

Modelling means identifying the structure of a skill that you observe someone else is particularly good at. If you can model the skill and the way the person achieves results, then you can learn to do it too.

The origins of NLP are in modelling. Richard Bandler and John Grinder modelled the language patterns of some of the great psychotherapists of the day and in doing so, not only unpicked the structure of language that influences others' thinking, but also developed the methodology to do the same with other skills.

The approach that Bandler and Grinder took was purely observational. This can be a very lengthy process, so we also use methods of interviewing the expert to find out more about what's going on in his or her head as they demonstrate the skill. Robert Dilts, the eminent NLP developer, trainer and practitioner, has contributed a lot to this more cognitive style of modelling.

If you're interested in learning how to model, there are some great books and courses that will show you how to do it. I'd like to share with you the results of two model-ling projects, so you get an idea of what can be achieved with this approach, and also to give you the benefit of two essential business skills.

Creating an air of authority

Working as a trainer and consultant, I'm privileged to meet a lot of senior people in a wide variety of organizations. Some people in senior roles have an air of authority that makes it obvious that they are in charge. Others don't. I became curious about what creates that air of authority, because I thought it would be a great thing to be able to do. So I started observing. And what I noticed surprised me, because it's really very simple to create an air of authority.

If you think it would help you at work, here's what you do. People who radiate an air of authority usually do two things:

1. Take their time
2. Take up space

Time: Taking time to do something or to answer a question or to begin a presentation gives a below conscious message that what you are about to do or say is worth waiting for. People who are nervous usually rush through everything, so doing the opposite shows you as confident and calm.

Simple ways to take your time:

- Do one thing at a time.

- After someone asks you a question, pause before answering.

- Believe that what you're about to say is worth waiting for and don't rush it.

- Stop if someone talks over you or causes a distraction.

Space: Even if you're not very tall or broad you can take up space with your belongings. When you take up space it demonstrates to people that you have a right to be there. People who feel uncomfortable in a situation often shrink away from it, so when you spread out it looks as if you're comfortable and that you belong.

 Simple ways to take up space:

- Relax. This will make you more likely to spread out, whereas if you are tense or nervous you will tend to make yourself smaller.

- In a meeting room, spread your belongings about in front of you; claim some territory (but don't encroach on your boss's space unless you want to challenge his authority).

- Stand fractionally further away from people than you would naturally.

- When speaking, make bold gestures with your hands.

Remembering names

I think that remembering names is an essential skill for all of us at work. Assuming you've been introduced to a person, when you use their name it is an acknowledgment of them and helps to create rapport. If you haven't been introduced, don't use their name by reading it off a security pass – some people will be offended and others will be startled.

Lots of people have difficulty remembering names and often the reason they can't retrieve a name from memory is that they didn't store it away in the first place. You can't retrieve something that isn't there. I'm blessed with a good memory and I've always been quite good at names, but I didn't always know how I did it.

The strategy for remembering names that I'm sharing here is the one I use and which was modelled from me by Craig Cummings as part of his Master Practitioner Certification. This was done by interview as well as observation – it would have been much harder to figure this one out by observation alone. Interestingly, having conscious knowledge of the process has made it more effective for me. Here's how I do it:

The foundation of this strategy is in the belief that it's important to remember people's names. Without that, the rest wouldn't happen. Because I believe this, when I'm about to meet someone new I make a conscious decision that I want to remember their name.

Belief/Value

- It's important to remember people's names. It makes people feel special and it demonstrates your connection with them.
- When I'm about to meet someone new I make a conscious decision that I want to remember their name.

Process

V – Look at the person's face. Concentrate.

A – Listen to the sound of their name.

V – Look at their name badge if they're wearing one, if not visualize one.

A – Repeat the name aloud, by saying, 'Pleased to meet you, John' or, if this is not appropriate, repeat the name in your head.

D – Check the correct spelling of the name, if possible.

K – Test until there is a feeling of confidence.

This process works because first of all it clears your mind of all distractions and focuses on the person whose name you want to remember. Then, by paying attention to sight, sound and spelling in turn, we create a neurological connection between the sight of the person's face, the sound of their name and the way it looks written down. If you use this process you'll be able to remember the name when you see the face, remember the face when you see the name written down, and also know how to pronounce the name correctly.

The 'testing' referred to is simply checking that I have retained the name. In a group situation I mentally go around the table saying each person's name to myself to check that I know it. I usually do this three times before I'm really confident that I know the names. (Now you know what my convincer pattern is!)

IF YOU REMEMBER ONE THING The key to remembering names is to pay attention to them. Decide before you meet someone new that you will store away their name. Then when you meet them, clear your mind of everything else and pay attention as you're introduced.

Using modelling to learn

Many of the really important skills in life you learned by modelling. How to walk, how to speak, how to eat and a host of other everyday activities. As a small child you couldn't ask, 'How do you do that?' – you just had to watch, experiment and learn.

You can still learn that way. If you notice someone who is really good at chairing a meeting, observe more closely exactly what they do. Then try it out for yourself and see what happens. You might have a colleague who's very good at dealing with upset customers. Notice what they say, their tone of voice, their facial expressions, and have a go for yourself.

There is a lot you can learn just by observing the people with the exceptional skills (if you're going to put in the effort to model someone, don't waste it modelling mediocrity!). There may also be opportunities when you can ask someone about what they do, but be prepared to be told, 'I don't know, I just do it' – remember that everything you do really well, you can do without thinking about it.

10. Language

The two biggest complaints I hear from employees in the companies I work with are 'Nobody told us' and 'Nobody asked us'. Communication is vitally important for business success and people need to feel that they are informed, involved, consulted and listened to. The larger the organization, the more difficult this is, and the more likely it is that individuals get the impression that their voice isn't being heard.

Most people think they're reasonably good at communicating and that they know how to use language. And most people are reasonably good at expressing their own thoughts in words. However, the purpose of communication is often much more than that. Especially at work, the purpose of communication is usually to influence someone.

If all you focus on is how to express your thoughts, you're probably not going to find it easy to influence others' opinions and attitudes, even if you have great rapport with them. The key to doing this is the direct connection between what you say (or write) and the internal representation that words create in the mind of your listener. Earlier on, we established that there are always multiple ways to describe the same thing. Notice the differences between:

'The project is late and overspent.'

'The project is behind schedule on delivery and ahead of schedule on spending.'

'The plan for this project was not realistic.'

'We will have to work more efficiently if we're going to meet the next project deadline and we need to review the budget.'

Each of the sentences above are describing the same thing, but do you notice that the way it's represented in your mind shifts as the words change? Imagine how you would react to each of these statements if you were the person's boss.

In choosing the 'best' way to express a thought or idea, we naturally consider who we're going to be speaking to. It's also helpful to consider your outcome.

 If you have a clear outcome you want to achieve with your words, it's much easier to decide how to structure your communication and to select the specific words to use.

Of course, it's rare that communication is a simple matter of crafting a message and delivering it. Most of the time we're involved in dialogue and group discussion. So listen to what other people say and you'll find out whether or not your words achieved their outcome. If not, it's time to have another go – rephrase your message and try again.

Most people would agree that there are two main things involved in communication: speaking and listening. However, just because these two things are going on, it does not always mean that communication occurs.

One of the reasons that communication fails is that, frequently, when we are having what passes for a conversation with someone else, we are either not fully listening or we make so many assumptions about what is being said that we do not learn anything new from what we are hearing. We simply use it as further support for our (already established) point of view.

We have already seen how we filter our experiences to support what we already believe to be true. The more deeply held a belief is, the harder it is to challenge it. So most of the time when we are engaged in conversation, we end up with an impression of what was said that is incomplete, inaccurate or both. Not all of this is a result of the listening – what we say is also filtered to make our conversation concise and easy to follow.

Therefore, there are three main skills in communication:

1. Asking questions
2. Making sense of what other people say
3. Directing other people to think in certain ways through your choice of words

(In fact, asking questions is a special case of directing other people's thoughts with your choice of language, but it's often helpful to think of it separately.)

When you use language purposefully, there are two main ways to get results: by elicitation and by installation.

Elicitation means asking questions to find out information about the structure and content of another person's map of the world. In elicitation we presuppose as little as possible and avoid making suggestions. Good, 'clean' elicitation will not change the person's map of the world.

The result of elicitation is that you find out something about the other person's thinking.

Installation means planting a thought or idea, a belief or a memory into a person's map of the world in such a way that they accept it as part of their map. Once this is done it can be very difficult to remove it and return the person's reality to the way it was before the installation. Hence it is important to realize *when* and *what* we are installing and to ensure it is in line with agreed outcomes.

The result of installation is that you change the other person's thinking.

As a general principle, find out as much as is necessary about the current attitude someone holds before you attempt to shift it.

Asking questions – the meta model

The meta model was developed by Bandler and Grinder as a result of their modelling of the great psychotherapist Virginia Satir. It is essentially a set of questions that can be used to dig a bit deeper into what people say.

When we make an internal representation of the outside world, it is composed of sensory information. The full linguistic description of our experience, outside of conscious awareness, is known as the **deep structure**. The deep structure cannot be communicated fully; it's like trying to describe a dream. The words we actually say in describing our experience are known as the **surface structure**. The deep structure is subjected to deletion, distortion and generalization to create the surface structure.

The meta model is usually divided into three parts that correspond with the three actions of the linguistic filters: deletion, distortion and generalization. The questions challenge incomplete or inaccurate statements and recover the meaning that has been lost to the filters. The result of using it is that you can help people to be more precise and specific in what they say, and sometimes help them to view their experience more constructively.

Deletion: There are several different kinds of deletion. One of the most common forms of deletion is heard in business in comments like 'Communication is terrible at present' or 'There's a complete lack of teamwork' or 'Performance is up this month.' What these statements have in common is that they have deleted information about what is actually being done and covered it up with an abstract noun. For example, the word 'communication' replaces details about is who is communicating what to whom (or not). To recover these details, simply ask, 'Who is not communicating about what?' Similarly, for the other examples you could ask, 'What work should the team be doing together?' or 'Who is performing well at what?' Simply ask. But since this is not the kind of question that people get asked regularly, you should take care in *how* you ask. It must be done with rapport; you must demonstrate that you are genuinely interested in the details, otherwise you risk sounding like the Spanish Inquisition!

Another common form of deletion is where the information about *who* is doing something is lost. For example, 'They don't listen to me' or 'She didn't invite me to the meeting' or 'They turned up late.' You may think you know who is being referred to, but sometimes it's a good idea to check. The ubiquitous 'they' is usually worth challenging, to find out whether or not the speaker actually knows whom they are referring to! In this case all you have to ask is: 'Who specifically?'

Sometimes this kind of deletion is paired up with a further type, known as an unspecified verb. This is when the details of *how* something was done are filtered off, as in 'He rejected my idea' or 'They upset us' or 'She ruined that project.' Again, we may think we know what is meant by each of these statements – and we may be wrong again! Find out more by asking, 'How specifically?' In the former case you may find that what 'he' said about the speaker's idea was he needed more time to consider it.

Of course, if you are going to ask these questions, there has to be a purpose. You will quickly lose a lot of goodwill from friends and colleagues if you start challenging every deletion you hear. It is probably most useful to use this kind of approach in interviews, coaching sessions, counselling and investigation of training and organisational development needs.

Distortion: The second area of the meta model deals with distortion. I should emphasize that this is unconscious distortion as a result of unconscious filters, not any wilful distortion of facts. One of the most common forms of distortion is 'mind-reading', where a person claims to know someone else's feelings or thoughts, as in 'You don't like me' or 'He's angry with me' or 'They won't agree.' Sometimes these judgements are made on valid information, but if you suspect that they are simply invented you can ask, 'How do you know?' Again, done with rapport and genuine interest, this can be a powerful question that could bring someone to a useful realization that they actually have no reason to assume hostility from another person (nor indeed to assume anything else).

Another form of distortion is when opinions and value judgements are expressed without being attributed to anyone. These may be heard in comments such as 'You should work hard all the time' or 'It's not a good idea to throw away those files' or 'It's best to keep an open mind.' None of these comments are presented as a subjective judgement, but rather as objective fact (all proverbs also come into this category). A statement like this may be preventing someone from thinking for himself. To challenge it, a good question to ask is 'Who says?' or 'According to whom?' There is, as with all meta-model questions, a risk of this sounding aggressive, so ensure that you are in rapport before asking.

A third kind of distortion comes when something is taken to have a meaning that may not be accurate, for example, 'I didn't get the promotion; the boss doesn't like me' or 'Sales are down; the product range is awful this season.' In neither case does the one statement necessarily imply the other, but the speaker is taking the two to be connected in this way. There are two ways that this can be challenged. First you can challenge the logic. For example, 'How does your not getting promoted mean that the boss doesn't like you?' or 'How does a drop in sales mean that there's something wrong with the product range?' The other way to challenge this would be to focus on the counter-example. For example, 'Are there other people whom the boss likes that didn't get promoted either?' or 'Have sales ever fallen when the product range was good?' Either way, to challenge this type of distortion will focus attention on faulty deductions and get back to what is actually known as fact. This can be really useful in problem solving and counselling situations where there seems to be no way forward and no choices. Challenging the distortions will usually open up some possibilities.

Generalization: The third area of the meta model deals with generalizations. Generalizations are useful because they give us the ability to spot patterns and to apply our experience in one situation to another similar one. However, they also form the basis of prejudice and narrow-mindedness, so it can also be useful to challenge them on occasion. The most common forms of generalization involve words like 'never', 'always', 'everyone', 'no one' and 'all'. For example, 'It always rains in Manchester' or 'Nobody ever listens to me' or 'All salespeople are extroverts.' Clearly these statements are untrue, but many people act as if their generalizations are fact and miss opportunities and possibilities because of it. The 'rule of thumb' here is that all generalizations are untrue – including this one! The way to challenge generalizations is simple: you echo back, 'Always?' or 'Nobody? Ever?' or 'All of them?' (with a suitable amount of incredulity). In doing this, you are again looking for the counter-example; your questions could alternatively be phrased as 'Is it ever fine in Manchester?' or 'Has anybody ever listened to you?' or 'Are there any salespeople who are introverts?'

I would urge particular care when challenging generalizations. Often when a person says something like 'Nobody ever listens to me' (especially if it's said with some feeling) it's an indicator that somebody in particular is not listening to something important right now. The person may be feeling undervalued or rejected and a smart response like 'What, nobody? Ever?' may not be the most tactful thing you can say. In that case, treat it like a deletion and ask, 'Who in particular is not listening to you?'

While there are other kinds of deletions, distortions and generalizations, the ones described above are some of the most common I have encountered in business. The overall effect of the questions I have suggested is to challenge the validity of the statement and encourage the speaker to be more precise, more specific in what they are telling you. With practice, you will be able to avoid some of the common frustrations of communications at work – such as preparing a report and being told afterwards, 'That wasn't what I wanted' or tailoring a presentation to suit what you had been told about the audience, only to discover that the information was incorrect. Using meta model questions won't entirely remove the subjectivity from a point of view, but it may bring you closer to understanding what is really being said – and what is really meant!

Presuppositions: making sense of what other people say

We've already explored some tools for making sense of what other people say. If you pay attention to the sensory language used, you can get an idea of a person's rep system preferences. If you listen for language associated with metaprogrammes, that will also help to clarify the *structure* of their thinking.

If you want to understand more about *what* someone thinks, then paying attention to the presuppositions they use will help.

Presuppositions are linguistic assumptions. We identify a presupposition in relation to what has been said. If a statement only makes sense if a second statement is true, then we can say that it presupposes the second statement.

For example, if I ask you, 'What colour is your car?' the question only makes sense if you have a car. We can therefore say that the question presupposes that you have a car. In this case the presupposition is unspoken, but nonetheless is part of the speaker's map of the world.

Sometimes the presupposition is part of what is said. For example, the observation, 'John's angry with me because I was late' presupposes that the cause of John being angry is the speaker's lateness. It also presupposes that John is angry. In conversation we may not immediately recognize them as presuppositions – especially if we share them!

The important thing in identifying presuppositions is to focus on the actual words used, not your own internal representation of what they mean.

Identifying presuppositions

Within any statement, there are likely to be several presuppositions. Some of these will be more obvious than others and, in terms of our responses, some will be of greater interest to us.

Take the following statement: 'I didn't apply to do the MBA because I didn't want to give up my social life.'

This statement contains the following presuppositions:

1. I could have applied to do the MBA
2. I didn't apply to do the MBA
3. I have a social life
4. It's not possible to do the MBA and also have a social life.

Another way of thinking of this is that the presuppositions are statements that can be logically inferred from what is said.

In the example above, the final presupposition is probably the one that is of most interest if you're in a conversation with the speaker. You could respond directly to that presupposition and say, 'So it's not possible for you to do the MBA and also have a social life?' It depends on your outcome for the conversation.

The value to you of looking at the presuppositions in what somebody says is that they give you an insight into what they're not saying but assume to be true, without you having to ask any questions.

Influencing another person's thinking using presuppositions

Not only is it useful to be able to recognize what is assumed by the speaker, it's also possible to make appropriate pre-suppositions in our own responses, thus creating internal representations in line with our outcomes.

When someone listens to you speaking they make internal representations based on the language you use. The presuppositions in that language can have a great impact on the internal representations they make and how they feel about what you have said. In other words, presup-positions have consequences. It is awareness of those con-sequences and of our own outcomes that lead to elegance and precision in communication.

The degree of influence you achieve by using language in this way will be determined partly by the degree of rap-port between you and the listener. It doesn't matter how subtle the language is, if you don't have rapport you're less likely to get the outcome you want.

TRY IT NOW! In each of the following examples, consider which language is most likely to achieve the stated outcome.

Outcome: to ensure that a colleague writes a particular letter.
1. Are you going to write that letter?
2. Are you going to write that letter before the meeting or after?

Outcome: to arrange a meeting with a busy senior colleague.
1. Can I have a meeting with you sometime next week?
2. When next week would be a good time for us to meet?

Outcome: to encourage a colleague who is about to make an important presentation.
1. You know that being nervous just means that you'll do a great job.
2. Are you nervous?

Outcome: to express frustration at lack of progress on a project.
1. You have let me down by missing that deadline.
2. I'm frustrated because we now have to work twice as hard to meet the next deadline.

Outcome: to persuade someone to make a decision now.
1. What else do you need to know to be able to make a decision?
2. What would it take for you to be able to decide right now?

Outcome: to 'sell' NLP to a colleague
1. If you were to learn NLP what would be most useful to you?
2. You ought to do NLP; it would really help you.

Outcome: to help a colleague understand a complex idea.
1. Shall I explain it again?
2. What's the one question you can ask me that will make it all clear for you?

Outcome: to challenge the approach to a task or project.
1. If this were not the best way of going about the job, how would you know?
2. Are you sure this is the best way to do it?

Outcome: to give feedback that someone may not be pleased to hear.
1. I know you're not going to like this but it's better that you know.
2. I know you are going to want to know how to improve the situation.

Outcome: to encourage a demoralized team.

1. Things can only get better.
2. Things can't get any worse.

Now that you've got the idea, you can only succeed in making your own communication more precise, more elegant and more effective.

Embedded commands

Another way of influencing another person's thinking is to use embedded commands. An embedded command is an instruction to do something that is delivered as part of a longer sentence, often a question. Native speakers of UK English do this a lot. It's common to hear people say things like 'Could you possibly shut the door?', 'Would you like to give me the figures now?' and 'Could I ask you to pass me that file, please?'

This is a pattern of speaking that most people regard as more polite than simply 'Shut the door', 'Give me the figures now' or 'Pass me that file.' However, in other languages and other countries this would be perfectly acceptable.

It's also common to hear people saying things like 'Don't worry about this. It's not a big problem' or 'It's urgent, but I can't ask you stay late to finish it' or 'It shouldn't take a long time. I don't think you'll miss lunch.'

Can you hear the way that these statements plant ideas in your mind that are actually the opposite of what the words literally mean?

Right now, this minute, don't think about what you're doing this weekend.

What came to mind, did you think about what you're doing at the weekend?

This is because at the simplest level of processing, it's practically impossible to represent 'don't' so we just represent the rest of the sentence: '…think about what you're doing at the weekend.'

So, even though at a conscious level we hear the 'don't', below consciousness, we're getting the opposite message. This is confusing and dilutes the impact of what is said.

Here's a quick exercise for you to practise spotting embedded commands in everyday language. For each sentence, underline the embedded command:

- Don't worry, it's not difficult.
- Don't waste your time doing that.
- That isn't the best way to do the job.
- Don't forget to bring the files.
- Don't start without me.

- You shouldn't use the front entrance.
- Don't laugh, but…
- Don't forget the security code.
- This isn't important, don't make it a priority.
- Don't even think about it!
- Don't think you can get away with doing that.
- It doesn't work if you do it that way.
- Don't forget to fill the car up with petrol.
- Don't lose sleep over it.
- Don't drop that!

Do you notice how the embedded command conflicts with the overall message? If you can edit this kind of conflict out of your own communication it will make a difference not just to your success in getting your message across but also to the way other people perceive you.

To avoid embedding conflicting commands in your language, focus on telling people what you want and how you want it rather than discussing what you don't want. This gives a much clearer message and also makes you sound more confident and reliable.

 If you do nothing else with this insight, scan important emails and presentations for embedded commands that you may have inadvertently included and which could detract from your carefully crafted message. Rephrase them to ensure they support your message completely.

The use of language is a vast subject within NLP and there is further reading you could do if you're interested. For the purposes of this book, though, let us finish by recapping the basic principles.

 The keys to effective use of language are:

- Know your outcome, not just for the conversation as a whole, but for every question you ask and every comment you make.

- Listen with your whole mind. Build up a clear second position. This will make it easier to tailor your communication to ensure that the person you are communicating with understands what you want them to understand.

- Notice whether the question you asked elicited the information you wanted. Notice the structure of the information.

- Be prepared to ask the same question in a number of different ways until you find out what you want to know.

A final word

NLP is not a science. It's not really an art either. It's definitely not a religion, although some people are evangelical about it. It's best thought of as a tool kit.

Like all good tool kits, there will be some bits you use a lot and some that rarely come out of the box. That's okay, you'll use the tools that are relevant and leave the ones that aren't.

This book has only been a brief introduction to some of the basic NLP techniques. Hopefully, by focusing on techniques that you can easily use in everyday work situations, it has already shown you ways in which these simple principles can become powerful tools in your life. Now it's up to you to go and experiment with what you have learned; to find out what works and what really makes a difference.

Just a suggestion: experiment with *all* the tools and find out how they work for you before you decide which ones are the best. That doesn't mean you have to use them all at once. Pick one or two and experiment with them until you feel confident that you know what they can be used for that's relevant for you.

You might pick one of the presuppositions of NLP and focus on it for a day. See the world through that presupposition as a filter and notice how it impacts your thinking and your results. Pick a different one on the next day, and see what happens.

Or you might start by paying attention to body language and by matching the behaviour of other people around you. Find out what it's like to do what they do. Experience second position.

You could simply begin by practising the well-formed outcomes pattern. Try doing five to ten per day for a week or two and notice how it affects your focus, effectiveness and motivation.

Of course, there is always more to learn, and the resources section that follows will give you some pointers as to where you could take your exploration of NLP next.

Whatever you do, if you begin with the intention of finding out how NLP can work for you, you'll quickly find ways to use NLP for work.

Resources

Further reading

Introducing Neuro-Linguistic Programming, John Seymour and Joseph O'Connor
Thorsons; Reissue edition (2003)
This book was one of the first NLP books written in the UK. It is a comprehensive look at the important concepts in NLP and in my opinion is much more than an introduction. It's written in a fairly formal style with good explanations of technical terms and an NLP glossary at the back. Perhaps not the most accessible cover-to-cover read, but a good reference text used by a lot of trainers and experienced practitioners.

NLP for Dummies, Kate Burton and Romilla Ready
John Wiley & Sons; 2nd edition (2010)
What I really appreciate in this book is the 'chunking' and layout. It has strong visual appeal and is easy to dip in and out of. It combines simple explanations of key concepts with fun activities and examples. As with most other NLP books, it covers a lot of information and would be a useful reference for practitioner training and beyond.

Business NLP for Dummies, Lynne Cooper
For Dummies (2009)
It has all the features that make the 'for dummies' books popular – clear chunking, examples, exercises and high visual appeal – and it's an intelligent guide to using NLP in business.

NLP at Work, Sue Knight
Nicholas Brealey Publishing; 3rd edition (2009)
If you are particularly interested in applying NLP in the business environment, then you will probably enjoy this book. It includes lots of anecdotes and examples of everyday business occurrences that will help you relate the subject to your own working life.

Words That Change Minds, Shelle Rose Charvet
Kendall/Hunt Publishing Co. (1997)
A classic for anyone who wants to master the art of adapting their communication style to suit the processing style of others. This book is a companion to the language and behaviour (LAB) profile and a good guide to the whole idea of metaprogrammes and how to use them. It is written in an informal and accessible style and does not presuppose any knowledge of NLP.

Performance Management, Michael D. McMaster
Metamorphous Press; 2nd revised edition (1993)
This is a thoroughly practical guide to applying NLP in a management context. It addresses the all-important issues of communication related to managing people, including planning, coaching and evaluation.

Persuasion Engineering, Richard Bandler and John La Valle
Meta Publications (1996)
This has to be the best book about NLP for salespeople! It's a wonderful exploration of communication, conversational

elicitation of values and beliefs, sales techniques and so on as well as being an entertaining history of some of Bandler's own experiences.

Me, Myself, My Team, Angus McLeod
Crown House Publishing (2000)
This book pulls together lots of aspects of NLP into a comprehensive guide to becoming an effective leader and coach. It's relevant to anyone who works with a team and can be applied whether or not you are the team leader.

NLP Workbook, Joseph O'Connor
Thorsons (2001)
I recommend this as a good third or fourth book on NLP. I wouldn't suggest it as a starting place, simply because it contains so much and some of the activities are quite complex. It is well structured and chunked into manageable pieces. It is comprehensive and thought provoking. The activities encourage the reader to explore the workings of their own mind and to challenge their own assumptions.

Modelling with NLP, Robert Dilts
Meta Publications; illustrated edition (1998)
This is a great book if you want to know more about the NLP modelling process and its applications. It defines the key communication and relational skills employed by effective leaders to achieve practical results in their working reality.

Online Resources

NLP Encyclopedia
www.nlpuniversitypress.com
The NLP Encyclopedia was created by Robert Dilts and Judith Delozier. It is a comprehensive reference library on everything related to NLP and it's free to use.

Use Your NLP
www.useyournlp.com
My online self-study programme that builds mastery of NLP one technique at a time.

NLP Conference
www.nlpconference.co.uk
The NLP Conference takes place in London in November each year. The website has all the details of the sessions and the speakers.

NLP Events
www.nlpevents.co.uk
This is a comprehensive listing of NLP training programmes and events taking place in the UK and Europe. The information is contributed by the training providers

ANLP
www.anlp.org
The Association for NLP provides a wealth of information for members and non-members too. There are articles,

recommended books, forums, course information and trainer information as well as professional standards, advice and networking.

Brilliant Minds
www.brilliantminds.co.uk
This is my main website, which includes articles, my blog, course information, coaching programmes, how to get me as a guest speaker and the opportunity to join my mailing list.

Success Strategies
www.successstrategies.com
I've mentioned Shelle Rose Charvet's excellent book, *Words That Change Minds*. This is her website where you can find her excellent free resource, 'Shelle's Top Tips'.

NLP University
www.nlpu.com
I've also referred to Robert Dilts several times in this book. You can find out more about his work at this site, which includes a massive archive of articles.

Index

adrenalin 20
agreement, expressing 115
anger 111–13
anxiety, communication and
    57–8
Ashby's law of requisite value
    73–5
association 136, 145–6
auditory preference 126–7
    communicating with 131
    engaging with in
        presentation 133–4
    language relating to 130
auditory submodalities
    144–5
auditory system 119
authority, creating air of
    166–8

Bailey, Rodger 36
Bandler, Richard 3, 165, 177
Battenberg, Jane 120
BBC World Service 79–80
behaviour
    appropriateness in some
        context of all 77–80
    distinction from person
        64–6

matching 101–5
pacing and leading
    114–15
present, as best choice
    available 70–3
purposefulness of every
    75–7
beliefs 28–31
    as filters 28–9, 31–6
    values vs. 29
body, and mind affect each
    other 80–3
body language 21–2
'body workers' 83
'breaking state' 70, 139
British Airways Engineering
    79

cats 76
change programmes 94–6,
    162–3
choice 153–4
    interventions should
        increase 84
Chomsky, Noam 44
Chopra, Deepak 82
circle of excellence
    68–70

communication 7, 173–5
  and anxiety 57–8
  cannot not happen 55–8
  consistent 58
  importance 173
  meaning in response it
    gets 58–61
  outcome and 174, 185,
    186–8
  skills 175
  written 59
  see also language
complaints, dealing with
  111–13
compromise, flexibility vs. 75
conscious thinking 16–18
content 2
  flexibility in 74–5
context
  appropriateness of
    behaviour in 77–80
  of outcome clear 91
  resources and 67–8
control
  flexibility and 73–5
  outcome under your 91
convincer demonstration
  157–61
convincer representational
  system 156–7
convincers 154, 155–62
  auditory 157
  automatic 157–8, 162

consistent 160–1, 163
  number of times 158–9,
    162–3
  period of time 159–60,
    163
  visual 157
core competence 78–80
counselling 180–1
counter-examples 180, 181
Cummings, Craig 168

decision-making strategies
  152–5
  stages 153–5
decisions 25–6
deep structure 44, 177
deletion 13–14, 177–9
digital preference 128–9
  communicating with 132
  engaging with in
    presentation 133–4
  language relating to 130
digital system 119
Dilts, Robert 165
dissociation 136, 145–6
distortions 13, 14, 177,
  179–81

'early adopters' 158, 162
ecology
  and categories 148
  of outcome 92–3
elicitation 176

embedded commands
188–91
'emotional baggage',
dealing with 8
emotions 20
awareness of 127
as clue to values 35–6
creation 20
monitoring 30–1
English language 26–7
experiences
creating our own 46–9
labelling 28
learning from 48

failure 61–4
feedback 61–4, 187
'filing codes' 146–7
filtering 1
filters, linguistic *see* linguistic
filters
first access 12
flexibility
compromise vs. 75
in content 74–5
in structure 74–5
and system control 73–5
friendship 29–30

Gandhi, Mahatma 64
generalizations 13, 15, 177,
181–2
gestures 105

goals, well-formed 94
Google 27
Grinder, John 3, 12, 165, 177
gustatory system 119, 120

habits 4–6
helpdesk system
implementation 95–6

impulse shopping 25
influence
rapport and 185
using embedded
commands 188–91
using presuppositions
185–8
information, managing 116
installation 176
intelligence, definition 75
internal representation 16
conscious 16
unconscious 16

James, William 74–5
Johnson, Denny Ray 120

kinaesthetic preference
127–8
communicating with 132
disguise of 128
engaging with in
presentation 133–4
language relating to 130

kinaesthetic submodalities 144–5
kinaesthetic system 119
kinaesthetics, sensory 12
Korzybski, Alfred 52

language 173–91
    embedded commands 188–91
    as filter 26–8
    keys to effective use 191
    see also communication; presuppositions
leading 109–10
    see also pacing and leading
linguistic 4
linguistic filters 13–16, 23–31, 45
linguistic responses 44–8
location, in visual field 145, 146

'making sense' 53–4, 55, 58
mapping across 147–9
maps of the world 1–2, 176, 183
    as not territories 52–4
memories 24–5
meta mirror 137–42
meta model 177–82
meta-messages 57
metaprogrammes 36–44

mind, and body affect each other 80–3
mind-body medicine 82
mind-reading 179
miscommunications 45
misunderstandings 45, 56, 59
modalities 143
modelling 3, 165–71
    creating air of authority 166–8
    definition 165
    remembering names 168–70
    using to learn 170–1
motivation 36–8, 153
    away from 36–8, 39–40, 153
    towards 36–7, 39, 40, 153

names, remembering 168–70
narrow-mindedness 181
neuro 4
neuro-linguistic programming *see* NLP
neurotransmitters 82
NLP
    definition 2
    origins 3
    as tool kit 193
    uses for work 6–8

NLP language and
communication model
11–49
diagram 11

objectives, well-formed 94
obstructiveness 71–3
olfactory system 119, 120
opinion, pacing and leading
115–16
options preference 41–3
outcome
and communication 174,
185, 186–8
context clear 91
ecology of 92–3
expressed in positive
89–90
secondary gain 88, 91–2,
94–6
in sensory-specific terms
90
team 94
under your control 91
well-formed 87–96
outcome frame 85–7
outcome thinking 85–96
keys to making it work
89–96

pacing 110
pacing and leading 110–17
behaviour 114–15

opinion 115–16
states 110–13
perceptual positions 135–42
associated 136
dissociated 136
first position 135
meta mirror 137–42
second position 135, 191
third position 135
performance appraisal
systems 65
physical health 23, 80–3
physiology 19–21
poltergeist 32
post-decision rationalization
25
prejudice 181
presuppositions 182–5
definition 183
identifying 184–5
influencing another
person's thinking using
185–8
presuppositions of NLP
51–84
all behaviour is
appropriate in some
context 77–80
all interventions should
increase choice 84
every behaviour has a
positive intention 75–7
map is not territory 52–4

205

presuppositions of NLP
(*contd.*)
  meaning of
    communication is in
    response it gets
    58–81
  mind and body affect
    each other 80–3
  part of any system which
    has most flexibility
    controls system 73–5
  people are doing the
    best they can with the
    resources they have
    available 70–3
  people have all the
    resources they need to
    succeed 66–70
  there is no failure, only
    feedback 61–4
  you are not your
    behaviour 64–6
  you cannot not
    communicate 55–8
problem frame 85–7
problem solving 180–1
procedures preference
  41–4
programming 4
proverbs 180
public speaking 66–7

questions, asking 177–83

rapport 97–109
  barriers to establishing
    101
  calibrating 99–100
  creating 100–4
  how it feels 107–8
  how it looks 108
  how it sounds 108–9
  and influence 185
  matching behaviour
    101–5
  matching voice 105–6
  matching words 106–7
  measuring 107
reassurance 154–5
reflex actions 20
relationships, addressing
  tension in 137–42
relaxing 167
remembering names 168–70
rep systems 119–34, 143
  communicating with
    different preferences
    130–2
  convincer 156–7
  indicators of preference
    125–9
  language relating to
    preferences 129–30
  preference quiz 121–5
  primary 120
  speaking to a group
    132–4

resources
  and context 67–8
  doing best you can with
    available 70–3
  having all you need 66–70

Satir, Virginia 177
Science and Sanity 52
secondary gain 88, 91–2,
  94–6
self-discipline 92
self-management 6–7
sensory preferences 125–9
  communicating with
    different 130–2
  language relating to
    129–30
  speaking to a group
    132–4
sensory-specific evidence 90
skills, learning new 7–8
space, taking up 167–8
states 111
  'breaking' 70, 139
  pacing and leading
    110–13
strategies 151–63
  change management
    94–6, 162–3
  definition 151
  see also convincers;
    decision-making
    strategies

stress 81
  reducing 155
structure 2, 177
  deep 44, 177
  flexibility in 74–5
  surface 44, 177
submodalities 143–9
  checklist 144–7
  mapping across 147–9
surface structure 44, 177

talking, to yourself 16–17
targets, well-formed 94
team outcomes 94
tidiness 126
time, taking 166–7
transformation, stages of
  12, 13
'trial period' 159

unconscious mind 17–19
unspecified verbs 178–9

VAKOG 12, 16
values 28–31
  beliefs vs. 29
  clashes of 34–6
  emotions as clue to
    35–6
  as filters 28–9, 33–6
  violation of 31, 112
visual preference 125–6
  communicating with 131

visual preference (*contd.*)
  engaging with in
    presentation 133–4
  language relating to 130
visual submodalities 144–5
visual system 119

voice, matching 105–6

well-being 23, 80–3
words
  matching 106–7
  meanings of 58–9

Notes

You can use the following pages to make your own notes on any of the exercises in the book.

Notes

Notes

Notes